D1206500

DESIGN, CONSTRUCTION, AND MONITORING OF LANDFILLS

DESIGN, CONSTRUCTION, AND MONITORING OF LANDFILLS

Second Edition

Amalendu Bagchi
Wisconsin Department of Natural Resources

A WILEY-INTERSCIENCE PUBLICATION

JOHN WILEY & SONS, INC.

NEW YORK · CHICHESTER · BRISBANE · TORONTO · SINGAPORE

This text is printed on acid-free paper.

Copyright © 1994 by John Wiley & Sons, Inc.

All rights reserved. Published simultaneously in Canada.

Library of Congress Cataloging in Publication Data:
Bagchi, Amalendu.
    Design, construction, and monitoring of landfills / Amalendu
Bagchi.—2nd ed.
        p.   cm.
    Rev. ed. of: Design, construction, and monitoring of sanitary
landfill. c1989.
    "A Wiley-Interscience publication."
    Includes bibliographical references and index.
    ISBN 0-471-30681-9 . alk. paper
    1. Sanitary landfills. I. Bagchi, Amalendu. Design,
construction, and monitoring of sanitary landfill. II. Title.
    TD795.7.B34   1994
628.4'4564—dc20   94-13853

Printed in the United States of America

10 9 8 7 6 5 4 3 2 1

PREFACE

It is heartening to write the preface to the second edition of a book. The first edition was well accepted both within the United States and internationally by teachers and professionals involved in design and construction of landfills. Response from the non-engineering community interested in the topic was also very good. Since the technology is changing rapidly I thought a revised edition would benefit the readers. I have tried to include new analytic techniques and new emerging concepts in this book. I hope the diverse opinions included in the book will help in creative thinking leading to the advancement of the state of knowledge as well as developing a better understanding of the technology as a whole.

Revisions have been made in all chapters. Significant additions were made to Chapters 3, 6, 7, and 8; minor additions were made to Chapters 9 and 10. I have tried to include all relevant information in the subject area of the book. All persons interested in various landfill related issues will find this revised edition helpful. As in the past, I appreciate readers' comments.

The opinions and conclusions in this book are mine and do not necessarily reflect the views and policies of the Wisconsin Department of Natural Resources.

The following persons have either provided significant comments on the first edition or have helped me with valuable information in preparing the manuscript for the second edition: B. Meredith Winn, Jr. (Westinghouse Environmental and Geotechnical Services, Inc.), Ajit Chowdhury (RMT Inc), and Hillary Inyang (Geoenvironmental Design and Research, Inc.). Their help is greatly appreciated. I also thank all the persons who provided comments on various topics included in the book.

I thank my daughter Sudeshna not only for her patience but also for her help in preparing the manuscript. Last but not the least, deepest gratitude to my wife Sujata for her encouragement and secretarial help.

AMALENDU BAGCHI

Madison, Wisconsin
February, 1994

PREFACE TO THE FIRST EDITION

The aim of this book is to provide a basic understanding of sanitary landfill related issues. Both theory and current practice on the subject are included so both students and practicing engineers can derive benefits from the book. Regulations, which vary in different parts of this country and in different parts of the world, have not been discussed. It is expected that a knowledge of the theory will help readers better address regulatory concerns.

The book deals with sanitary landfill technology applicable to both hazardous and nonhazardous waste disposal. The emphasis in this book is on *why* certain things are done rather than on *what* is to be done. A basic knowledge of geology and civil engineering is necessary to fully understand the design principles and construction guidelines. However, a basic understanding of science will be enough for readers interested in some basics of landfill design and monitoring. Although civil engineers and hydrogeologists are primarily responsible for the design, construction, and monitoring of landfills, a second group (e.g., lawyers, environmental control managers in various industries, and city and county administrators) is involved in the management of landfills. The book will provide some help to the second group of persons as well.

The need for a book such as this was felt during conversations with professionals and teachers involved in the profession. Significant information is available in the literature on the topics of most of the chapters. It was a difficult task to condense the information into the length of some of the chapters. Topics that are needed for design, construction, and monitoring are emphasized. None of the chapters is a summary of all the information on the topic of that chapter. References are cited for readers interested in more information. The book will provide a well-organized approach for understanding landfill technology as a whole.

Chapters 2 through 8 provide information primarily for design and Chapters 9 and 10 provide information for construction and monitoring, respectively. Chapter 11 provides information regarding operation and Chapter 12 provides information on economic analysis. Laboratory demonstrations and field trips may be used to supplement certain topics.

I offer my regards to the following three special teachers in my life: Subroto Ghosh, the mathematics teacher in high school who created my interest in science, and A. N. R. Char and T. B. Edil, who provided the necessary stimulation that helped me learn geotechnical engineering.

The opinions and conclusions in this book are mine and do not necessarily reflect the views and policies of the Wisconsin Department of Natural Resources.

The following persons provided critical reviews of one or more chapters: Professor Robert K. Ham, Professor Tuncer B. Edil, and Anik Gangully (University of Wisconsin, Madison); Kenneth J. Quinn (Warzyn Engineering Inc.); Garry T. Griffith (Nekoosa Packaging); Dennis Sopeich (Dane County Public Works Department); Barbara B. Gear, Dennis P. Mack, and Julian D. Chazin (Wisconsin Department of Natural Resources). In addition, many colleagues and friends provided valuable input. Janet Christopher performed the word processing tasks for parts of the book. Their assistance is greatly appreciated.

I thank my daughter, Sudeshna, for her patience during the preparation of the manuscript. Finally, I express deepest gratitude to my wife, Sujata, for her encouragement, patience, understanding, and secretarial help.

AMALENDU BAGCHI

Madison, Wisconsin
August, 1989

CONTENTS

DESIGN, CONSTRUCTION, AND MONITORING OF LANDFILLS

1 Introduction

Land disposal of waste has been practiced for centuries. In the past it was generally believed that leachings from waste are completely attenuated (purified) by soil and groundwater and hence contamination of groundwater was not an issue. Thus, disposal of waste on all landforms (e.g., gravel pits and ravines) was an acceptable practice. However, with increasing concern for the environment in the late 1950s landfills came under scrutiny. Within a short period of time several studies [California Water Pollution Control Board (CWPCB), 1954, 1961; Apgar and Langmuir, 1971; Garland and Mosher, 1975; Kimmel and Braids, 1974; Walker, 1969] showed that landfills do contaminate groundwater. Although percolation of leachings from chemical industry waste to groundwater aquifers was considered unsafe, leachings from nonchemical industrial waste and municipal waste was considered less harmful. As a result waste was divided into two categories: hazardous and nonhazardous. In many countries separate regulations were developed for these two types of wastes. Although collection of leachate from nonhazardous waste was not mandated, leachings from hazardous waste were required to be collected. For nonhazardous waste the emphasis was on transforming waste dumps into "sanitary landfills."

As a result of these different attitudes toward the effect of leachate on groundwater pollution, two separate design concepts evolved: natural attenuation (NA) type landfills (Fig. 1.1) and containment type landfills (Fig. 1.2). In NA type landfills leachate is allowed to percolate into the groundwater aquifer. The design guidelines for NA type landfills included a minimum allowable thickness of the unsaturated zone, a depth to bedrock, a distance to the nearest home with a private well, and so on. In containment type landfills the design concept consisted of constructing a low permeability liner to restrict leachate from percolating through the base of the landfill and a pipe system to collect leachate generated within landfills. The second category of landfills, also known as engineered or secured landfills, was mandated for disposal of hazardous waste. In addition to the trench-type landfill configuration in which waste is disposed below grade (Fig. 1.1 and 1.2), there are two other configurations in use. The second configuration may be termed as Canyon Fill landfills in which waste is disposed in sloping canyons and swales (Fig. 1.3). The third configuration may be termed as At Grade landfills in which the waste is disposed at or slightly below grade within the cofines of berms (Fig. 1.4). Although most of the design issues are the same for all

1

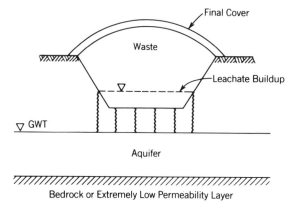

FIG. 1.1. Natural attenuation landfill.

types of landfills, there are special issues for each landfill type (e.g., slope stability of Canyon Fill landfills) that should receive due consideration.

Further research indicated that soil cannot attenuate all the contaminants leached even from nonhazardous waste (Bagchi, 1983), no matter how thick the underlying unsaturated zone is or how high the cation-exchange capacity of the soil is (refer to Sections 5.2 and 5.3 for a detailed discussion). Thus, subsequently, restrictions were imposed on NA type landfills. Currently NA type landfills are totally banned in some places. Certain changes also took place in the design of hazardous waste landfills. Initially these landfills were required to have a single liner. However, because of the possibility of leakage through a single liner, especially if the liner material is clay, a second liner (Fig. 1.5) was mandated. In many instances a synthetic membrance was required as the first liner. Thus, at present, land disposal of waste is no longer a simple practice.

1.1   COMMENTS ON REGULATORY REQUIREMENTS

Landfill siting is a fairly involved process. Statutes governing standards for waste management have been adopted in many countries. These statutes allow a regulatory agency to adopt minimum standards for landfills.

Prior to planning a landfill it is necessary to obtain all pertinent information about the units of government that may exercise control over different operational issues of a landfill. For instance, for containment type landfills leachate is collected and treated either on site or in a treatment plant off-site. In many instances the landfill management is handled by one section and the treatment facility management is handled by a separate section of the regulatory body. Therefore, the planner must become familiar with the requirements of both these bureaus or sections. A landfill also has to meet several locational criteria (see Section 2.1.2); separate sections of the regulatory agency may

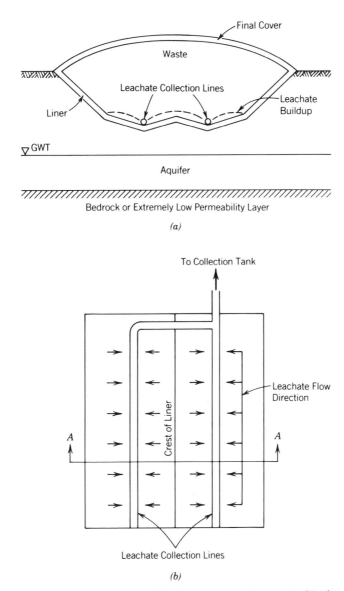

FIG. 1.2. Containment landfill: (a) Cross-section A–A; (b) plan.

exercise control over each of these locational criteria. Construction of a flow diagram indicating regulatory requirements and the name of the bureau involved in each operation is helpful in obtaining all the necessary approvals from different bureaus. Local governments sometimes exercise control over refuse collection, storage, hours of operation, and so on. In some regions local or regional authorities exercise control over "waste flow" (or how and

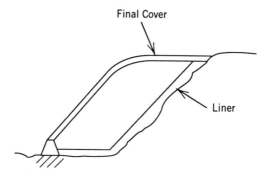

FIG. 1.3. Canyon fill landfill.

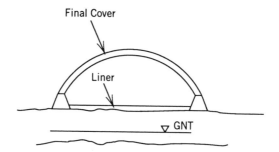

FIG. 1.4. At grade landfill.

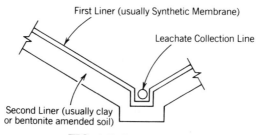

FIG. 1.5. Double liner.

where the waste generated within the region is to be handled). This type of "waste flow" control becomes an important issue if an incinerator is located within that region and the waste generated within that region is not high. It may become essential to direct a minimum waste volume to the incinerator so that it can be operated economically. Because landfill technology is evolving and the regulations are changing, it is important to remain up to date about both.

1.2 SCOPE AND ORGANIZATION OF THE BOOK

Rather than discussing the regulatory requirements in detail, this book concentrates on current theory and practice regarding landfills. Attempts have been made to discuss the pros and cons of different concepts so the designer can make the correct decision in each case. Landfill siting is usually a three-step process. The first step includes choosing several sites using locational criteria. The second step is to conduct a feasibility study of potential sites developed in step 1 and to determine which site is best for the landfill. A detailed discussion on feasibility study is beyond the scope of this book. However, some guidelines regarding the site-selection process is included in Chapter 2. The third step is to develop a detailed design for the site identified in a feasibility study. This book primarily deals with the issues confronted by the engineers involved in the process. (Note: usually hydrogeologists and engineers work as a team on landfill projects. This book is primarily aimed at engineers, although hydrogeologists will find it informative.) The book is written primarily for practicing engineers and engineering students; however, it will be very helpful to any person interested in learning the fundamentals of landfill technology. Attempts have been made to provide both theory and current practice mainly regarding landfill design and construction. Fundamental concepts of landfill monitoring are included so that engineers can interact with hydrogeologists as a team. The contents of each chapter are briefly discussed below.

Chapter 1 includes a brief history of changes in attitudes toward landfilling of waste. Because in most countries landfills are regulated by government, comments are included regarding approval by local regulatory agencies. The scope and organization of the book are also included.

Chapter 2 discusses how to select a site. Both preliminary site selection and the final selection process are discussed briefly.

Chapter 3 discusses how leachate and gas are generated in a landfill. Approaches for assessing leachate quality and quantity and the typical leachate quality of various wastes are included.

Chapter 4 discusses how to characterize waste. Characterization of both hazardous and nonhazardous waste is discussed separately.

Chapter 5 discusses natural attenuation processes and the design approach used for natural attenuation landfills.

Chapter 6 discusses the design of containment type landfills for both hazardous and nonhazardous waste. The advantages and disadvantages of natural attenuation and containment type landfills are discussed.

Chapter 7 discusses the various materials used for landfill construction.

Chapter 8 discusses the design of several landfill elements. The design of gas venting systems and how to retrofit an existing natural attenuation landfill to function as a containment type landfill are also included.

Chapter 9 discusses construction-related issues and the tests usually performed for quality control purposes.

Chapter 10 discusses fundamental concepts regarding performance monitoring. A detailed listing of items usually monitored in a landfill is included.

Chapter 11 discusses landfill operation and long-term care needed for maintaining a sanitary landfill.

Chapter 12 discusses how to estimate the costs of construction, operation, final closure, and long-term monitoring.

Landfill technology is multidisciplinary. It requires application of principles from hydrogeology, civil engineering, chemistry, mathematics, and material science. Although each concept is discussed in simple language, some knowledge of the above disciplines is assumed for readers interested in an in-depth study (i.e., students and professionals). Landfill technology is developing and is fairly complex. Correct, unambiguous answers to all issues are not yet available. Some points made may be in dispute and some discussions may be incomplete. Hopefully these uncertainties can be resolved in the future. Since publication of the first edition research efforts have been reported regarding new ways of disposal and landfill space use. In Europe most countries now require that municipal waste be treated prior to disposal. Currently treatment includes incinerations and composting. However, concepts of "reactor landfills" are also being studied. Landfill mining, which essentially consists of digging up old landfills and recovering usable materials, is no longer a dream; several landfills have already been mined. Although, these items are outside the perview of landfill design, brief discussion and references on the subjects are included to help readers in their search for alternatives to landfilling.

Landfill design and construction-related issues are discussed in reasonable detail. In addition, references are cited for readers interested in follow-up reading. Attempts have been made to include detailed design steps whenever possible, but justice could not be done in some cases. Some comments regarding research needs and some worked out examples are included for classroom use.

A list of references is included at the end of the book.

2 Site Selection

A landfill site has to meet several locational and geotechnical design criteria and be acceptable to the public. A preliminary list of potential sites is developed satisfying the first two criteria. For this purpose, usually a circle indicating a "search radius" (the maximum distance the waste generator is willing to haul the waste) is drawn on a road map of the region, keeping the waste generator (a city or industry) at its center (Fig. 2.1). The "search radius" depends on economics of waste hauling. Hauling of waste is one of the high cost items in landfill operation. So it is essential to keep the cost as low as posible. One should start with a small search radius and enlarge it if needed. A methodology for estimating the optimal service region for a landfill has been proposed by Wenger and Rhyner (1984). Once a landfill site is identified, this methodology may be used to identify the waste generation points that can use the landfill at a reasonable cost. Both the landfill operating cost and hauling cost are taken into consideration while developing the optimal service region using this approach. This approach is very helpful in identifying the communities that should participate in a municipal waste landfill located in a region. Search radius may have to be increased if potential sites cannot be located within the search area. If more than one waste generators is involved (e.g., several cities within a county) then a compromise location acceptable to the waste generators is used as the center. A discussion with the waste generator(s) regarding the search radius and center of the search area (especially if more than one generator is involved) must be undertaken before site selection begins. A landfill site may be owned by an individual, a company, or a public body (e.g., municipality). Since acceptibility to the public is crucial to the landfill siting process, the citizens to be affected should be informed regarding the site selection process as early as possible. Locational criteria dictated by the regulatory body must also be studied. A close look at the flexibility of these locational criteria is helpful. For instance, in the United States a landfill can never be sited within a critical habitat area; however, some compromise may be available for landfills located near a highway. The process of selecting a landfill site is complex and it involves three major issues: data collection, locational criteria, and preliminary assessment of public reactions. In the following sections each issue is discussed separately.

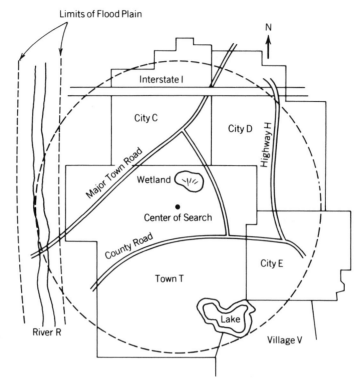

FIG. 2.1. Search radius for siting a landfill.

2.1 DATA COLLECTION

Several maps and other information need to be studied to collect data within
the search radius. The following information is needed: topographic map(s),
soil map(s), land use plan(s), and transportation plan(s) (discussed in Sections
2.1.1 to 2.1.8), and waste type and waste volume (discussed in Sections 2.1.9
to 2.1.15). Brief discussions on each of these items follows.

2.1.1 Topographic Maps

The topography of the area indicates low and high areas, natural surface
water drainage pattern, streams, and wetlands. A topographic map will
help find sites that are not on natural surface water drains or within a
wetland. Debate exists regarding locating a landfill on a groundwater recharge or discharge zone, an issue to be settled with the local regulatory
agency.

2.1.2   Soil Maps

These maps, primarily meant for agricultural use, will show the types of soil near the surface. Although these maps are partly useful for natural attenuation type landfill siting, they have very little use for containment type landfills.

2.1.3   Land Use Plans

These plans are useful in delineating areas with definite zoning restrictions. There may be restrictions on the use of agricultural land or on the use of forest land for landfill purposes. These maps are used to delineate possible sites that are sufficiently away from localities and to satisfy zoning criteria within the search area.

2.1.4   Transportation Maps

These maps, which indicate roads and railways and locations of airports, are used to determine the transportation needs in developing a site. If clay needs to be hauled from a distant source for constructing liner for a site, then these maps are used to estimate the hauling distance. Allowable axle loads on roads leading to a potential site must be studied to find out whether any road improvement will be necessary. Restriction may exist regarding locating a municipal waste landfill within a certain distance of an airport to minimize "bird hazard" for airplanes.

2.1.5   Water Use Plan

These maps are usually not readily available. However, once potential areas are delineated the water use in those areas must be investigated. A plan indicating the following items should be developed: private and public wells indicating the capacity of each well, major and minor drinking water supply line(s), water intake jetty located on surface water bodies, and open wells. A safe distance (365 m or more) should be maintained from all drinking water sources. A minimum distance between an existing well and a proposed landfill site may be specified by the regulatory agency.

2.1.6   Flood Plain Maps

These maps are used to delineate areas that are within a 100 year flood plain. For hazardous waste landfills a 500 year flood plain may be used. Landfill siting must be avoided within the flood plains of major rivers. A landfill may be constructed near an intermittent stream if additional protection measures (e.g., levee) are implemented.

2.1.7 Geologic Maps

These maps will indicate geologic features and are very important for glaciated regions. A general idea about soil type can be developed from a glacial geologic map. They are also very helpful in identifying clay borrow sources. In nonglaciated regions they may be used to identify predominantly sandy or clayey areas.

2.1.8 Aerial Photographs

Aerial photographs may not exist for the entire search area. Once a list of potential sites is developed aerial photographs or preferably a photogrametric survey of each of the potential sites may prove to be extremely helpful. Surface features such as small lakes, intermittent stream beds, and current land use, which may not have been identified in earlier map searches, can be easily identified using aerial photographs.

2.1.9 Waste Type

The first thing to identify is whether the waste is hazardous or nonhazardous. Regulations are significantly different for these two types of waste. If the waste is nonhazardous then the designer should know whether it is municipal or industrial waste. Municipal wastes are highly mixed types of wastes whereas industrial wastes are usually either monotypic or a mixture of two or three different waste streams with identifiable characteristics. Characterization of the waste (refer to Chapter 4 for details) should be performed to develop proper landfill design.

2.1.10 Waste Volume

The volume of industrial waste (hazardous or nonhazardous) can be easily estimated by studying the previous disposal records. For a new plant the estimate of waste volume should be made from the waste generation rate of similar types of industries elsewhere. Although municipal waste generation rates vary widely, 0.9–1.8 kg (2–4 lb) per person per day is a reasonable estimate; 650–815 kg/m^3 (40–50 lb/ft^3) is a reasonable range of bulk unit weight for municipal garbage. An estimation of population during active life of the landfill should be done; the estimated population in each year is then multiplied by the waste generation rate to obtain the waste volume in each year. For industrial waste the bulk unit weight, if not readily available, should be determined by laboratory testing.

2.1.11 Landfill Volume

Landfill volume is estimated by adding the daily, intermediate (if used) and final cover volume to the waste volume. Daily cover is mandatory for most municipal garbage landfills. If soil is used as daily cover, then a waste to daily cover ratio of 4 : 1 to 5 : 1 by volume is a reasonable estimate. The intermediate and final cover volume can be estimated from the thicknesses of these covers.

2.1.12   Availability of Landfill Equipment

Although this item is not involved in developing the landfill plan, these data should be gathered at the planning stage. Special hauling trucks are sometimes required for hauling sludges, especially if they have a high liquid content or are hazardous.

2.1.13   Recycling and Incineration Options

A study regarding possible recycling or incineration of all or part of the waste should be undertaken. Recycling of a waste, if technology is available, is sometimes mandated by regulatory authorities. Recycling and incineration, although technically feasible in some cases, may not be an economically acceptable option. Therefore both the technical and economic feasibility of recycling and incineration should be checked. Incineration will reduce the landfill volume (and may change the design if the ash is found to be hazardous) needed for the disposal of ash. From an environmental standpoint, every effort should be made to recycle waste as much as possible and then arrangements should be made to incinerate the rest if it is technically and economically feasible.

2.1.14   Existing Disposal Option

The available landfill volume within a reasonable haul distance should be studied. The cost of disposing of waste in existing landfill could be less than developing and operating a new landfill. There are certain hidden costs of operating a landfill (e.g. monitoring of groundwater wells, payment for bond purchase if required by regulation) that are sometimes overlooked. A list of landfills around the proposed site will also be helpful in emergency situations (e.g., the landfill could not be built by the target date due to litigation).

2.1.15   Funding

The cost of developing a landfill is quite high. Funds for the initial investigation, preparation of the report, landfill construction, and so on must be obtained for planning purposes. A proper estimate for each stage of the proposal preparation and the flow of necessary funds must be studied. Discussion regarding the availability of funds with the person(s) dealing with the budget is essential. Funding is a very critical issue for facilities owned publicly or by giant corporation.

2.2 LOCATIONAL CRITERIA

Usually a landfill cannot be sited within a certain distance of the following: lakes, ponds, rivers, wetlands, flood plain, highway, critical habitat areas, water supply well, and airports. In addition, landfill siting is not allowed in areas in which a potential for contamination of groundwater or surface water bodies exists. No information is available as to how the distances mentioned in the following subsections were arrived at, however, these distances are widely accepted. Usually permission from a regulatory agency is required if a proposed landfill site does not meet the locational criteria. If it is absolutely essential to site a landfill within the restricted zone(s) then permission from the regulatory agency should be sought at an early date. In the absence of any regulatory requirements regarding safe distances, the following distances may be used.

2.2.1 Lake or Pond

No landfill should be constructed within 300 m (1000 ft) of any navigable lake, pond, or flowage. This distance may be reduced for a containment type landfill. However, because of concerns regarding runoff of waste contact water, a surface water monitoring program should be established if a landfill is sited less than 300 m from a lake, pond or flowage.

2.2.2 River

No landfill should be constructed within 90 m (300 ft) of a navigable river or stream. The distance may be reduced in some instances for nonmeandering rivers but a minimum of 30 m (100 ft) should be maintained in all cases.

2.2.3 Flood Plain

No landfill should be constructed within a 100 year flood plain. Regulations may require a more restrictive flood plain (e.g., 500 year flood plain) siting criteria. A landfill may be built within the flood plains of secondary streams if an embankment is built along the stream side to avoid flooding of the area. However, landfills must not be built within the flood plains of major rivers.

2.2.4 Highway

No landfill should be constructed within 300 m (1000 ft) of the right of way of any state or federal highway. This restriction is mainly for aesthetic reasons. A landfill may be built within the restricted distance if trees or berms are used to screen the landfill site.

2.2.5   Public Parks

No landfill should be constructed within 300 m (1000 ft) of a public park. A landfill may be constructed within the restricted distance if some kind of screening is used. A high fence around the landfill and a secured gate should be constructed to restrict easy entry of unauthorized persons in the landfill.

2.2.6   Critical Habitat Area

No landfill should be constructed within critical habitat areas. A critical habitat area is defined as the area in which one or more endangered species live. It is sometimes difficult to define a critical habitat area. If there is any doubt then the regulatory agency should be contacted. Siting a landfill within critical habitat areas is not suggested.

2.2.7   Wetlands

No landfill should be constructed within wetlands. It is often difficult to define a wetland area. Maps may be available for some wetlands, but in many cases such maps are absent or are incorrect. If there is any doubt then the regulatory agency should be contacted. Disturbance of wetlands should be avoided.

2.2.8   Airports

No landfill should be constructed within 3048 m (10,000 ft) of any airport. This restriction is imposed to reduce bird hazard. Birds are attracted to landfills where food is available (in general, municipal landfills fall in this category). Permission from the proper authority (in most cases the airports are under the administration of a central agency) should be sought if the proposed landfill site is within the restricted zone.

2.2.9   Water Supply Well

No landfill should be constructed within 365 m (1200 ft) of any water supply well. It is strongly suggested that this locational restriction be abided by at least for down gradient wells. Permission from the regulatory agency may be needed if a landfill is to be sited within the restricted area.

2.3   PRELIMINARY ASSESSMENT OF PUBLIC REACTIONS

The public should be informed regarding the possibility of siting of a landfill in their area as soon as a list of potential sites is developed. The public is less suspicious and more open to discussion if they are informed by the owner rather than getting the news from other sources. Public education regarding the dangers and benefits of a landfill should be undertaken. A preliminary assessment of public opinion regarding all the sites in the list is essential.

A site may be technically and economically feasible yet may be opposed heavily by the public. The "not in my back yard" (NIMBY) sentiment is high initially, however, with proper discussion it can be overcome in some cases. Early assessment regarding how strong the NIMBY sentiment is, can significantly reduce the time and money spent on obtaining the final permit for a landfill site. In many instances residents around a proposed site cooperate if the owner's representative listens to concerns of the area residents and considers those concerns in designing and monitoring a site. Noise, dust, odor, increases in traffic volume, and reduction in property value concern the area residents more than the fear of groundwater contamination.

2.4   DEVELOPMENT OF A LIST OF POTENTIAL SITES

After studying the information discussed in Sections 2.1 to 2.3, areas having potential for site development should be delineated. A road map may be used to show the potential sites that are technically acceptable and satisfy the regulatory locational criteria. The sites should be ranked based on public reactions. This first list of potential sites should then be discussed with the owner(s) of the site. It is a good idea to develop stronger public relations in the areas that are identified as prime candidates for landfill development. Preliminary geotechnical investigation should be undertaken at each of the potential sites where public opposition is low. If possible three or more sites should be studied initially. After studying the geotechnical information a potential list of two or more sites should be developed. A report indicating geotechnical information, conceptual design, and discussion regarding locational criteria should be developed for each of the potential sites. These reports should then be discussed with the regulatory authority, the area residents of the potential sites, and the municipality in which the site is to be located.

2.5   FINAL SITE SELECTION

Based on the discussion of the reports mentioned in Section 2.4, one or two sites are selected finally. A detailed investigation for each of these sites needs to be undertaken to develop a feasibility report. In some states/countries, requirements of a feasibility report are clearly spelled out in administrative code(s) issued by the regulatory agency. These codes should be obtained and followed when writing a feasibility report. Two major items of this report with which engineers get involved are on-site geotechnical investigation and borrow source investigation. These two items are discussed below.

2.5.1   On Site Geotechnical Investigation

The purpose of geotechnical investigation is primarily to obtain data to study the different soil stratum present at the site and to prepare a groundwater map for the site.

Subsoil Investigation. Continuous soil samples need to be collected to determine soil stratification. Mechanical properties (strength and consolidation characteristics) are not a prime concern in subsoil investigation for landfills, although consolidation characteristics of highly organic (e.g., peat) clay layer(s) and strength characteristics of suspected collapsible soil should be studied carefully. Permeability of the soil layers (both laboratory permeability of undisturbed soil samples and field permeability) should be studied carefully. For major projects both horizontal and vertical permeability should be investigated. Ratios of horizontal to vertical permeability of undisturbed soil samples between less than one to seven were reported for several clay types (Mitchell, 1956). Olsen (1962) predicted a ratio of up to 20 for kaolinite and 100 for other clay types for totally horizontal particle orientation. He used a tortuous flow path model for this analysis.

In addition to the items mentioned in the above paragraph, the following items are of interest in soil investigation for a landfill project: Atterberg limits of fine grained soils, grain size distribution of soil samples, existence of fracture in the clayey layer, thickness of each stratum, depth to bedrock, identification of bedrock (this can be done by studying a geologic map of the area), natural moisture content, and degree of saturation of the clayey strata. The last two items should be studied to discover the probable location of the groundwater table (GWT) in the clayey stratum and the existence of perched GWT. This will also help in installing groundwater monitoring wells to define GWT. An experienced soil engineer or hydrogeologist should be present during the entire soil investigation program. A detailed borelog should be prepared, which should include comments and observations.

Differences of opinion may exist regarding the total number of borings required to define the soil stratigraphy and groundwater condition at a site. A clearer concept of the soil formation process (geology) at the site is helpful in developing a strategy. In the absence of specific requirements regarding the number of borings the following guidelines may be followed:

1. The borings should be distributed in such a way that it covers an area at least 25% larger than the proposed waste limits.
2. Five borings should be done for the first 2 hectares (ha) (five acres) or less and two additional borings for each additional hectare. The boring should be well distributed over the entire area. Refer to Section 2.6.2 for suggested number of groundwater wells.
3. The borings should extend at least 7.5 m (25 ft) below the proposed base of the landfill.

As far as landfill design goes, soil borings help to identify the soil type(s) and bedrock depth and the depth and thickness of usable groundwater aquifer. In some sites the bedrock or the aquifer may be too deep. In those sites regional geologic data may be used to assess the bedrock and water table depth. One or two borings should extend at least 2 m (6 ft) into the aquifer/ bedrock for verification purposes.

Knowledge of local geology and hydrogeology is essential for proper planning of a soil boring program. It must be borne in mind that the above-mentioned numbers of borings provide a guideline only. The exact number of boreholes and groundwater table wells may be far in excess of the minimum suggested number. The reverse may also be true in some cases, that is, far fewer borings may be sufficient to define the soil stratigraphy of a site completely. Usually there is a lump sum cost for bringing the drill rig on site. This cost can be saved if all the necessary borings can be done when the rig is on site.

Seismic Hazard Investigation. In sites located within the seismic impact zone additional investigations to find seismic design parameters should be undertaken. The two mechanisms that can cause damage to landfills are (1) strong ground motion and (2) displacement of the ground below or adjacent to the landfill base due to movement along a fault. The strong ground motion may cause liquefaction of subsoil leading to substantial settlement of the landfill base.

The magnitude of an earthquake is measured by the Richter scale, which was developed by Charles Richter in the early 1900s. In general, earthquakes of 5.0 and above on the Richter scale can cause significant damage to structures. It should be noted that humans cannot sense earthquakes below 2.0 on the Richter scale. The following parameters need to be assessed for use in seismic design: Maximum Horizontal Acceleration (MHA) and Maximum Horizontal Velocity (MHV). Although vertical motions are also observed during earthquakes leading to a momentary reduction in weight, such vertical motions are not important for bulky soil structures such as landfills and dams.

Design earthquake ground motions can be found from existing seismic hazard maps (note: these maps are available from appropriate government organizations) or by conducting site-specific seismic hazard analysis. Note, however, that published maps usually provide high values of MHA and MHV. On the other hand, site-specific seismic data collection and analysis are time consuming and costly. It is suggested that values obtained from existing maps be used for sites with low or moderate seismic activity. For sites located in high seismic activity regions, a site-specific seismic study should be undertaken.

Certain soil types (e.g., saturated loose sand) are prone to liquefaction. So the liquefaction potential of soils from various strata below the landfill base should be studied carefully during subsoil investigation. It is also recommended that investigations regarding the existence of faults below a proposed site should be undertaken for sites located within a seismic impact zone (note: see Section 8.6). Such investigations of the existence of faults are costly. A stepped approach may be used, which will help avoid unnecessary cost and time involved in the investigation. The suggested steps are (Weiler et al., 1993):

1. Review of the published seismic data.
2. Review of subsurface exploration data to determine if a fault exists.
3. A geologic reconnaissance survey of the area.
4. Review of the regional seismological and geological history.
5. Geophysical investigation utilizing one or more of the following: seismic refraction/reflection, gravimetric survey, and magnetic survey.
6. Angular boring.
7. Test trenching to search for evidence of recent faulting.

Usually the first four steps will provide needed information regarding the possibility of the existence of a fault. Steps 5 through 7 are undertaken only when strong evidence regarding the existence of a fault is found in the first four steps.

2.5.2   Borrow Source Investigation

A conceptual design of the site should be developed based on the subsoil investigation report and the waste type. This conceptual design should then be discussed with the regulatory agency to find out whether they agree with the design. Material requirements for natural attenuation type landfills are minimal whereas those for containment type sites are significant. In a containment type landfill, liner design is a major issue. The allowable type of material for both liner and cap will dictate borrow source investigation. If the liner material is clay and the drainage blanket is clean sand then identification of borrow sources for these two materials will become critical to the construction of the landfill. If the recommendations given in Section 2.1 are followed then initial data regarding the availability of these materials may already be known. The next step is to embark on a detailed investigation regarding the properties of these materials. The following sections discuss guidelines for investigation that are to be followed for each type of geologic materials used in landfill construction.

2.5.2.1   Clay. Clay may be used as primary or secondary (if a double liner is proposed) liner material. Test pits and borings are used to find the vertical and horizontal extent of the potential clay borrow source. The total number of test pits and borings required needs to be discussed with the regulatory agency. Five test pits and five borings (a total of 10) for the first 2 ha (5 acres) or less and two test pits or borings for each additional hectare are reasonable. The locations of the test pits and borings should be well distributed on a uniform grid pattern. Logs should identify the geologic origin, testing results, soil classification (using a system acceptable to the technical community of the region), and visual description of each major soil layer. The layer or layers of soil should be identified based on test pit logs and boring logs. It is a good idea to avoid the use of a clay layer for liner construction that is less than 1.5 m (5 ft) in thickness. It may become extremely difficult to procure clay from such a thin layer. Grain size distribution curves and Atterberg limits should be developed for at least two to three samples obtained from each potential clay layer(s).

Differences of opinion may exist regarding specifications for clay, which when compacted in the field will provide a low permeability (1 $\times$ 10^{-7} cm/sec or less) layer; in the absence of guidelines the following specifications may be used to identify a clay borrow source that can provide a low permeability liner: liquid limit between 20 and 30%; plasticity index between 10 and 20%; 0.074 mm or less fraction (P200 content: 50% or more; clay fraction (0.002 mm or less size): 25% or more. Grain size distribution significantly influences the permeability of soil. Thus, a soil that has a classical "inverted S type" particle size distribution can be easily compacted to achieve low permeability whereas a soil with a very high clay content but with poor particle size distribution may not be easily compacted to achieve low permeability. Five representative samples for the first 4 ha (10 acres) or less and one additional sample for each additional 2 ha (5 acres) or less should be used to develop modified proctor curves. Each proctor curve should be developed based on a minimum of five points. Thus, there will be at least five modified proctor curves for the first 4 ha (10 acres) and one additional modified proctor curve for each additional 2 ha (5 acres) of clay borrow area. It is essential to study the relationship between compaction, moisture content during compaction, and recompacted permeability of the clay layer(s). For this purpose permeability of some of the modified proctor samples should be studied to find out the range of permeability at different moisture content and compaction. Literature indicates (Mitchell, 1976) that soils compacted wet of optimum moisture will develop lower permeability than the ones compacted low of optimum. Compaction at wet of optimum moisture provides a better kneading effect in the field. Refer to Sections 7.1.8 and 9.2.1 for additional information.

2.5.2.2 Sand. Sand is used as the drainage blanket and sometimes as a protective layer over a clay cap. The primary property needed for the drainage layer is high permeability (1 $\times$ 10^{-2} to 1 $\times$ 10^{-3} cm/sec). Usually clean sand containing a maximum of 5% below the 0.074 mm fraction (P200) is capable of providing high permeability. Coarse sand can be easily washed to satisfy the permeability criteria.

The number of test pits and borings used to identify a clay borrow source is also applicable for sand borrow source identification. Borrowing of sand from a sand layer less than 1.5 m (5 ft) thick should be avoided for the same reason described in the above section. As far as the test goes only two types of tests need to be done on representative samples. Two or three samples from each test pit/boring should be collected and tested individually for grain size distribution and permeability at 80–90% relative density (Note: permeability of clean sand does not depend heavily on compaction.)

2.5.2.3 Silty Soil. A 60- to 90-cm (2–3 ft) or more silty soil layer is preferred as the protective layer over the barrier (clay) layer in the landfill final cover to protect the barrier layer from freeze–thaw and desiccation effects (Note: freeze–thaw cycles and desiccation cracks increase permeability of compacted clay.) The number of test pits and boring necessary to identify a silty soil borrow source can be the same as suggested for clay. No specification is generally assigned to the soil in that layer but it is preferred that the soil be a silty loam.

2.5.2.4 Topsoil. Topsoil is used on the final cover of a landfill. Usually the area stripped for the construction of a landfill has topsoil that can be stock-piled for future use. However, if there is not enough topsoil at the site, additional borrow sources should be identified. Local agricultural experts/horticulturists may be contacted for this purpose. Sometimes an additional nutritional requirement and pH adjustment are necessary before a soil can be used for planting. Although detailed nutritional testing is not necessary at the feasibility stage, it is essential that a topsoil source be identified if needed.

2.5.2.5 Leachate Collection Pipes and Synthetic Membrane for Liner. Supplier(s) for leachate collection pipes and synthetic membrane should also be identified if synthetic membrane is to be used for liner and/or final cover construction. A list of available products should be obtained for future use.

2.6 PREPARATION OF FEASIBILITY REPORT

The purpose of a feasibility report is to determine whether a particular site has the potential for use as a landfill for the disposal of a particular waste type. The submittal requirements for this type of report are clearly indicated by regulatory agencies in some state/country. A familiarity with the technical information requirements and other necessary legal steps (e.g. notifying the town board in which the landfill is proposed) for obtaining a permit is essential. Detailed discussion of the items is beyond the scope of this book. However, the following sections will summarize some of the major issues generally associated with a feasibility report, some of which have already been discussed in earlier sections of this chapter.

2.6.1   Geotechnical Information

A detailed description of the geology of the area including a bedrock map or depth, distribution of unconsolidated units, and surface topography must be included. The topography of the area should be mapped using 30- to 60-cm (1–2 ft) contour interval.

2.6.2   Hydrogeology

The depth to groundwater, direction of groundwater flow, vertical and horizontal hydraulic gradients, existence of any groundwater divide, and the depth to usable aquifer should be investigated. To obtain this information, groundwater wells must be installed. The total number of wells necessary to define the hydrogeology of a site must be resolved with the regulatory agency. In the absence of a specific requirement the following guidelines may be used: two water table wells/hectare (2.5 acre) for the first 2 ha (5 acres). At least at 25% of the groundwater table locations a second well should be installed to form a piezometer nest to find the vertical gradient. All water table wells and piezometers must be developed properly. The purpose of developing a well is to remove fine particles, drill cuttings, and so on, and clean the well screen to ensure a proper inflow of water. Water quality in a properly developed well remains stable. The well development process usually involves pumping two to three "well volumes" of water from the well. Because of this water removal the water level remains unstable for a period of time; recovery time depends on the fineness of the medium in which the well is placed. Recovery time is maximum for clayey soil and minimum for sandy soil. It may be noted that the water level in a well fluctuates within the year (seasonal fluctuation) and also fluctuates on a long term basis. Therefore the water level should be measured for at least 1 year to find the seasonally high water table.

The existing water quality should also be established prior to disposing of waste in a landfill. Therefore water samples from all the wells and piezometers should be collected and tested for a list of parameters appropriate for the waste type. It is recommended that at least four (preferably eight) water samples collected at least one month apart from each well be tested to form some idea about the background (existing) water quality at the site.

2.6.3   Environmental Impact

This may become an important issue for some landfills. The landfill may impact flora, fauna, groundwater, surface water, ambient air quality, and so on. A detailed discussion of each is beyond the scope of this book. Thorough discussions should be held with the regulatory agency to find out the exact requirements in each case.

2.6.4 Conceptual Design

A conceptual design of the landfill should be presented in the report. Decisions must be made regarding subbase level, volume of landfill, material and thickness of liner, leachate collection system design, final contour, final cover design, routing of surface water, leachate treatment (on site or off site), and final use of the landfill. For major landfills some engineering analysis may need to be performed at this stage.

2.6.5 CONTINGENCY PLAN

A discussion should be included in the report regarding actions to be taken in case the groundwater at the site is impacted. This may not be a serious issue for smaller landfills but for a large hazardous waste landfill a contingency remedial action plan is worth investigating at an early stage. The analysis may show that remedial action at the proposed site may be very expensive or technically not feasible. Such findings may require a more stringent liner and leachate collection system design or may lead to discarding the proposed site.

3 Leachate and Gas Generation

The subjects of Chapters 3 and 4 are interrelated. Both chapters should be read to understand the importance of leachate characteristics in landfill design. The main aim of this chapter is to discuss leachate and gas generation as they relate to landfill design. The quality and quantity of leachate generated during the active life and after closure of a landfill are important in managing a landfill. In addition, the quality of leachate is an important issue for leachate treatment.

Leachate is generated as a result of the percolation of water or other liquid through any waste and the squeezing of the waste due to self weight. Thus, leachate can be defined as a liquid that is produced when water or another liquid comes in contact with waste. Leachate is a contaminated liquid that contains a number of dissolved and suspended materials. Part of the precipitation (snow or rain) that falls on a landfill reacts (both physically and chemically) with the waste while percolating downward (Fig. 3.1). During this percolation process it dissolves some of the chemicals produced in the waste through chemical reaction. The percolating water may also dissolve the liquid that is squeezed out due to weight of the waste (e.g., squeezing out of pore liquid of papermill sludge). Many studies were conducted to determine the role of microbial activity in decomposing municipal waste and subsequent leachate formation (Rovers and Farquhar, 1973; Caffrey and Ham, 1974; CWPCB, 1961). In a municipal waste, methane, carbon dioxide, ammonia, and hydrogen sulfide gases are generated due to anaerobic decomposition of the waste. These gases may dissolve in water and react with the waste or dissolved constituents of the percolating water. For instance, carbon dioxide combines with water to form carbonic acid, which then dissolves minerals from the waste [American Public Works Association (APWA), 1966]. Several other chemical reactions also take place releasing a wide range of chemicals, depending on the waste type. The percolating water plays a significant role in leachate generation. It should be noted that even if no water is allowed to percolate through the waste, a small volume of contaminated liquid is always expected to form due to biological and chemical reactions. The concentration of chemical compounds in such liquid is expected to be very high. The percolating water dilutes the contaminants in addition to aiding its formation. The quantity of leachate increases due to the percolation of water, but at the same time the percolating water dilutes the concentration of contaminants. Both quality and quantity of leachate are important issues for landfill design.

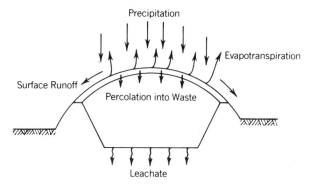

FIG. 3.1. How leachate is generated.

In general, to minimize leachate generation the present concept is to construct final cover on a landfill as soon as the waste reaches the designed final grade. However, increasingly it has been suggested that the final cover construction be delayed. The advantage of prompt construction of final cover is a significant reduction in leachate quantity within a short period of time after the construction of final cover. The disadvantage of prompt construction of final cover is that the leachate that will need treatment is likely to be produced for several years after closure. Although the current thought is that leachate that needs treatment will not be generated 40 years after landfill closure, there are no definite data to support the theory. In a recent study on an old landfill, sufficient amount of biodegradable matter was identified 40 years after disposal (Suflita et al., 1992). This uncertainty regarding the length of time for which leachate treatment will be needed has economic and legal ramifications. The advantage of not constructing a final cover promptly after reaching final grade is that the contaminants from the waste are expected to flush out sooner. Although data from actual landfills are not available, results from three large test cells (25,000 to 35,000T of municipal waste) reported by Lechner et al. (1993) show that biodegradable matter reduced significantly within 21 months after disposal. The disadvantage of not constructing a final cover promptly is that the gas produced from a landfill cannot be collected efficiently; odor from such an "open landfill" is also expected to pose problems. In many countries in Europe only treated waste (i.e., after incineration or some form of biodegradation) is allowed to be land disposed. Enhanced aerobic or anaerobic biodegradation prior to land disposal needs further study for successful application in actual landfills.

3.1   FACTORS THAT INFLUENCE LEACHATE QUALITY

Various factors influence leachate quality. In general, leachate quality of the same waste type may be different in landfills located in different climatic regions; landfill operational practices also influence leachate quality. The following subsections discuss the basic reasons why such variations are observed.

3.1.1 Refuse Composition

Variation in refuse composition is probably at a maximum in municipal waste and at a minimum in industrial waste. Because of this variation in refuse composition, the quality of municipal leachate varies widely (Garland and Mosher, 1975; Lu et al., 1985). In general, quality variation is higher for putrescible wastes than for nonputrescible waste.

3.1.2 Elapsed Time

Leachate quality varies with time. In general the overall quality of leachate generated in year 1 will be less strong than that generated in subsequent years. Leachate quality reaches a peak value after a few years and then gradually declines. Figure 3.2 shows an idealized relationship of leachate quality with time (Ham, 1980; Ham and Anderson, 1974; Pohland, 1975). For an actual landfill leachate the quality variation is not as smooth, although distinctive zones of upward and downward trends can be observed if quality variation is plotted with time. All the contaminants do not peak at the same time and the time versus concentration variation plots of all contaminants from the same landfill may not be similar in shape.

3.1.3 Ambient Temperature

The atmospheric temperature at the landfill site influences leachate quality. The temperature affects both bacterial growth and chemical reactions. Sub-zero temperatures freeze some waste mass, which reduces the leachable waste mass and may cause inhibition of some chemical reactions. There are no reports of freezing an entire landfill during winter in cold regions, although chunks of frozen mass may be found in landfill. There are several studies on landfill leachate temperature (Chian and Dewalle, 1977; Fungaroli and

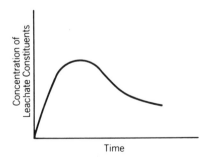

FIG. 3.2. Idealized variation of leachate quality with time.

Steiner, 1979; Wigh, 1979; Leckie et al., 1979) but no study is reported relating ambient temperature with leachate quality.

3.1.4   Available Moisture

Water plays a significant role in biodegradation and subsequent leaching of chemicals out of a waste. Leachate quality from waste disposed in a wet climate is expected to be different from the leachate quality of the same waste disposed in a dry climate.

3.1.5   Available Oxygen

The effect of available oxygen is notable for putrescible waste. Chemicals released due to aerobic decomposition are significantly different from those released due to anaerobic decomposition. The anaerobic condition in a landfill develops due to frequent covering of waste with soil (daily/weekly cover) or with fresh waste. The supply of oxygen starts to become depleted as soon as the waste is covered (either with soil or with more waste). A predominantly anaerobic condition develops in thicker refuse beds.

3.2   FACTORS THAT INFLUENCE LEACHATE QUANTITY

Like quality, leachate quantity is also dependent on weather and operational practices. The following subsections discuss the basic reasons why such variations are observed.

3.2.1   Precipitation

The amount of rain and snow falling on a landfill influences leachate quantity significantly. Precipitation depends on geographical location.

3.2.2   Groundwater Intrusion

Sometimes landfill base is constructed below the groundwater table. In these landfills groundwater intrusion may increase leachate quantity.

3.2.3   Moisture Content of Waste

Leachate quantity will increase if, because of is own weight, the waste releases pore water when squeezed (e.g., sludges). Unsaturated waste continues to absorb water until it reaches field capacity (a water saturation state). So dry waste will reduce leachate formation. However, it must be noted that in actuality, channeling causes water to flow through the waste without being absorbed by the waste; thus, the water absorption is much less than that predicted by laboratory or small-scale field studies. Codisposal of sludge or liquid waste in municipal landfills will increase the leachate quantity in a landfill.

3.2.4   Final Cover Design

Leachate volume is reduced significantly after a landfill is closed and finally covered because of two reasons: vegetation grown in the topsoil of a final cover reduces infiltratable moisture significantly by evapotranspiration and the low permeability layer reduces percolation. A properly designed final cover will reduce postclosure leachate quantity significantly.

3.3   ASSESSING PROBABLE QUALITY OF LEACHATE

Assessing leachate quality is difficult, and is more so for putrescible waste. Both laboratory and field scale studies have been done mostly on municipal waste. It should be mentioned that leachate quality predicted by laboratory tests may vary widely from the actual leachate obtained from a matured landfill. Before expanding any further on the test details it is essential to understand why assessment of leachate quality is important for landfill design purposes. There are basically four reasons for assessing leachate quality at an early stage: (1) to identify whether the waste is hazardous, (2) to choose a landfill design, (3) to design or gain access to a suitable leachate treatment plant, and (4) to develop a list of chemicals for the groundwater monitoring program. To assess the leachate quality of a waste, the normal practice is to perform laboratory leachate tests wherever possible and to compare the data with the quality of actual landfill leachate, if available. Difficulty arises when field data are not available for a particular waste type. In such cases it is better to take a conservative approach while performing/designing laboratory leachate tests on the waste.

Three approaches are available for assessing leachate quality: (1) laboratory test, (2) field study using lysimeters, and (3) predictive modeling. The laboratory tests are not applicable to biodegradable wastes in which bacteria play an important role.

3.3.1   Laboratory Tests

Several laboratory procedures are available: the water leach test, the standard leachate tests, toxicity characteristic leaching procedure (TCLP), and synthetic precipitation leachate procedure.

Water Leach Test. ASTM water leach test (D 3987-85) is used to predict leachate quality under laboratory conditions. The test data may not provide a representative quality of a "field leachate." In this method usually 70 g of waste is mixed with 1400 ml (or a ratio of 1:20) of water (meeting the specification of D1193 of ASTM) in a 2-liter watertight container (note: a container with a venting mechanism should be used for samples in which gases may be released; it should be noted that such venting may affect the concentration of the volatile components in the extract) and agitated for approximately 18 hr at 18°–27°C. The agitation is done using a motor with an axial rotation of 29 rpm (Fig. 3.3). The extract is analyzed for specific constituents using an available standard method. The solid liquid ratio and the rpm of the motor may significantly influence the leachate quality. The method is not suitable for use with organic, monolithic, and solidified wastes.

Standard Leach Test. This test (Ham et al., 1979) consists of two alternative mixing series. Procedure R is intended to indicate the maximum quantity of contaminants likely to release. Procedure C is intended to indicate the probable maximum concentration of contaminants in the leachate. In both tests either distilled water or a "synthetic leachate" can be used as a leaching medium. The distilled water medium is intended to model a monofill scenario and the "synthetic leachate" medium is intended to model a codisposal (with municipal waste) scenario. The synthetic leachate consists of a complex mixture of organic and inorganic chemicals to model a municipal solid waste leachate. This leachate must be used in an anaerobic test environment. The leaching is similar to the ASTM procedure. However, a different agitation method and solid-to-liquid ratio are used. In method C new waste is added

FIG. 3.3. Shaking apparatus. (Courtesy of RMT Inc., Madison, WI.)

TABLE 3.1. Toxicity Characteristic Contaminants and Levels

Contaminant	Regulatory Level (mg/liter)
Arsenic	5.0
Barium	100.0
Benzene	0.5
Cadmium	1.0
Carbon tetrachloride	0.5
Chlordane	0.03
Chlorobenzene	100.0
Chloroform	6.0
Chromium	5.0
o-Cresol	200.0
m-Cresol	200.0
p-Cresol	200.0
Cresol	200.0
2,4-D	10.0
1,4-Dichlorobenzene	7.5
1,2-Dichloroethane	0.5
1,1-Dichloroethylene	0.7
2,4-Dinitrotoluene	0.13
Endrin	0.02
Heptachlor (and its epoxide)	0.008
Hexachlorobenzene	[3]0.13
Hexachlorobutadiene	0.5
Hexachloroethane	3.0
Lead	5.0
Lindane	0.4
Mercury	0.2
Methoxychlor	10.0
Methyl ethyl ketone	200.0
Nitrobenzene	2.0
Pentachlorophenol	100.0
Pyridine	[3]5.0
Selenium	1.0
Silver	5.0
Tetrachloroethylene	0.7
Toxaphene	0.5
Trichloroethylene	0.5
2,4,5-Trichlorophenol	400.0
2,4,6-Trichlorophenol	2.0
2,4,5-TP (Silvex)	1.0
Vinyl chloride	0.2

TABLE 3.2. Comparison of EP Toxicity and TCLP Leaching Tests

Topic	EP	TCLP	Comments
1. Extraction device	Rotary mixer or tumbler	Tumbler at 30 ± 2 rpm	—
2. Structural integrity	Test with falling weight for monolithic waste	*All* samples to be crushed, ground, or broken to pass through ³/₈-in. sieve	Could change leachability of waste (surface area to liquid ratio)
3. Filtering	Use of 0.45-μm filter	Use of 0.6- to 0.8-μm filter	More particulates will pass and may change results from EP test
4. Leaching period	24–28 hr	18 hr	—
5. Leaching method	Manual pH adjustment	All leachant added at start of test	—
6. Leachant media	0.5 *N* acetic acid	Two media	
		0.5 *N* acetic acid	Acetic acid for highly alkaline waste only
		Acetate buffer at pH 5	Acetate buffer may change leachability of some constituents
7. Temperature	Room temperature	22 ± 3°C	Need to control temperature
8. Highly alkaline waste	Procedure same for all wastes	Screening procedure to be used to identify highly alkaline wastes and leachant to be used	Adds time to test
9. Quality control requirements	All analyses by standard additions	Standard additions only required where accuracy is <50% or >150% or when test results are within 20% of regulatory level	More reasonable approach but no time saver
10. Leaching bottles	Plastic or glass	Teflon, glass, or ZHE	—
11. ZHE (zero headspace extractor)	Not used	Used for volatiles	Separate device and leaching procedure
12. Laboratory analysis of leachates	Smaller number of chemicals	Many more chemicals to test	Analytical problems anticipated due to wide range of matrices

After Duranceau (1987); courtesy of Madison waste conf. committee.

during the test to maximize contaminant levels in the extract. Currently this test is not used widely.

TCLP Test. The toxicity characteristic leaching procedure (TCLP) was developed by USEPA. Table 3.1 lists the 40 chemicals for which the TLCP test can be used and their toxic concentration level. The TCPL test uses two leaching media: pH 5 acetate buffer and 0.5 *N* acetic acid. The choice of leaching medium depends on whether the waste is classified as highly alkaline. The 0.5 N acetic acid is to be used for highly alkaline waste and the pH 5 acetate buffer is recommended for all other waste types. A comparison of the EP toxicity test (a test procedure used by USEPA that was subsequently discontinued by the agency) with the TCLP test is included in Table 3.2 (Duranceau, 1987). The toxic concentration levels for EP toxicity tests were chosen arbitrarily, whereas the toxic concentration levels for the TCLP test are based on toxicological health data and groundwater models. A device known as a Zero-Headspace Extractor (ZHE) is used to capture volatile compounds released during the test (Fig. 3.4).

Synthetic Precipitation Leachate Procedure. The extraction fluid used in the method is dictated by the site location and whether leachability of cyanide and volatiles needs to be determined. The following three types of fluids are used in this method: (1) Fluid #1: a 60/40 wt% mixture of sulfuric and nitric acids is mixed with reagent water (ASTM type 2 or equivalent) until the pH is 4.2 ± 0.5. The fluid is used to determine the leachability of waste disposed

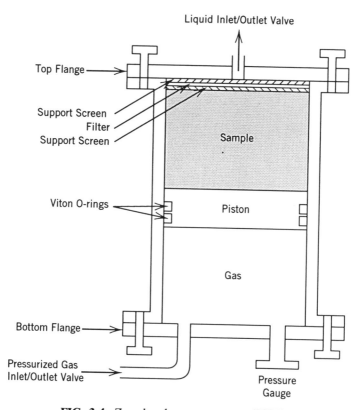

FIG. 3.4. Zero-headspace extractor (ZHE).

in a site east of the Mississippi River in the United States. (2) Fluid #2: the composition of the fluid is the same as fluid #1 except that the pH is adjusted to 5.00 ± 0.05. The fluid is used to determine leachability of waste in a site west of the Mississippi River in the United States. (3) Fluid #3: this is reagent water (ASTM type II or equivalent) that is used to determine cyanide and volatiles leachability from a waste. Figure 3.5 shows a flow chart used for the test.

3.3.2 Leachate Study Using Lysimeters

Rectangular or cylindrical lysimeters are used to study refuse leaching. Lysimeter studies can be conducted indoors or outdoors. Several researchers used lysimeters to study leaching of municipal refuse (Quasim and Burchinal, 1970b; Rovers and Farquhar, 1973; Pohland, 1975; EMCON Associates, 1975; Wigh, 1979; Walsh and Kinman, 1979; Fungaroli and Steiner, 1979; Ham, 1980). Both leachate quality and quantity can be monitored by using lysimeter study. Large lysimeters (15–60 m long × 6–9 m wide × 2.5–3 m deep) constructed outdoors and exposed to natural climatic conditions will provide valuable data if studied for a long period of time (5–6 years minimum). Refuse is not stabilized in a short period of time and hence data from a short-term lysimeter study, even if it is constructed outdoors, will not provide reliable results.

3.3.3 Predictive Modeling

Two approaches to model leachate composition have been reported in the literature. In one approach an attempt is made to quantify the physical, chemical, and biological processes. In the other approach an attempt is made to develop empirical equations to predict leachate concentration with time. Quasim and Burchinal (1970a) attempted physical process modeling to predict the concentration of leachate constituents. The experimental and theoretical concentrations of leachate constituents were fairly close. Straub and Lynch (1982a,b) proposed a model to predict both leachate quantity and concentration of both inorganic and organic constituents. It is essentially a one-dimensional kinetic model. Demetracopoulas et al. (1986) proposed a mathematical model that is solved numerically. It predicts a hydrograph-like contaminant concentration history at the bottom of the landfill that showed good qualitative agreement with the measured concentration history. Revah and Avnimeleik (1979) developed an empirical relationship to model the variations of concentration with time for the following leachate constituents: organic carbon, TKN, NH_3, NO_3, Fe, Mn, and volatile acids.

Leachate modeling has not developed enough to predict field leachate quality. However, the usefulness of leachate modeling must not be underestimated.

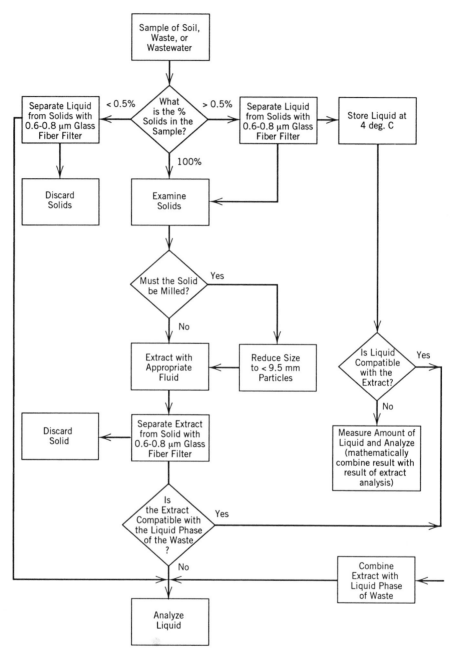

FIG. 3.5. Synthetic Precipitation Leachate Procedure Flow Chart.

3.4 ESTIMATION OF LEACHATE QUANTITY

Leachate quantity depends heavily on precipitation, which is difficult to predict. The preclosure and postclosure leachate generation rates in a landfill vary significantly and the methods used to calculate them are also different. An estimation of the preclosure leachate generation rate is needed to determine the spacing of the leachate collection pipe at the base of the landfill, the size of the leachate collection tank, and the design of an on site/off site plant for treating the leachate. Based on a field study of 13 municipal landfills in north West Germany Ehrig (1983) reported the following leachate generation rate, which depends on type of compactor used for compacting the waste: 15–25% of annual precipitation if steel wheel compactors are used; 25–50% of annual precipitation if crawler tractors are used. An estimation of the postclosure leachate generation is needed primarily to determine the long-term care cost (refer to Section 12.1.3 for details). The leachate generation rate is higher during the active life of the landfill and is reduced gradually after construction of the final cover. The following sections discuss how to estimate the preclosure and postclosure leachate quantity.

3.4.1 Preclosure Generation Rate

Leachate is generated primarily as a result of the precipitation and squeezing out of pore liquid in waste disposed in the landfill. Decomposition of putrescible waste mass can also release water/liquid. In a study conducted in a California landfill, the leachate generated due to decomposition from water was reported to be 0.5 in./ft of waste (CWPCB, 1961). For practical design purposes the volume of leachate generated due to decomposition from water is negligible. Surface run-on water may also cause an increase in leachate quantity (Lu et al., 1985); however, in a properly designed landfill surface water should not be allowed to run on into the waste. So this issue is also not addressed here. However, if for an existing landfill surface run-on water is unavoidable, then the volume of run-on water must be estimated using principles of hydrology to calculate the volume of leachate. The preclosure leachate generation rate is guided by Eq. (3.1). However, in reality a model needs to be used to predict the preclosure leachate generation rate (refer to Section 6.5).

$$L_v = P + S - E - AW \tag{3.1}$$

where L_v = preclosure leachate volume
$\quad S$ = volume of pore squeeze liquid
$\quad P$ = volume of precipitation
$\quad E$ = volume lost through evaporation
AW = volume lost through absorption in waste.

It is difficult to estimate S, E, and AW in a real landfill. Discussion of all the variables except precipitation follows.

Leachate Volume due to Pore Squeeze. When a layer of sludge is disposed in a landfill, the liquid within the pores of the sludge layer is released due to the self weight of the sludge and the weight of the layers above it. The pore water is released essentially because of the consolidation of sludge. Both primary and secondary consolidation can take place. Although secondary consolidation may be high for putrescible waste, partly due to the creep of fibers and partly due to the microbial decomposition of organic matter present in the waste, the total volume of liquid drained due to pore squeeze is not expected to be high. Usually primary consolidation accounts for the majority of the pore squeeze liquid, which can be predicted reasonably well using laboratory values (Charlie and Wardwell, 1979). Charlie and Wardwell (1979) developed a mathematical relationship between the leachate generation rate and the primary consolidation properties of sludge.

Terzaghi's one-dimensional consolidation theory was used in developing the relationship. Although doubt was expressed regarding the use of Terzaghi's consolidation theory in predicting the settlement (Bagchi, 1980), studies conducted by Mar (1980) on municipal digested sludge supported Charlie and Wardwell's (1979) approach of predicting the leachate generation rate. Usually the following laboratory testing is used to predict leachate generation from sludge: the sludge is placed in a mould (usually a proctor's mould) and pressure is applied on the sludge that is equal to the anticipated maximum weight of the sludge in the field. The pressure is applied for several days and the settlement at the end of the period is recorded. It is assumed that the settlement is solely due to release of pore liquid. Based on this assumption the field leachate volume is estimated for the entire sludge volume to be disposed in the landfill. Solseng (1978) proposed a different approach for estimating pore squeeze liquid using standard consolidation data. Pore squeeze liquid from mechanically pressed sludges will be negligible because the pressure applied to the sludge mechanically is much higher than the pressure on the sludge after disposal. So the landfill designer needs to know whether the sludge is mechanically pressed prior to disposal. Estimation of leachate quantity in a codisposal situation (e.g., disposal of sludge and municipal waste in the same landfill) is difficult. The absorbing capacity of the municipal waste will influence the volume of leachate generated.

Loss of Leachate due to Evaporation. Precipitation moisture or the moisture already present in a landfill may evaporate under favorable conditions. Evaporation depends on factors such as ambient temperature, wind velocity, difference of vapor pressure between the evaporating surface and air, atmospheric pressure, and the specific gravity of the evaporating liquid. A 1% decrease in evaporation rate due to each 1% rise in specific gravity of the evaporating liquid has been reported (Keen et al., 1926; Fisher, 1927; Penman, 1948; Veihmeyer and Henderickson, 1955; Chow, 1964). Soil tends to bind water molecules by an attractive force that depends on the moisture content of the soil and its characteristics. The evaporation rate of unsaturated soils is almost constant over a range of moisture content of the soil. A shallow surface layer of soil (approximately 10 cm for clays and 20 cm for sand) will continue to evaporate until the layer reaches a permanent wilting point (the point at which the moisture content of the soil prevents the soil from supplying water at a sufficient rate essentially due to intermolecular surface tension). Evaporation from deeper soil is negligible (Chow, 1964). The water budget method, energy budget method, and mass transfer techniques have been used to predict evaporation from open water bodies (Viessman et al., 1977). As discussed above, evaporation opportunity depends on the availability of water. It is 100% from saturated soil but nearly zero from dry soil. Evaporation from open water bodies can be measured directly by pan evaporation. In the United States usually an unpainted galvanized iron pan 4 ft in diameter and 10 in. in height is used in determining pan evaporation. The pan is mounted 12 in. above the ground, on a wooden frame. The evaporation observed from the pan is multiplied by a factor of 0.67–0.81, known as the pan coefficient, to determine evaporation from large open water bodies such as lakes (Linsley and Franzini, 1972). The estimate must be based on long-term observations to avoid significant error. The average evaporation from an active landfill surface will be much lower than pan evaporation because of unsaturated conditions.

Loss of Leachate due to Absorption in Waste. Waste may absorb some moisture before allowing it to percolate through. Theoretically once the field capacity of the refuse is reached all precipitation that falls on the waste will show up as leachate. The field capacity of the waste is defined as the maximum moisture content that waste can retain against gravitational forces without producing a downward flow of liquid. However, moisture absorption by waste is not uniform. A high heterogeneity of waste mass exists in a landfill; as a result channeling of precipitated water occurs in a landfill. The absorptive capacity of the waste depends on the composition of the waste. A detailed study of water absorption capacity of waste components was conducted by Stone (1974). The study indicated that field capacity of any refuse can be estimated with reasonable accuracy if the relative percentage of each waste component is known. The initial moisture content and field capacity of municipal solid waste as reported in several studies are summarized in Table 3.3 (Rovers and Farquhar, 1973; Walsh and Kinman, 1979, 1981; Wigh and Brunner, 1981; Fungaroli and Steiner, 1979). The data in Table 3.3 indicate that on average, a field capacity of 33 cm/m (4 in./ft) is reasonable for municipal solid waste. From Table 3.3 the average initial moisture content of municipal solid waste can be assumed to be 12 cm/m (1.5 in./ft). Thus, on average, a municipal waste can absorb an additional 21 cm/m (2.5 in./ft) of moisture. However, in actual field situations absorption of moisture to

TABLE 3.3. Summary of Field Capacity of Waste

S1 Number	Data Source	Wet Density (lb/yd³)	Dry Density (lb/yd³)	Initial Moisture Content (in./ft)	Field Capacity (in./ft)
1	Rovers et al. (1973)	530	Not available	1.92	3.62
2	Walsh et al. (1979)	808	526	2.0	3.82
3	Walsh et al. (1981)	798	520	1.98	4.85
4	Wigh (1979)	658	510	1.0	4.4
5	Fungaroli (1979)	563	476	0.62	4.1

the full field capacity is reduced due to channeling. Sludges are mostly saturated, hence reduction in leachate volume due to absorption may be neglected for sludges. The field capacity would be very low for a sandy nonputrescible waste (e.g., foundry sand). Thus, loss of moisture due to absorption depends on waste type—a point that should be borne in mind while estimating preclosure leachate generation rate. Sequencing of waste placement to allow maximum moisture absorption can reduce leachate quantity.

3.4.1.1 Computer Model. Two computer-based models are available for predicting preclosure leachate generation rate (Schroeder et al., 1984; Bagchi and Ganguly, 1990). A study reported by Mbela et al. (1991) for four Wisconsin landfills indicates that error range in predicting preclosure leachate generation rate using the Help model (Schroeder et al., 1984) is between 84.1 and 196.7%. The error range using a model reported by Bagchi and Ganguly (1990) for two of the four landfills studied by Mbela et al. (1991), is between −65.2 and −7.7%. Reasonable accuracy in prediction of preclosure leachate generation on a daily basis is needed in the following two situations: (1) if the leachate is to be treated in relatively small municipal or industrial waste water treatment plants; and (2) if a pretreatment or onsite treatment plant is needed for treatment of the leachate.

3.4.2 Postclosure Generation Rate

After the construction of the final cover only the water that can infiltrate through the final cover percolates through the waste and generates leachate. Five approaches are available to predict the long-term leachate generation rate: the water balance method, computer modeling in conjunction with water balance method, empirical equation, mathematical modeling, and direct infiltration measurements. Descriptions of each of these methods and summary comments are included in the following sections.

3.4.2.1   Water Balance Method.   Up to early 1980s the water balance method was used to predict the long-term leachate generation rate. In simple terms the water balance equation can be written as

$$L_v' = P - ET - R - \Delta S \tag{3.2}$$

where L_v' = postclosure leachate volume
   P = volume of precipitation
   ET = volume lost through evapotranspiration
   R = volume of surface runoff
   ΔS = volume of soil and waste moisture storage
When precipitation falls on a covered landfill, part of it runs off the surface (R) and part of it is used up by vegetation (ET). The remaining part infiltrates the cover (Fig. 3.6), but part of it is held up by soil and waste (ΔS). The water balance method is applicable only for landfills in which a relatively high permeable layer of soil is used as final cover. A significantly lesser amount of water will infiltrate into a landfill if it is covered with a low permeability clay layer or synthetic membrane.

3.4.2.1.1   Evapotranspiration.   Evapotranspiration is a term that combines evaporation and transpiration. Evaporation, discussed in detail in Section 3.4.1, is the loss of water that occurs from the soil surface. Transpiration on the other hand is the loss of water from the soil due to uptake by plants and its subsequent partial release to the atmosphere. Because of the difficulties in measuring the two items separately, they are measured as one item and termed evapotranspiration. Since the goal in a water budget is to "predict" the future leachate generation rate, potential evapotranspiration rather than actual evapotranspiration is of interest to the designer. Essentially two methods are available for predicting potential evapotranspiration.

USE OF AN EMPIRICAL RELATIONSHIP.   The rate of transpiration is approximately equal to the pan evaporation rate from a free water surface reduced by the pan evaporation coefficient, provided plants continue to get sufficient water (Chow, 1964; Linsley and Franzini, 1972). Because the type of vegetation greatly influences evapotranspiration (Foth and Turk, 1943), the approach may over- or underestimate potential evapotranspiration.

EMPIRICAL/THEORETICAL   APPROACHES.   Several   empirical/theoretical equations are available for estimating the potential evapotranspiration rate (Veihmeyer, 1964). A brief description of the equations used to predict monthly/daily evapotranspiration rates is given below.

Blaney–Morin Equation.   This equation, proposed in 1942, essentially predicts evapotranspiration empirically using percentage daytime hours, mean monthly temperature, and mean monthly relative humidity. The equation takes into account the seasonal consumptive use of several irrigated crops.

Thornthwaite Equation. This equation, originally proposed in 1944, uses an exponential relationship between mean monthly temperature and mean monthly heat index. This method for predicting evapotranspiration was further developed by providing additional tables necessary for calculation (Thornthwaite and Mather, 1957). The relationship is based on studies conducted mostly in the central and eastern United States. The method is widely used to predict evapotranspiration from landfill cover. A detailed discussion on how to use the method for estimating leachate production in landfills is provided by Fenn et al. (1975).

Penman Equation. This is a theoretical equation based on absorption of radiation energy by ground surface. The values of variables used in the equation can be obtained from graphs and tables found elsewhere (Veihmeyer, 1964). Daily evapotranspiration can be calculated using this equation. This method is also widely used to predict evapotranspiration from landfill covers.

Blaney–Criddle Equation. This is a revised form of the Blaney–Morin equation that does not consider the annual mean relative humidity used in the Blaney–Morin equation.

As mentioned earlier, of all these equations Thornthwaite's and Penman's equations are most widely used. Thornthwaite's equation requires extensive use of tables that were developed mostly from observations in the central and eastern United States. However, because the studies were performed at different locations, the effect of latitude on evapotranspiration can be accounted for by using this approach. Thornthwaite and Mather (1955) recognized that evapotranspiration is dependent on root zone and vegetation type and may vary 400-fold from one location to another. They therefore issued a caveat regarding the use of the tables for precise estimates. Since the tables were developed mostly from observation stations in the United States, its worldwide application may not be useful, a fact recognized by the authors of the method.

3.4.2.1.2 Surface Runoff. Approaches for estimating surface runoff are different for water and snow. Surface runoff for water is discussed under the headings "Field Measurement" and "Empirical Relationship." For snow, the infiltration rather than runoff from snow melt is estimated, which is discussed under the heading "Snow Melt."

FIELD MEASUREMENT. For field measurement of surface runoff a test plot needs to be fenced to collect the runoff from the enclosed area. A precipitation gauge must be located next to the fenced area to measure precipitation at definite intervals of time (but not more than an hour apart). Several areas, with different slopes but each with the same type of topsoil and vegetation as proposed for the landfill, must be studied. The need for this type of study cannot be justified because runoff from different soils and slopes can be predicted fairly accurately using empirical relationships described in the next section. However, if the designer is certain that at the location where the landfill is to be sited, use of empirical equations will not provide reasonable runoff estimates, then use of the field measurement technique to estimate surface runoff is justified.

EMPIRICAL RELATIONSHIP. These relationships were essentially developed from extensive field measurements. Several methods are available for surface runoff measurements (Chow, 1964; Varshney, 1979). However, only the two methods widely used in the United States are discussed. F.P.S. units are used for both methods.

Rational Method. The following equation is used to calculate peak surface runoff (*R*) in ft^3/sec.

$$R = CIA_s \qquad (3.3)$$

where I = uniform precipitation rate in inches
A_s = area of the landfill surface in acres
C = runoff coefficient.
The surface runoff can be predicted fairly accurately if a proper value of C is chosen. Different sets of values for C for different surface conditions are available [American Society of Civil Engineers (ASCE), 1960; Chow, 1964; Perry, 1976; Salvato et al., 1971]. Of these sets of values the ones by Salvato et al. (1971) were part of a landfill study. The rational method cannot account for the relationship between duration of precipitation and runoff, antecedent soil moisture content, frequency of precipitation, and permeability of the cover material.

Example 3.1 (in F.P.S. units)

Calculate the surface runoff for a 10.5-acre landfill. Based on precipitation data, the 10-year 24-hr storm intensity is found to be 2.7 in./hr; the landfill has a cover that consists of the following layers: 1 ft of sand over the waste, 2 ft of recompacted clay, 2.5 ft of silty sand, and 6 in. of topsoil. The landfill has good vegetative cover and the top slope varies between 2 and 5%.

The surface runoff is over a sandy loam with grass cover; the surface slope is 2–5%. From Table 3.4 the value of C is between 0.15 (sandy soil with a 2–7% slope) and 0.22 (heavy soil with a 2–7% slope). Assume an average value of 0.18. [Note: For this case, values of C obtained from other sources (Chow, 1964; Perry, 1976; Salvato et al., 1971) vary between 0.3

TABLE 3.4. Runoff Coefficients for Storms of 5- to 10-Year Frequency

S1 Number	Description of Area	Runoff Coefficients
1	Unimproved areas	0.10–0.30
2	Lawns: sandy soil	
	Flat, 2%	0.05–0.1
	Average, 2–7%	0.1–0.15
	Steep, 7%	0.15–0.2
3	Lawns: heavy soil	
	Flat, 2%	0.13–0.17
	Average, 2–7%	0.18–0.22
	Steep, 7%	0.25–0.35

After American Society of Civil Engineers (ASCE) (1960).

and 0.45.] It is a good idea to minimize surface runoff while predicting leachate volume and to maximize surface runoff when designing storm water drainage systems.

For the example landfill, $C = 0.18$, $I = 2.7$ in./hr, $A_s = 10.5$ acres; $R = 0.18 \times 2.7 \times 10.5 = 5.1$ ft^3/sec.

Curve Number Method. The curve number method proposed by the Soil Conservation Service of the United States is used to predict surface runoff from agricultural land [Soil Conservation Service (SCS), 1975]. In addition to rainfall volume, soil type, and land cover, the method accounts for land use and antecedent moisture conditions. The antecedent moisture condition is first divided into three groups based on season (dormant and growing) and a 5-day total antecedent rainfall in inches. Soil is grouped into four different types based on ability to cause runoff (e.g., clayey soil has high runoff potential and sand or gravel has low runoff potential; all other soil types are classified between these two extremes). The land use and land cover are then determined. The weighted curve number is then established using tables. The direct runoff can then be estimated for different rainfall using Eq. (3.4):

$$R_i = \frac{\{W_p - 0.2[(1000/CN) - 10]\}^2}{W_p + 0.8[(1000/CN) - 10]} \tag{3.4}$$

where R_i = surface runoff in inches, W_p = rainfall in inches, and CN = curve number.

Example 3.2

A landfill has a surface area of 4.3 acres and has a 2-ft sandy silt final cover with 6 in. of topsoil. The landfill has poor pasture cover. Assume that the 10-yr 24-hr storm intensity for the area is 2.65 in./hr, which occurred in a dormant season. The total 5-day antecedent rainfall was 0.45 in. Estimate surface runoff using the curve number method.

1. From Table 3.5 the antecedent moisture condition = AMC I.
2. From Table 3.6 the hydrologic soil group is C.
3. From Table 3.7 CN = 86 for AMC II.
4. From Table 3.8 obtain CN for AMC II to CN for AMC I = 72.
5. From Eq. (3.4) for CN = 72 and W_p = 2.65 in. the direct runoff = 0.8 in.

3.4.2.1.3 Snow Melt Infiltration. In many areas leachate generated as a result of infiltration during snow melt is significant. The majority of snow melt usually occurs in early spring. Infiltration from snow depends on the condition of the ground (frozen or unfrozen), ambient temperature and its duration (snow melt will depend on whether a temperature of 32°F and above prevails for 1 day or several days), radiation energy received (more snow melt occurs on sunny days than on a cloudy day), rainfall during snow melting (rainfall accelerates the snow melting process), and so on. Because of the variables involved it is difficult to predict snow melt runoff or infiltration. Two methods are usually used: the degree day method and the U.S. Army Corps of Engineers equation. Because it is simpler, only the degree day method is discussed here. A detailed discussion of the U.S. Army method can be found elsewhere (Lu et al., 1985; Chow, 1964).

DEGREE DAY METHOD. The following equation is used to estimate snow melt infiltration (SCS, 1975)

$$SM = K(T - 32°F) \tag{3.5}$$

where SM = potential daily snow melt infiltration in inches of water
K = constant that depends on the watershed condition
T = ambient temperature above 32°F
$T - 32°$ is the number of degree per day. The total snow melt infiltration predicted must not exceed the total water equivalent of precipitated snow (note: 1 in. of water = 10 in. of snow).

TABLE 3.5. Antecedent Moisture Class for 5-Day Rainfall

| S1 Number | 5-Day Antecedent Rainfall (in.) | | Moisture Condition Class |
	Dormant Season	Growing Season	
1	<1.1	>2.1	III
2	0.5–1.1	1.4–2.1	II
3	<0.5	<1.4	I

After Soil Conservation Service (SCS) (1972).

TABLE 3.6. Soil Groups Relevant to Landfill Cover Design

S1 Number	Description of Soil	Soil Group
1	Soils having moderate infiltration rates when thoroughly wetted; moderately coarse to moderately fine textured soil	B
2	Soils having slow infiltration rates when thoroughly wetted; moderately fine to fine textured soil	C
3	Soils having very slow infiltration rates when thoroughly wetted; chiefly clayey soils with low permeability	D

After Soil Conservation Service (SCS) (1972).

Example 3.3 (in F.P.S. units)

Estimate the snow melt infiltation from an 18-in. snow pack during spring. The average daily air temperatures for the next 5 days were 33°, 34°, 29°, 31°, and 36°F.

From Table 3.9, the value of $K = 0.02$. (Note: assume low runoff to maximize leachate production.)

$$18 \text{ in. of snow} = 18 \times (1/10) = 1.8 \text{ in. of water}$$

$$\begin{aligned}
\text{Total expected infiltration} &= 0.2 \times [(33 - 32) \\
&\quad + (34 - 32) + (36 - 32)] \\
&= 0.14 \text{ in.}
\end{aligned}$$

TABLE 3.7. Runoff Curve Number for Different Soil Groups and Land Use Conditions Relevant to Landfill Cover Design

S1 Number	Land Use	Agricultural Practice	Hydrologic Condition[a]	Soil Groups B	C	D
1	Fallow	Contoured	Poor	79	84	88
		Contoured	Good	75	82	86
		Contoured, terraced	Poor	74	80	82
		Contoured, terraced	Good	71	78	81
2	Pasture or range	Contoured	Poor	57	81	88
		Contoured	Fair	59	75	83
		Contoured	Good	35	70	79

After Soil Conservation Service (SCS) (1972).

[a] Poor hydrologic conditions mean heavily grazed, with no mulching a surface or less than 50% of the area covered with plants. Fair hydrologic conditions mean moderately grazed with plant cover or 50–75% of the area. Good hydrologic conditions mean lightly grazed with plant cover on more than 75% of the area.

TABLE 3.8. Runoff Curve Numbers (*CN*) Relevant to Landfill Cover Design

	CN for Antecedent Moisture Condition	
CN for Condition II	I	II
90	78	96
88	75	95
86	72	94
84	68	93
82	66	92
80	63	91
78	60	90
76	58	89
74	56	88
72	53	86
70	51	85
68	48	84
66	46	82
64	44	81
62	42	79
60	40	78
58	38	76
56	36	75
54	34	73
52	32	71
70	31	70
48	29	68
46	27	66
44	25	64
42	24	62
40	22	60
38	21	58
36	19	56
34	18	54

After Soil Conservation Service (SCS) (1972).

(Note: All temperatures below 32°F are to be neglected because snow will not melt if the temperature is 32°F or below.

3.4.2.1.4 Soil Moisture Storage. Part of the infiltrating water will be stored by the soil. Only part of this stored water is available for use by vegetation. Soil moisture storage capacity is expressed as

$$\Delta S = \text{Field capacity} - \text{wilting point} \qquad (3.6)$$

TABLE 3.9. Values of K for the Degree Day Equation Relevant to Landfill Cover Design

S1 Number	Watershed Condition	K
1	Average heavily forested area	
	North facing slopes	0.04–0.06
2	South facing slopes	0.06–0.08
3	High runoff potential	0.03

After Soil Conservation Service (SCS) (1972).

Soil moisture storage depends on soil type, state of compaction, and thickness of the soil cover. Lutton et al. (1977) published data for soil moisture storage for different types of soil. The effect of state of compaction is not accounted for in this study.

3.4.2.1.5 Comments on the Water Balance Method. The preceding paragraphs included discussion on how to estimate evapotranspiration, surface runoff and soil storage. Postclosure leachate volume production is calculated by subtracting evapotranspiration, surface runoff, and soil and waste moisture storage from the precipitation for each day. Use of a computer facilitates calculation. Test cell data and other field verification of the water balance method indicate that the margin of error is very high. Wigh (1979) and Wigh and Brunner (1981) found a difference of 43% between the leachate volume predicted using average climatic data and the actual leachate collection. In general, test cell data for the verification of the water balance method are in good agreement (Fungaroli and Steiner, 1979; Walsh and Kinman, 1981). SCS engineers (1976) collected leachate generation data from five existing sites located in five different geographic areas. Twenty-five difference methods were used to predict leachate generation; different combinations of methods were used to estimate surface runoff, infiltration, and evapotranspiration. Thus, in all, the study reports on 125 individual cases (25 methods × 5 sites). The data indicate that in 54 cases (42.4%) the leachate generation was underestimated and in 71 cases (56.8%) the leachate generation was overestimated. The average error for the 25 methods varied between 83 and 1543%. So the study shows that use of the water balance method for predicting the leachate generation rate can be highly erroneous. The error range for predicting the leachate generation rate using the HELP model (Schroder et al., 1984) is between −96 and +449% (Peyton and Schroeder, 1988). In the past the water balance method was used widely for estimating the leachate generation rate and more than 100 approaches are available for its calculations. A study is available which compares all the approaches and identifies a method for estimating the long-term leachate generation rate (Lu et al., 1985). Dass et al. (1977) and Perrier and Gibson (1980) conducted a parametric study to determine the sensitivity of the predicted leachate volume. These studies show that the sensitivity is moderate to high for all of the parameters involved in the water balance equation. Therefore, caution should be exercised in using the water balance method for predicting the long-term leachate generation rate.

The following are the drawbacks of using the water balance method alone for estimating postclosure leachate generation:

1. The method does not take into account the permeability of the barrier layer, which is an essential design feature for minimizing leachate generation.
2. In an attempt to maximize evapotranspiration, a designer may use a very thick vegetative layer without realizing that the root system of the chosen vegetation may not penetrate the entire thickness of the layer. In addition, one may note that approximately 80% of the moisture uptake from soil by a root system is performed by only the top few centemeters of roots of grass/shrub type vegetation (Foth and Turk, 1943) commonly used in landfill cover. So the total length of root should not be used for estimating evapotranspiration. Thus increasing the vegetative layer beyond a certain depth will not increase evapotranspiration and thus will not reduce infiltration by increasing evapotranspiration.

3.4.2.2 Computer Modeling in Conjunction with the Water Balance Method. Several computer models available for predicting the postclosure leachate generation rate. The HELP model developed by Schroeder et al. (1984) uses the water balance equation but also considers permeability of the barrier layer to predict infiltration into the landfill. In the model, apportionment of infiltrating water after evapotranspiration, into vertical percolation through the barrier layer, and horizontal runoff through the barrier layer are estimated (Fig. 3.6). A simulation study by Peyton and Schroeder (1988) indicated that the HELP model can predict leachate generation fairly accurately. However, it is interesting to note that in the study the input of model variables was based on judgment, making the validity of the prediction questionable. In addition, the model has to share the inaccuracies associated with the water balance method. Miller and Mishra (1989) also raised doubts about several aspects of the verification process. Of most importance were the inability of the model to simulate infiltration under unsaturated condition and the effect of macropores (due to desiccation cracks) on the infiltration rate. In their response Peyton and Schroeder (1989) indicated that the HELP model computes unsaturated hydraulic conductivity using an equation proposed by Campbell (1974) and that the most appropriate use of the model is for comparison of designs rather than prediction of quantities. It may be

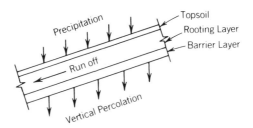

FIG. 3.6. Apportionment of precipitation through landfill cover.

mentioned, however, that the HELP model is used widely in the United States.

In addition, three computer-based models are also available for predicting postclosure leachate generation rate (Bagchi and Ganguly, 1990; Demetraco-poulas et al., 1984; Scharch et al., 1985). The models by Demetracopoulos et al. (1984) and Scharch et al. (1985) are modifications of Wong's model (Wong, 1977); in these models infiltration through the cover soil is an input that is estimated using a water balance method. Both models can predict the total postclosure leachate generation rate in a year but the error is high for an estimate of the daily leachate generation. The model reported by Bagchi and Ganguly (1990) needs field data regarding the postclosure leachate gener-ation rate from a closed landfill located within the same climatic region for calibrating the model (refer to Section 6.6 for a detail discussion of these models).

3.4.2.3 Empirical Equation. Gee (1986) proposed an empirical relationship for predicting percolation through the top cover:

$$\text{Percolation} = \exp[K + \beta_1 \ln W_p + \beta_2 \ln(W_i/W_f) \\ + \beta_3 \ln \gamma_d + \beta_4 \ln \alpha + \beta_5 C_c] \quad (3.7)$$

where W_p = rainfall in inches
$\quad \alpha$ = slope of the landfill surface (%)
$\quad W_i$ = initial soil moisture content of the cover (%)
$\quad W_f$ = field capacity of the soil (%)
$\quad \gamma_d$ = soil density in lb/ft^3
$\quad C_c$ = coefficient of curvature of the soil
β_1 to β_5 = constants.

The accuracy of the equation was judged against field observations. From the comparison of the equation with Thornthwaite's water balance method, the HELP model, and field data,it was concluded that the equation can predict percolation through the landfill cover fairly accurately.

3.4.2.4   Mathematical Model. Korfiatis et al. (1984) proposed a mathematical model using the theory of unsaturated flow. The model was tested in the laboratory. Comparisons of predicted and measured leachate quantity showed reasonable agreement.

3.4.2.5   Direct Infiltration Measurements. Published data are available regarding the average long-term percolation rate in soil in the northern United States. A 16-year average yearly percolation rate (in Ohio) in a silt loam was observed to be 17% (range 13–27%) (cited in Foth and Turk, 1943). Percolation into solid waste test cells was reported by Ham (1980). The average yearly percolation was about 18% for a 7-year period. Of this 18%, approximately 80% took place during the spring. Thus, an average yearly percolation rate of 20% may be used for rough estimation of the long-term leachate generation rate.

3.4.2.6   Summary Comments on Long-Term Leachate Generation. Prediction of the postclosure leachate generation rate is a difficult task. So far as landfill management goes, the long-term rate is needed to establish a fund for long-term maintenance of the landfill (refer to Section 12.2 for more details). The full advantage of evapotranspiration is not available until a stable vegetative growth is established over the entire landfill, which takes at least 1 to 2 years. The leachate volume remains high at least for those 1 or 2 years after closure. So the money set aside for long-term leachate treatment will be insufficient if evapotranspiration is taken into consideration to predict the postclosure leachate generation rate in the first few years. Thus, prediction based on models that use evapotranspiration becomes more of a theoretical exercise than of any practical use, at least for the first few years after closure. The infiltration rate may be estimated roughly by assuming a 20–30% of precipitation as infiltration from landfill cover to calculate the fund requirements for long-term care. The fund can be adjusted in subsequent years based on actual field data. This approach will provide a better safeguard against a low long-term care fund. As an alternative the author's model (refer to Section 6.6) which uses a field calibration technique may be used for better prediction of postclosure leachate generation rate. However, if regulatory requirements dictate the use of a water balance method for predicting the postclosure leachate generation rate then long-term climatic records (20–30 years) should be used for the estimation. The computer models mentioned in Section 3.4.2.2 are useful for parametric studies in determining a better final cover design option and for comparing landfill designs.

3.5   TYPICAL LEACHATE QUALITY OF VARIOUS NONHAZARDOUS WASTES

The ranges of concentrations of different parameters in leachate of various nonhazardous wastes are included in Tables 3.10 to 3.15. The aim is to provide some ideas about leachate quality for different waste types. As discussed earlier, leachate quality for any waste type is not unique. Hence the constituents and their concentrations are not recommended for automatic use for landfill design, but are provided here essentially to form a data base. The leachate quality for incinerator ash needs particular attention because the concentration of lead and cadmium may exceed the permissible level in some cases and therefore the ash may become hazardous (Sopcich and Bagchi, 1988; U.S. Congress, 1989).

3.6   LEACHATE TREATMENT

As discussed above, landfill leachate is highly contaminated liquid that cannot be discharged directly into surface water bodies. Studies have shown that discharge of raw municipal leachate into streams impacts aquatic life and causes degradation of water quality (Hansen, 1980; Nutall, 1973; Cameron and McDonald, 1982).

The following traditional techniques used for waste water systems are

TABLE 3.10.   Typical Leachate Quality of Iron Foundry Waste (Composite Sample)

S1 Number	Parameter	Concentration (mg/liter) (except as indicated)
1	Arsenic	0.113
2	Barium	<0.2
3	Cadmium	<0.01
4	Chloride	139
5	Total chromium	<0.05
6	COD	150
7	Cyanide	<0.05
8	Fluoride	1.60
9	Iron	<0.03
10	Lead	<0.005
11	Manganese	<0.01
12	Mercury	<0.0002
13	Nickel	<0.04
14	Phenols	0.118
15	Selenium	0.030
16	TDS	1990
17	Sulfate	5.1
18	Zinc	<0.01
19	pH	12.3 units

TABLE 3.11. Range of Concentration of Different Parameters in Leachate of Municipal Incinerator Ash

S1 Number	Parameters	Range of Concentration (mg/liter except as indicated)
1	Aluminum	2.3–88.8
2	Arsenic	0.005–0.218
3	Barium	0.055–2.48
4	Boron	0.42–3.2
5	Cadmium	<0.001–0.3
6	Calcium	21–3200
7	Chromium	<0.002–1.53
8	Cobalt	0.007–0.04
9	Copper	<0.005–24
10	Iron	<0.01–121
11	Lead	<0.0005–2.92
12	Magnesium	0.006–41
13	Manganese	0.103–22.4
14	Mercury	<0.00005–0.008
15	Molybdenum	<0.03
16	Nickel	<0.005–0.412
17	Potassium	3.66–4300
18	Selenium	0.0025–0.037
19	Silver	<0.001–0.07
20	Sodium	11.5–7300
21	Strontium	0.07–1.03
22	Tin	0.005–0.013
23	Zinc	0.002–0.32
24	Chloride	32.6–305
25	Fluoride	0.1–3.39
26	Hardness	49–742
27	Nitrate–nitrogen	0.011–0.59
28	Phosphate	0.16–0.43
29	Specific conductivity	253–1,874 μmho/cm
30	Sulfate	105–4900
31	Alkalinity	60.9–243
32	pH	8.47–9.94 units
33	Benzaldehyde	ND–0.008
34	Biphenyl	ND–0.051
35	Dimethyl propane diol	ND–0.120
36	Dioxins (ng/liter)	
	total	0.06–543
	2,3,7,8-TCDD	0.025–1.6
37	Ethyl hexyl phthalate	ND–0.08
38	Furans, total (mg/liter)	0.04–280
39	Hexa tiepane	ND–0.082
40	PCBs (ng/μl)	<1
41	Sulfonylbis sulfur	ND–0.011
42	Thiolane	ND–0.400

Based on Bagchi and Sopcich (1989) and U.S. Congress (1989).

TABLE 3.12. The Range of Concentration of Different Parameters in Leachate of Papermill Sludge

S1 Number	Parameter	Range of Concentration (mg/liter) (except as indicated)
1	pH	5.4–9.0 units
2	TDS	289–9,810
3	TSS	80–320
4	Conductivity	70–14,370 μmho/cm
5	Alkalinity	174–5,500
6	Hardness	682–6,600
7	BOD	36–10,000
8	COD	4–43,000
9	Sulfate	0.9–550
10	Sodium	9–4,500
11	Calcium	5.5–2,400
12	Aluminum	0.008–18
13	Chloride	1–1,200
14	Iron	<0.1–950
15	Zinc	<0.018–0.03
16	Color	1,315–38,300 color units
17	Turbidity	NR[a] turb. units
18	Phenols	0.0011–4.5
19	Tannin-lig	13–90
20	Kjeldahl-nitrogen	34.5–385
21	Ammonia-nitrogen	<0.1
22	Nitrate	<0.1–15
23	Nitrite	<0.01–0.018
24	Sulfite	4–64
25	Sulfide	ND[a]
26	Phosphate	0.11–0.58
27	Total volatile solids	211–483
28	Total fixed solids	144–266
29	Barium	0.011–1.1
30	Bromide	ND[a]
31	Cadmium	0.006–0.02
32	Chromium	0–0.15
33	Cobalt	0.005–0.014
34	Copper	<0.01–0.21
35	Lead	0.037–0.1
36	Magnesium	3.8–6,000
37	Manganese	0.1–200
38	Mercury	<0.01–7 μg/liter
39	Nickel	<0.005–0.024
40	Potassium	140
41	Selenium	75
42	Tin	<0.1
43	Titanium	0.04
44	Vanadium	<0.01
45	TOC	1,350
46	Silicon	<3
47	Phosphorous	0.65
48	Arsenic	0.029
49	Cyanide	0.017

After Benson (1980). [a] NR, not reported; ND not detected.

TABLE 3.13. The Range of Concentration of Different Parameters in Leachate of Coal Burner Fly Ash

S1 Number	Parameter	Range of Concentration (mg/liter except as indicated)
1	Aluminum	0.85–1.7
2	Antimony	<0.02
3	Arsenic	0.135–0.41
4	Barium	<0.1
5	Boron	1.8–2.3
6	Cadmium	0.01
7	Calcium	60–22
8	Chromium	0.03–0.29
9	Cobalt	<0.01–0.02
10	Copper	<0.01–0.04
11	Germanium	<3.0
12	Iron	0.07–0.24
13	Lead	<0.01–0.04
14	Magnesium	1.1–4.3
15	Manganese	0.01–0.05
16	Mercury	<0.009
17	Molybdenum	0.29–3.8
18	Nickel	0.03–0.06
19	Potassium	20–29
20	Rubidium	<0.09
21	Selenium	0.05–0.18
22	Silica	5.1–51.0
23	Sodium	9.0–50.0
24	Strontium	<0.04–0.99
25	Sulfur	53.3–222
26	Tin	<1.0
27	Titanium	<0.5
28	Uranium	<0.005
29	Vanadium	0.26–0.92
30	Zinc	0.02–0.04
31	Alkalinity ($CaCO_3$)	37–50
32	COD	6–16
33	Chloride	<1.0–1.7
34	Conductivity	409–1213 μmho/cm
35	Dissolved solids	390–1240
36	Hardness ($CaCO_3$)	216–596
37	pH	7.83–9.05 units
38	Phosphorus	0.04–0.08

TABLE 3.14. The Range of Concentration of Different Parameters in Leachate of Municipal Waste[a]

S1 Number	Parameter	Range of Concentration (mg/liter except as indicated)
1	TDS	584–55,000
2	Specific conductance	480–72,500 μmho/cm
3	Total suspended solids	2–140,900
4	BOD	ND–195,000
5	COD	6.6–99,000
6	TOC	ND–40,000
7	pH	3.7–8.9 units
8	Total alkalinity	ND–15,050
9	Hardness	0.1–225,000
10	Chloride	2–11,375
11	Calcium	3.0–2,500
12	Sodium	12–6,010
13	Total Kjeldahl nitrogen	2–3,320
14	Iron	ND–4,000
15	Potassium	ND–3,200
16	Magnesium	4.0–780
17	Ammonia-nitrogen	ND–1,200
18	Sulfate	ND–1,850
19	Aluminum	ND–85
20	Zinc	ND–731
21	Manganese	ND–400
22	Total phosphorus	ND–234
23	Boron	0.87–13
24	Barium	ND–12.5
25	Nickel	ND–7.5
26	Nitrate-nitrogen	ND–250
27	Lead	ND–14.2
28	Chromium	ND–5.6
29	Antimony	ND–3.19
30	Copper	ND–9.0
31	Thallium	ND–0.78
32	Cyanide	ND–6
33	Arsenic	ND–70.2
34	Molybdenum	0.01–1.43
35	Tin	ND–0.16
36	Nitrite-nitrogen	ND–1.46
37	Selenium	ND–1.85
38	Cadmium	ND–0.4
39	Silver	ND–1.96
40	Beryllium	ND–0.36
41	Mercury	ND–3.0
42	Turbidity	40–500 Jackson units

Based on McGinley and Kmet (1984), Lu et al. (1981), and Tharp (1991).

[a] Several bacteria and fungi species and several priority pollutants are found in the leachate.

TABLE 3.15. Range of Concentration of Different Parameters in Leachate of Construction/Demolition Waste[a]

S1 Number	Parameter	Range of Concentration (mg/liter except as indicated)
1	pH	6.5–7.3
2	Specific conductance (μmho/cm)	2,920–6,850
3	BOD, 5 day	100–320
4	COD	3,080–11,200
5	Total dissolved solids	2,412–4,270
6	Total suspended solids	1,000–43,000
7	Total organic carbon	76–1,080
8	Chloride	125–240
9	Calcium	148–578
10	Magnesium	92–192
11	Sodium	256–1,290
12	Potassium	118–618
13	Carbonate	0
14	Bicarbonate	2,090–7,950
15	Sulfate	<40
16	Fluoride	<0.1–0.4
17	Nitrate	4–13
18	Nitrite	—
19	Ammonia (N)	30–184
20	Phenolphthalein alkalinity ($CaCo_3$)	0
21	Alkalinity ($CaCo_3$)	1,710–6,520
22	Hardness ($CaCo_3$)	597–1,516
23	Phosphorus	2.5–3.89
24	Oil and grease	18–47
25	Iron	29–172
26	Filtered iron	0.24–11
27	Manganese	1–4.9
28	Boron	1.4–3.9
29	Cyanide	<0.10
30	Phenol	0.7–2.99
31	Nickel	0
32	Arsenic	0.017–0.075
33	Barium	1.5–8.0
34	Cadmium	0.02–0.03
35	Chromium	0.1–0.25
36	Hex chromium	0.18–4.92
37	Copper	0.14–0.49
38	Lead	0.22–2.13
39	Mercury	<0.002–0.009
40	Selenium	<0.001
41	Silver	<0.01–0.03
42	Zinc	1.7–8.63

Based on Norstrom et al. (1991).

[a] Leachate quality depends on type of demolition debris disposed in the landfill.

also used for treating landfill leachate: biological treatment (aerobic and anaerobic biological stabilization) and physical/chemical treatment (precipitation, adsorption, coagulation, chemical oxidation, and reverse osmosis). As indicated earlier, both short-term and long-term variability of leachate characteristics are expected. The variation of leachate characteristics makes the design of a treatment system difficult. Leachate from municipal waste landfills can have quite high BOD concentration and significant concentrations of metals and trace organics. The concentrations of chemicals in leachates of other types of waste (both hazardous and nonhazardous) are usually significantly high. Leachate from each landfill is unique; however, some generalization can be made for leachate based on waste type and landfill location. Thus, a general approach of treatment for a particular waste type from all landfills located in a region appears to be reasonable. Although bench scale studies may be needed for leachate from major landfills (especially from hazardous waste), certain general trends are being observed regarding leachate treatment. The following comments on the subject are based on the current trend in leachate treatment technology [Boyle and Ham, 1974; Carlson and Johansen, 1975; Cook and Foree, 1974; Uloth and Mavinic, 1977; Steiner et al., 1979; Stegmann, 1979; Rebhun and Galil, 1987; Chian and Dewalle, 1977; American Society of Civil Engineers/Water Pollution Control Federation (ASCE/WPCF), 1977; Cadena and Jeffers, 1987; Kremer et al., 1987; Metcalf & Eddy, Inc., 1979; Yong, 1986; Lange et al., 1987; Meidl and Peterson, 1987; Hoffman and Oettinger, 1987; McShane et al., 1986; Ying et al., 1987; Osantowski et al., 1989; Ehrig, 1984].

1. Generally landfill leachate is treated either in an on-site leachate treatment plant (common for extremely large municipal waste landfills and most hazardous waste landfills) or in an off-site existing waste water treatment plant (common for most nonhazardous waste, which also includes municipal waste landfills). In some instances pretreatment of the leachate (in an on-site or off-site treatment plant) is done and the effluent is discharged in an existing waste water treatment plant.

2. In many instances a combination of biological and physical/chemical processes is utilized for treating leachate.

3. Leachate with a high organic content is best treated with a biological process, whereas leachate with a low organic content is best treated with a physical/chemical process.

4. To avoid shock to a treatment plant leachate should be slowly introduced into the treatment stream. Necessary leachate storage at the treatment plant should be arranged if a slow introduction is envisioned. This type of storage is important where the available capacity of the treatment plant is low and the leachate is hauled to the treatment plant using trucks.

5. The projected variation of leachate quality and quantity with time (daily, seasonal, and long-term) needs to be communicated to the waste

water treatment plant designer/operator who is responsible for designing/maintaining the effluent quality of the treatment plant.

6. Investigation regarding treatability of leachate from a proposed landfill should be undertaken at an early stage. Current practice is to treat municipal waste leachate and most other types of nonhazardous waste leachate in a municipal waste water treatment plant. In many instances leachate from an industrial waste landfill is treated in the waste water treatment plant of the same industry. A detailed bench scale study is generally undertaken for hazardous waste landfills; in many instances an on-site pretreatment is used and the effluent is discharged in an existing waste water treatment plant.

7. Both granular and powdered activated carbon are found to be useful in removing organic compound from leachate.

8. A reverse osmosis technique is also used for leachate treatment (Longman, 1990).

Figure 3.7 shows a flow diagram of a leachate treatment system that combines physical, chemical, and biological processes. The flow diagram also includes air stripping, an option useful for Volatile Organic Compound (VOC) removal.

Recirculation of municipal leachate is promoted as a method for treating leachate. Although leachate recirculation reduces BOD and COD concentration, the concentration of metals and chloride increases (EMCON Associates, 1975; Robinson and Maris, 1985; Stegmann, 1979). It is argued that recirculation will reduce leachate volume due to increased evaporation and absorption in the waste. Problems such as reduction in permeability of the cover, perching of leachate, and odor have been reported by researchers (McGinley and Kmet, 1984; Robinson and Maris, 1985; Lechner et al., 1993). It appears that recirculation of municipal waste leachate may be successful initially but not in the long run. No data are available on recirculation of leachate from other waste types. Although research may be undertaken to study leachate recirculation for other waste types, recirculation does not appear to be a viable option for leachate treatment.

Land disposal for municipal leachate, as a means for treatment, has also been studied by some researchers (Chan et al., 1978; Bramble, 1973). Symptoms of toxicity were observed on plants grown in fields irrigated with municipal waste leachate (Menser and Winant, 1980). The concentrations of chemicals in leachate of most waste types are expected to be high. Leachate from those waste types, which needs to be disposed of in containment type landfills, will impact the groundwater if land disposed. Thus, in general, land disposing of leachate as a means of treating it is not a logical approach. If leachate from a landfill in which only one type of waste has been disposed is observed to be of such quality that it can be landspread or directly discharged into surface water bodies, then one may reconsider the need for disposal of the waste in a landfill.

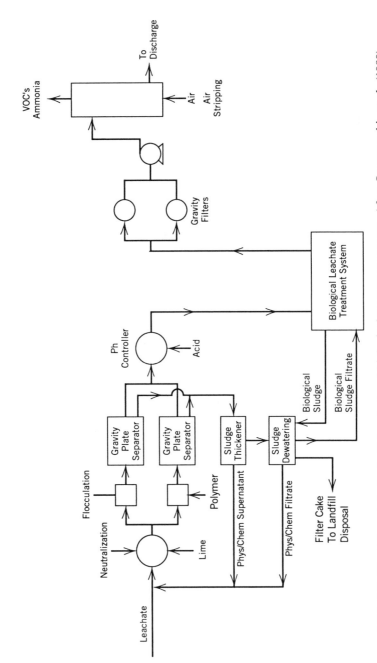

FIG. 3.7. The integrated physical/chemical and biological treatment system. After Osantowski, et al. (1989).

3.7   GAS GENERATION

Although gas generated within a few waste type landfills may be negligible (e.g., foundry waste), most waste type is expected to generate a significant quantity of gas. It should not be assumed that only putrescible waste can generate gas. Gas generation from nonputrescible waste should be studied carefully prior to designing a landfill for the waste. The quality of gas depends mainly on the waste type. As with leachate, the quality and quantity of landfill gas vary with time. The discussion on quality and quantity of gas that follows pertains mainly to municipal waste landfills.

There are usually three distinct but closely related stages of anaerobic digestion of biodegradable waste (e.g., municipal waste): (1) hydrolysis and fermentation due to bacterial activities; (2) acetogenesis and dehydrogenation; and (3) methanogenesis (Kayhanian et al., 1991). The first stage is dominated by the formation of acidic and propionic acid due to bacterial fermentation. In this stage the pH of the landfill leachate drops significantly along with a simultaneous rise of COD and BOD. In the second stage the soluble materials are oxidized to low-molecular-weight organic acids. Hydrogen gas is produced in this stage. In the third stage methane fermentation takes place, which primarily leads to the formation of methane and carbon dioxide; however, small amounts of other gases such as hydrogen sulfide, hydrogen, and nitrogen also form in this stage.

Within a relatively short period of time a stable methane phase is reached whereby gas composition (consisting mainly of methane, carbon dioxide, and a low percentage of nitrogen) and gas production rate remain constant

TABLE 3.16. Typical Composition of Stabililized Municipal Landfill Gas (Waste Volume <0.35 million m^3)

S1 Number	Parameter	Percentage or Concentration
1	Methane	30–53%
2	Carbon dioxide	34–51%
3	Nitrogen	1–21%
4	Oxygen	1–2%
5	Benzene	ND–32[a] ppm
6	Vinyl chloride	ND–44[a] ppm
7	Toluene	150[a] ppm
8	t-1,2-Dichloroethane	59[a] ppm
9	$CHCl_3$	0.69[a] ppm
10	1,2-Dichloroethane	19[a] ppm
11	1,1,1-Trichlorethane	3.6[a] ppm
12	CCl_4	0.011[a] ppm
13	Trichloroethane	13[a] ppm
14	Perchloroethane	19[a] ppm

[a] Maximum concentration obtained from a survey of 20 landfills. ND, not detected.

over a longer period of time. Finally, the biological activity decreases, leading to a gradual decrease in gas production, which may be looked at as a fourth stage of anaerobic decomposition (Lechner et al., 1993). A detailed discussion on refuse decomposition can be found elsewhere (Barlaz and Ham, 1993). The time dependency of the percentage of methane is critical for landfill gas recovery and reuse projects. The typical quality of gas generated in municipal waste landfills is included in Table 3.16 (EMCON Associates, 1980; Kester and Van Slyke, 1987; Wood and Porter, 1987; Kalfka, 1986). It may be noted that percentage/concentration of various parameters depends on age and composition of waste in a landfill.

The quantity of gas generated depends on waste volume and time since deposition in landfill. Gas production may be increased by adding sewage sludge or agricultural waste, removal of bulky metalic goods, and use of less daily and intermediate cover soil. The methane production rate ranges from 1.2 to 7.5 liters/kg/year (0.04–0.24 ft^3/lb/year) (EMCON Associates, 1980).

If gas is expected to be generated from a landfill then proper arrangements should be made for venting/extraction and subsequent treatment (where necessary). Whether gas should be vented/extracted from a landfill is sometimes argued by design professionals. The following issues should be considered before deciding not to vent gas from the landfill.

1. Gas pressure: Some estimate regarding gas pressure should be made. The estimated gas pressure should be low enough so that it will not cause any rupture of the landfill cover. If the waste is expected to generate gas due to biodegradability and/or other physical/chemical processes then venting of the gas should be recommended.

2. Stress on vegetation: The effect of the gas diffused through the cover on the vegetation should be studied. Stress may cause vegetation to die, which in turn will lead to increased erosion of the final cover.

3. Toxicity of the gas: The toxicity of the landfill gas should be studied. Release of the gas, by diffusion, through the final cover is unavoidable. The rate, concentration of release, and toxicity of the gas will determine whether such diffusional release will violate any air quality criteria.

4. Location of the landfill: The diffused gas may pose a health risk to the population residing in the immediate vicinity of the landfill.

LIST OF SYMBOLS

L_v = preclosure leachate volume
S = volume of pore squeeze liquid
P = precipitation volume
A = effective area of the landfill
FC = field capacity of the waste

E	=	volume lost through evaporation
L_v'	=	postclosure leachate volume
ET	=	volume lost through evapotranspiration
R	=	surface runoff volume
ΔS	=	soil and waste moisture storage volume
I	=	uniform precipitation rate in inches
A_S	=	area of the landfill surface in acres
C	=	runoff coefficient
SM	=	potential daily snow melt infiltration in inches of water
K	=	constant that depends on the watershed condition
T	=	ambient temperature above 32°F
R_i	=	surface runoff in inches
W_p	=	rainfall in inches
α	=	slope of the landfill surface (%)
W_i	=	initial soil moisture content of the cover
W_f	=	field capacity of the soil (%)
γ_d	=	soil density in lb/ft^3
C_c	=	coefficient of curvature of the soil
β	=	constant
CN	=	curve number

4 Waste Characterization

Waste characterization must be undertaken prior to designing a landfill. In general the characteristics of wastes may vary within a type of industry (e.g., waste from different papermills may not have the same characteristics) and may vary over time as a result of change in the industrial process. Therefore, it is a good practice to characterize a new waste and repeat the characterization if a process change occurs. Characterization of municipal garbage is usually not performed, because it is extremely difficult to perform tests on the waste and many studies have already been done to characterize the waste. However, since the composition of municipal waste may vary widely across a country (e.g. metropolitan areas versus small towns, industrial versus nonindustrial cities) and in different parts of the world, waste characterization studies of municipal garbage should be undertaken wherever possible. A typical range of major components of municipal garbage in the United States is indicated in Table 4.1. The range of the components is compiled from several studies conducted in the United States (Glaub et al., 1983; Wigh, 1979; Walsh and Kinman, 1981; EMCON Associates, 1975; Fungaroli and Steiner, 1979). The range of components has changed slightly due to recycling efforts. In general, waste characterization is done to address the following issues.

1. Whether the waste is hazardous.
2. Whether the waste can be landfilled.
3. Probable leachate constituents (necessary for judging liner compatibility, treatment plant design, and groundwater monitoring program design).
4. Volume rate of waste generation.
5. Physical properties of the waste necessary for the design of a landfill.
6. Physical properties of the waste necessary for the operation of a landfill.
7. Identification of safety precautions to be observed by landfill operators and inspectors.
8. Identification of waste reduction alternatives.

Therefore both the physical and chemical properties of the waste must be determined to address the above eight issues. The detailed steps for waste characterization for each source can vary widely; therefore a general guideline is provided that must be further developed on a case-by-case basis.

TABLE 4.1. Typical Range of Major Components of Municipal Garbage in the United States

S1 Number	Major Components	Range (% of wet weight)
1	Food waste	4.4–15.3
2	Garden waste	12.5–24.2
3	Glass	6.5–10.9
4	Metals (iron and aluminum)	4.0–9.0
5	Moisture	27.1–35.0
6	Other combustibles	1.6–12.1
7	Other noncombustibles	1.8–11.1
8	Paper	41.6–53.5
9	Plastics	0.76–5.7

4.1   GENERAL GUIDELINES FOR WASTE CHARACTERIZATION

The guidelines provided in this section are primarily for industrial waste. For characterizing municipal garbage, study can be conducted either by collecting samples from curbside or from garbage trucks arriving at a landfill. The first step in characterizing a waste is to study the material flow. A correct material flow diagram must be developed. Each final waste stream that is disposed in a landfill should be identified with a number. Most industries practice inhouse recycling of waste; these waste recyclings should be noted carefully. Many times these recycled wastes are wrongly identified as "waste stream" and characterized. Only those waste streams that are leaving the building need to be characterized. A typical (partial) flow diagram along with the waste identification number is shown in Fig. 4.1.

The raw material safety data sheet that identifies the chemicals used in the process needs to be studied. This will provide information on the probable constituents of the waste and thereby help narrow the list of parameters for future chemical analysis.

4.1.1   Sampling

Once the sampling points are identified in the flow diagram, the next step is to decide sampling frequency and volume of sample. A defective sampling plan will not provide the true characteristics of the waste. The sampling should be done in such a way that the test data are representative and indicate variability of the waste (USEPA, 1986).

The methods and equipment used for sampling waste materials will vary with the form and consistency of the waste. Different sampling protocols are used for sampling different waste types. The following is a list of standards

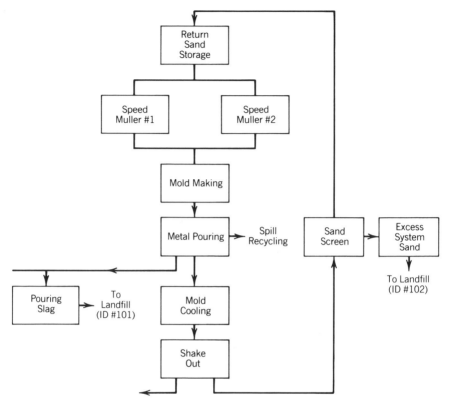

FIG. 4.1. Typical flow diagram used for identifying waste sampling points.

tests that may be used in the absence of a regulatory directive on sampling protocol:

1. ASTM standard D140-70 for extremely viscous liquid (ASTM).
2. ASTM standard D346-75 for crushed or powdered waste (ASTM).
3. ASTM standard D420-69 for soil or rock-like waste (ASTM).
4. ASTM standard D1452-65 for soil type waste (ASTM).
5. ASTM standard D2234-76 for fly ash type waste (ASTM).

Sludges may be treated as viscous liquids for the purpose of sampling.

Waste characteristics may vary due to the variability in raw material input or the production process. For instance, the municipal garbage fed into incinerators may vary daily or seasonally. A papermill may vary raw chemicals input depending on the end product in each production period (e.g., different colored paper at different times of the day or days of the week).

Such variation must be taken into consideration when deciding sampling frequency. Random sampling provides a better assessment of waste characteristics. If samples are composited, care must be taken to ensure that the composite is representative of the original (e.g., for an incinerator a composite of samples collected in the same season is allowable but a composite of samples collected over several seasons is not recommended). If two or more waste types are disposed of simultaneously in a landfill then a composite of each waste in the same ratio (volume or weight) in which it is disposed in the landfill is admissible. Individual samples are preferred over composite samples. However, sometimes to obtain a more representative sample, a composite of several samples is practiced (e.g., hourly samples of fly ash from an incinerator may be collected and composited to obtain one sample for the whole day).

4.1.2   Chemical Tests

Two types of chemical tests are done on each waste stream: bulk chemical analysis and the leach test.

Bulk Chemical Analysis. The aim of bulk chemical analysis (also termed total analysis) is to determine the chemical makeup of the waste. In general, bulk analysis involves solubilization of waste constituents to the greatest possible extent and then identification of them so that the sum total of all constituents is 99.99% of the original total bulk weight of the sample. No universal procedure for all waste or chemicals is available, but there are several test guidelines [Water Pollution Control Federation (WPCF, 1981; ASTM D-2795-84 and E-886-82; American Society for Testing and Materials (ASTM), 1986; USEPA, 1986]. Study of the raw material input to the process would help in finalizing the list of chemicals for which tests are to be run. Tables 4.2 and 4.3 include parameters of concern and Table 4.4 includes recommended lists of parameters for different waste types.

The lists are by no means final and should be modified whenever necessary. The lowest detection limits, as permissible by current technology, should be used in both bulk chemical analysis and the leach test. (Limit of detection is the lowest concentration at which a chemical species can be detected and limit of quantification is the lowest concentration at which a chemical species can be quantified.)

The advantages and disadvantages of bulk chemical analysis are discussed below. The advantages are as follows:

1. Sources of error are few because the number of variables to control the test are minimal.
2. Bulk chemical analysis may be the only way of determining the contamination potential from an unstable waste.

TABLE 4.2. Health-Related, Public Welfare-Related, and Indicator Parameters

Public-health-related parameters	*Public welfare-related parameters*
Aldicarb	Chloride
Arsenic	Color
Bacteria, total coliform	Copper
Barium	Foaming agents MBAS
Benzene	(methylene-blue active substances)
Cadmium	Iron
Carbofuren	Managanese
Chromium	Odor
Cyanide	Sulfate
1,2-Dibromoethane	Total dissolved solids (TDS)
1,2-Dibromo-3-chloropropane (DBCP)	Zinc
p-Dichlorobenzene	
1,2-Dichloroethane	*Indicator parameters*
1,1-Dichloroethylene	
2,4-Dichlorophenoxyacetic acid	Alkalinity
Dinoseb	Biochemical oxygen demand (BOD_5)
Endrin	Boron
Fluoride	Calcium
Lead	Chemical oxygen demand (COD)
Lindane	Magnesium
Mercury	Nitrogen series
Methoxychlor	Ammonia nitrogen
Methylene chloride	Organic nitrogen
Nitrate + nitrite (as N)	Total nitrogen
Selenium	Potassium
Silver	Sodium
Simazine	Specific conductance
Tetrachloroethylene	Total hardness
Toluene	Total organic carbon (TOC)
Toxaphene	Total organic halogen (TOX)
1,1,1-Trichloroethane	
1,1,2-Trichloroethane	
Trichloroethylene	
2,4,5-Trichlorophenoxypropionic acid	
Vinyl chloride	
Xylene	

3. Bulk chemical analysis will provide the total contaminant content.
4. The test data provide a baseline for comparing the waste with other waste or natural material.
5. It provides a basis for studying and ranking the leachate constituents obtained.

TABLE 4.3. Priority Pollutants Suggested for Bulk Chemical Analysis and Leach Test

Metals, cyanide, and total phenols

Antimony
Arsenic
Beryllium
Cadmium
Chromium
Copper
Lead
Mercury
Nickel
Selenium
Silver
Thallium
Zinc
Cyanide
Phenols, total

Dioxin

2,3,7,8-Tetrachlorodibenzo-*p*-doxin

GC/MS fraction—volatile compounds (purgeable)

Acrolein
Acrylonitrile
Benzene
bis(chloromethyl)ether
Bromoform
Carbon tetrachloride
Chlorobenzene
Chlorodibromomethane
Chloroethane
2-Chloroethylvinyl ether
Chloroform
Dichlorobromomethane
Dichlorodifluoromethane
1,1-Dichloroethane
1,2-Dichloroethane
1,1-Dichloroethylene
1,2-Dichloropropane
1,2-Dichloropropylene
Ethylbenzene
Methylbromide
Methylchloride
Methylene chloride
1,1,2,2-Tetrachloroethane

TABLE 4.3. *Continued*

Tetrachloroethylene
Toluene
1,2-*trans*-Dichloroethylene
1,1,1-Trichloroethane
1,1,2-Trichloroethane
Trichloroethylene
Trichlorofluoromethane
Vinyl chloride

GC/MS fraction—acid compounds (*acid extractable*)

2-Chlorophenol
2,4-Dichlorophenol
2,4-Dimethylphenol
4,6-Dinitro-o-cresol
2,4-Dinitrophenol
2-Nitrophenol
4-Nitrophenol
p-Chloro-*m*-cresol
Pentachlorophenol
Phenol
2,4,6-Trichlorophenol

GC/MS fraction—base/neutral compounds
(*base/neutral extractable*)

Acenaphthene
Acenaphthylene
Anthracene
Benzidine
Benz[*a*]anthracene
Benzo[*a*]pyrene
Benzo[*b*]fluoranthene
Benzo[*g,h,i*]perylene
Benzo[*k*]fluoranthene
Bis(2-chloroethoxy)methane
Bis(2-chloroethyl)ether
Bis(2-chloro-*iso*-propyl)ether
Bis(2-ethylhexyl)phthalate
4-Bromophenyl phenyl ether
Butyl benzyl phthalate
2-Chloronaphthalene
4-Chlorophenyl phenyl ether
Chrysene
Dibenz[*a,h*]anthracene
1,2-Dichlorobenzene
1,3-Dichlorobenzene
1,4-Dichlorobenzene

TABLE 4.3. *Continued*

3,3-Dichlorobenzidine
Diethyl phthalate
Dimethyl phthalate
Di-*n*-butyl phthalate
2,4-Dinitrotoluene
2,6-Dinitrotoluene
Di-*n*-octyl phthalate
1,2-Diphenylhydrazine
Fluoranthene
Fluorene
Hexachlorobenzene
Hexachlorobutadiene
Hexachlorocyclopentadiene
Hexachlorethane
Indeno[1,2,3-*c*,*d*]pyrene
Isophorone
Naphthalene
Nitrobenzene
n-Nitrosodimethylamine
n-Nitrosodi-*n*-propylamine
n-Nitrosodiphenylamine
Phenanthrene
Pyrene
1,2,4-Trichlorobenzene

*GC/MS fraction—pesticides and polychlorinated
biphenyls (PCBs)*

Aldrin
α-BHC
β-BHC
γ-BHC
δ-BHC
Chlordane
4,4'-DDT
4,4'-DDE
4,4'-DDD
Dieldrin
α-Endosulfan
β-Endosulfan
Endosulfan sulfate
Endrin
Endrin aldehyde
Heptachlor
Heptachlor epoxide
PCB-1242
PCB-1254

TABLE 4.3. *Continued*

PCB-1221
PCB-1232
PCB-1248
PCB-1260
PCB-1016
Toxaphene

After Weston (1984).

BHC: benzene hexachloride; DDT: dichlorodiphenyltrichloroethane; DDE: dichlorodiphenyldichloroethylene; DDD:
dichlorodiphenyldichloroethane.

TABLE 4.4. Parameters Recommended for Bulk Chemical Analysis and the Leach Test for Different Waste Types

Waste Type	Chemical Substances/Parameters
Alum mud	Parameters listed in Table 4.2 plus aluminum, berillium, hydrogen sulfide, pH, and sulfide
Coal fly ash	Parameters listed in Table 4.2 plus cobalt, molybdenum, nickel, pH, strontium, thallium, and vanadium
Foundry waste	Parameters listed in Table 4.2 plus aluminum, formaldyhyde, molybdenum, nickel, pH, phosphorus, phenol, and tin (Note: If other chemicals are identified in the raw materials then those should be included in this list)
Hazardous waste (all sources)	Parameters listed in Table 4.2 plus priority pollutants listed in Table 4.3 plus other chemicals identified in the raw materials used in the process
Municipal waste	Parameters listed in Tables 4.2 plus priority pollutants listed in Table 4.3 plus aluminum, antimony, beryllium, molybdenum, nickel, pH, total phosphorus, total suspended solids, thallium, and tin
Municipal solid waste incinerator ash (both bottom ash and fly ash)	Parameters listed in Table 4.2 plus aluminum, cobalt, strontium, tin, phosphate, and pH
Papermill sludge	Parameters listed in Table 4.2 plus priority pollutants listed in Table 4.3 plus aluminum, bromide, cobalt, phosphate, phosphorus, phenols, pH, tin, titanium, total suspended solids, and vanadium
Other non-hazardous waste (minimum recommendation)	Parameters listed in Table 4.2 plus other chemicals identified in the raw materials used in the process

The disadvantages of bulk chemical analysis are as follows:

1. Although bulk chemical analysis is aimed at a complete analyses of the waste, in many instances it does not identify 99.99% of the waste components.
2. The highly aggressive nature of the test is not representative of the field situation. The leachable concentration of a component in a waste will seldom equal the total concentration of the component in the waste.

Leach Test. Prior to deciding on a leaching media, the situation under which the waste is to be landfilled should be investigated. For monofills water leaching is acceptable; however, for codisposal with municipal waste, acid leaching should be performed. A synthetic leachate may also be used as a leaching medium for the codisposal scenario. The list of chemicals to be tested in the elutriate should be the same as for bulk chemical analysis. However, if bulk chemical analysis indicates a very low percentage of certain chemicals that are not expected to be dissolved in the leaching medium, then those chemicals may be deleted from the list of chemicals to be used for the leaching test. A discussion on available leach tests is included in Section 3.3.1.

4.1.3   Physical Tests

The physical tests to be performed depends on the landfill design and the available knowledge about the physical behavior of the waste. The following list includes the physical properties necessary for landfill design; the reader has to use judgment to choose the most appropriate tests: Compacted bulk density or unit weight, specific gravity, grain size distribution, permeability, consolidation characteristics, Atterberg limits, and static and dynamic strength characteristics (i.e., angle of internal friction and cohesion). Of the above items specific gravity may not be necessary in most cases. In addition to its design-related use, compacted bulk density is necessary to estimate the tipping fee for disposing of waste in a landfill; determination of Atterberg limits is not necessary for most waste types. Interface friction angle with construction materials (e.g., synthetic membrane) may be needed for stability design.

Standard soil testing methods can be used for testing soil type waste, however, difficulty arises with putrescible waste and sludges (especially those that release gas under pressure). The problem is further multiplied if the waste is heterogeneous (e.g. municipal waste). No standard test methods are available for such waste at present. The physical properties of waste are found by using standard geotechnical equipment and testing procedures with minor modification (Wardwell and Charlie, 1981; Lowe and Andersland, 1981; Zimmerman et al., 1977; Rao et al., 1977; Andersland and Mathew, 1973; Hagerty et al., 1977; Kulhawy and Sangrey, 1977; Somogyi and Gray,

1977; Pervaiz and Lewis, 1987). Doubts have been expressed regarding the use of standard geotechnical procedures for testing sludge (Bagchi, 1980; Wardwell and Charlie, 1981). This is an area in which further research may be done.

4.2   IDENTIFICATION OF HAZARDOUS WASTE

Prior to developing any detailed physical and chemical testing program, the first step should be to determine whether the waste is hazardous. Dawson and Mercer (1986) have provided a good discussion on how to define hazardous wastes and how they are defined in different countries. A discussion with the regulatory agency regarding its definition of hazardous waste is suggested. In general, two approaches are taken in the United States to define hazardous waste: by listing and by identification of characteristics. A third approach using toxicological standards for defining hazardous waste is under consideration. The list is subject to change. A solid waste containing any of the hazardous constituents is considered hazardous unless proven otherwise by the disposer.

In summary the steps involved in identifying a hazardous waste are as follows:

1. Determine whether it is already listed as a hazardous waste.
2. If not then characteristic identification tests are to be performed to determine whether the waste is hazardous due to any characteristics detailed in Section 4.2.2.

It may be noted that radioactive waste and infectious waste (often termed biohazardous waste) are not included in this discussion and, therefore, any reference to hazardous waste in this book would always mean nonradioactive, noninfectious hazardous waste.

4.2.1   Listing

A list of chemicals whose health hazard is already known is developed. Wastes that contain any of these compounds are characterized as hazardous. A list of such hazardous waste compounds is included in Appendix I.

4.2.2   Characteristics Identification

The following four characteristics are used to identify hazardous waste.

Corrosivity. If the waste is aqueous, has a pH less than 2 and greater than 12.5, and corrodes plain carbon steel (carbon content of 0.2%) at the rate of 6.35 mm or greater per year at 55°C, then the waste is characterized as corrosive waste.

TCLP Toxicity. The TCLP test is performed on the waste. Refer to Section 3.3.1 for more details.) If the extract contains any of the substances listed in Table 3.1 at a concentration equal to or greater than the respective value given in that table, then the waste is considered as TCLP toxic waste.

Ignitability. If the waste is liquid other than an aqueous solution containing less than 24% alcohol by volume, and has a flash point of less than 60°C as determined by ASTM test D-93-79 or D-93-80 (ASTM), it is characterized as ignitable waste. A nonliquid waste (other than gaseous) is also characterized as ignitable if it causes fire at 0°C and at a pressure of 1 atmosphere through friction, absorption of moisture, or spontaneous chemical changes or burns vigorously creating a fire hazard. For gaseous waste the ignitability is determined by ASTM test D-323 (ASTM).

Reactivity. A waste is characterized as reactive if it is normally unstable and undergoes violent change without detonating or reacts violently with water, or forms a potentially explosive mixture with water or generates significant quantities of toxic gas when mixed with water endangering human health or the environment.

4.3 RESTRICTION ON LAND DISPOSAL OF HAZARDOUS WASTE

In general hazardous wastes should not be disposed of in a landfill without pretreatment. Although compatibility tests may demonstrate the suitability of the proposed liner material for the waste type, it is good practice to "stabilize" the waste so the mobility of the hazardous constituents is reduced. Waste that produces toxic fumes due to contact with water or other waste in the landfill, or waste with flash points >140°F, should not be disposed of in landfills. In addition, waste having the following characteristics should not be disposed of in landfills (Stanczyk, 1987).

1. Waste with high percentages of volatile organic content.
2. Waste with high percentages of aromatic, halogenated and nonhalogenated compounds.
3. Waste with high percentages of metallics, especially arsenic, cadmium, lead, mercury, and selenium.
4. Waste with high percentages of cyanide and sulfide.
5. Powdery hazardous waste that may cause dust problems in and around the landfill.
6. Waste with very low shear strength that may preclude construction of a final cover for the landfill.
7. Waste with high percentages of liquid that may generate too much leachate in the landfill.

The regulatory agency may impose additional criteria for restricting land disposal of waste. A discussion with the regulatory agency on this issue is recommended. Currently in the United States, there is a ban on disposal of hazardous waste in landfills. Hazardous waste must be stabilized (i.e., rendered nonhazardous) prior to disposal.

4.4   IDENTIFICATION OF NONHAZARDOUS WASTE

As previously indicated, complete waste characterization will include determination of both the physical and chemical properties of the waste. For known nonhazardous waste usually only the leach test and some physical tests are performed. However, it is good practice to perform at least the TCLP test to determine whether the waste is TCLP toxic. The regulatory agency should be contacted to determine the test requirements for each waste type. Scattered information on the physical and chemical properties of different waste types is available in the literature. Since waste characterization is rather waste specific, a summary of all available data is not included.

5 Natural Attenuation Landfills

The design concept for natural attenuation (NA) type landfills consists of allowing the leachate to percolate through the landfill base with the expectation that the leachate will be attenuated (purified) by the unsaturated soil zone beneath the landfill and by the groundwater aquifer. In the past only NA type landfills were used for disposal of all types of waste. At that time it was thought that the soil in the unsaturated zone is capable of completely attenuating the leachate. This concept of attenuation by soil has changed significantly. Presently only nonhazardous wastes are disposed of in NA type landfills. Recent studies indicate that even small NA type landfills (waste volume up to 50,000 yd^3) may impact groundwater (Friedman, 1988). Whether such an impact on groundwater is to be considered as severe depends on the prevailing groundwater rules in the area in which the landfill is to be located. Currently in some countries of the world (e.g., West Germany) and in some states of the United States (e.g., Wisconsin), NA type landfills are not allowed regardless of volume or waste type. Therefore, before designing a NA type landfill, the designer should discuss the issue with the local regulatory agency.

Two types of filling methods are used in operating NA type landfills: the area method and trench method. In the area method an entire area is excavated to the subbase grade and filled up from one end (Fig. 5.1). In the trench method, individual trenches are excavated, filled, and covered progressively (Fig. 5.2). The area method would require less land for disposing the same volume of waste compared to the trench method. However, in the area method of filling, because precipitation comes in contact with the entire waste area, the quantity of leachate generated is higher and leachate quality is worse compared to the trench method of filling. Therefore, in choosing between the two types of filling, a designer has to strike a balance between land availability and extent of allowable groundwater impact. A third option of filling has evolved that reduces the leachate generation rate and at the same time needs less disposal area (Fig. 5.3). In this filling method, the landfill area is progressively excavated on one side while the area that has reached final grade on the other side is capped. Usually the subbase has a downward slope in the direction in which excavation proceeds. A small (0.6–0.9 m (2–3 ft) berm is constructed at a suitable distance so that noncontact water collected between the berm and the end of excavation is pumped out using a small pump. However, if a pump is used for pumping noncontact water, the operator must be made aware of the fact that the waste limit must be

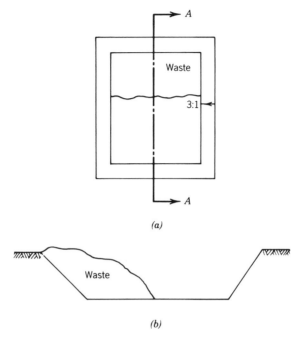

(a)

(b)

FIG. 5.1. Area method of landfilling: (a) plan; (b) section A–A.

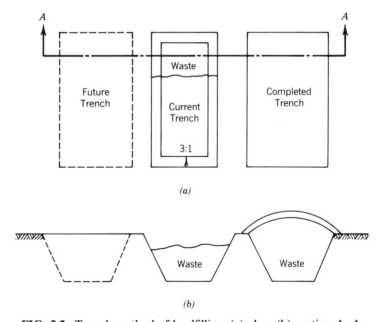

(a)

(b)

FIG. 5.2. Trench method of landfilling: (a) plan; (b) section A–A.

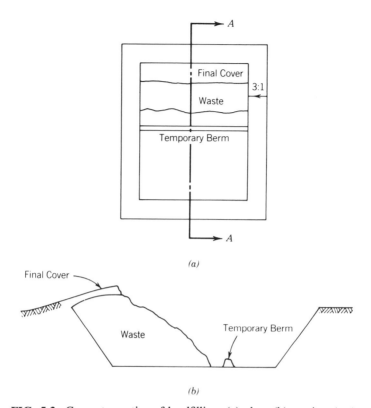

FIG. 5.3. Current practice of landfilling: (a) plan; (b) section A–A.

inside the small berm to avoid discharging contaminated water into surface water bodies. The designer should contact the regulatory agency prior to using a pump to discharge noncontact water into surface water bodies.

The trench filling method works best when a relatively small volume of waste is disposed of in a short time period (several days) and the interval between such disposals is high (6 months or more). The area method works best when a daily flow of waste volume is expected; the third option of landfill mentioned in the previous paragraph may also be used in this situation.

5.1 NATURAL ATTENUATION PROCESS

Figure 5.4 shows a cross section of an ideal NA type landfill. Generalized soil stratigraphy of an ideal NA type landfill includes a clayey stratum [which should consist either of ML or ML-CL type soil per the Unified Soil Classification System (USCS) followed in the United States] directly below the landfill base, which overlies a sandy aquifer. The bedrock underlying

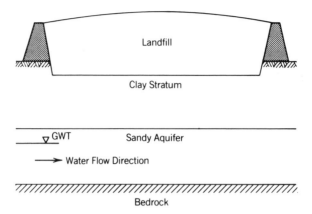

FIG. 5.4. Soil stratification for an ideal natural attenuation landfill.

the sandy aquifer should be quite deep [ideally 15–18 m (50–60 ft) or greater below the water table].

Attenuation of leachate occurs in two stages. In the first stage, soil in the (initially) unsaturated zone reacts with the leachate constituents and attenuates the leachate in part. The second stage of attenuation occurs in the groundwater equifer (Fig. 5.4).

5.2 MECHANISMS OF ATTENUATION

Natural attenuation may be defined as a process by which the concentration of leachate parameters is reduced to an acceptable level by natural processes. Based on this definition, the following mechanisms of attenuation are identified: (1) adsorption, (2) biological uptake, (3) cation- and anion-exchange reactions, (4) dilution, (5) filtration, and (6) precipitation reactions (Bagchi, 1983). Except for dilution, the other mechanisms can be operative in the unsaturated zone. All the mechanisms except for biological uptake can be operative in the aquifer.

5.2.1 Adsorption

Adsorption is the process by which molecules adhere to the surface of individual clay particles. Because of the difficulty in distinguishing adsorption from exchange reactions (explained in Section 5.2.3) experimentally, it is sometimes referred to as adsorption-exchange reactions (Philips and Nathwani, 1976; Minnesota Pollution Control Agency, 1978; Roberts and Sangrey, 1977). However, there is a basic difference between adsorption and exchange reactions. Adsorption will cause a decrease in the total dissolved solid (TDS) in the leachate, whereas exchange reactions will not. Therefore adsorption may be considered to really attenuate leachate, whereas exchange reactions will simply change the type of ions present in the exfiltrate (the liquid generated after leachate percolation through the unsaturated zone). The adsorption reaction is pH dependent and to determine the adsorption capacity of a particular clay, adsorption isotherms must be established experimentally. An adsorption isotherm is a plot of the amount of ion adsorbed versus the concentration of the ion in the solution. In addition to the theoretical difficulties of adsorption analysis, there is a general lack of data concerning the adsorption by clay minerals of individual ions from a solution of various ions. Much of the available information pertains to adsorption for specific ions, although some studies have included analyses of adsorption isotherms for the leachate–soil system (Griffin and Shimp, 1976; Griffin, 1977). Although some generalization of leachate quality for municipal waste can be made for a region, in most cases, leachate quality is significantly site specific. Thus, in the absence of experimental results using the on-site soil and leachate obtained from the waste to be disposed of at a particular site, site-specific quantitative analysis cannot be performed.

The opposite of adsorption—desorption—occurs in many systems (Gebhard, 1978). The isotherm based on a young system may not be true for an aged adsorbate-solid complex. Hence, even the adsorption isotherm obtained experimentally should be used carefully for site-specific design.

5.2.2 Biological Uptake

Biological uptake is a mechanism by which microorganisms either break down or absorb leachate constituents and thereby attenuate leachate. Microbial growth in soil systems can have a tremendous impact on leachate attenuation initially. Processes that soil microorganisms either perform or mediate include the following (Fuller, 1977; Wood et al., 1975):

1. Breaking down carbonaceous wastes.
2. Production of carbon dioxide and subsequent formation of carbonic acids.
3. Production of various organic acids.
4. Using up available oxygen supplies and creation of an anaerobic environment.
5. Participation in metal ion reactions.
6. Oxidation or reduction of inorganic compounds.
7. Transformation of cyanide to mineral nitrogen compounds and then denitrification of the compound to nitrogen gases.
8. Methylation of metals and metalloids.
9. Production of complex organic compounds that react with leachate constituents.

10. Production of large and small organic molecular species on which leachate constituents can be absorbed.

11. Production of small sized organic debris that can infiltrate pore spaces and thereby reduce soil permeability.

Mineralization is the process by which elements or organic matter, microbial tissues, and organic complexes are converted into an inorganic state. Biological immobilization is considered to be the reverse of mineralization. Trace and heavy metals are incorporated into microbial tissues and the mobility is controlled by cells or cell tissues. For elements that are relatively immobile in soils such as inorganic complexes, incorporation into cell materials may be thought of as a mechanism by which they can migrate as minute particles and cell materials when the tissues die and decay. Movement of phosphorus in an organic form is an example of this phenomenon (Hannapel et al., 1964). Thus, the presence of microbes beneath a landfill may be beneficial for attenuating some leachate constituents but could be detrimental for mobilizing others. Organic complexes in typical municipal landfill leachates probably could immobilize many of the trace metal parameters by precipitation under aerobic conditions (Fuller, 1977). However, under anaerobic and acid conditions, attenuation of metals by microbes will be less effective, especially if the pH drops to 3 or less.

Biological uptake in biodegradable (putrescible) waste landfills is basically through anaerobic bacteria. Methane gas generation is an indicator of the anaerobic biological activity. It is known that methane gas generation attains a peak volume and then decreases. The rate of depletion of the generation of gas is related to the rate of decay of the microorganisms. The microorganisms in a landfill can be only heterotropic (surviving on a food source) not autotrophic (capable of food synthesis through the use of sunlight). Since the strength of the landfill leachate decreases with time, causing a decrease in the availability of food, the activity of the microorganisms is bound to decrease. Therefore, it is obvious that the chemical fixation of pollutants is not permanent. As the biological population dies, pollutants once fixed in microbial cells could be released through a mineralization process. It is possible to study the decay rate and subsequent pollutant release of landfill microorganisms using principles of biokinetics; however, this has not been reported in the literature. Because of the uncertainties involved in the decay rate and the fact that the biological uptake is not permanent, it is prudent not to use biological uptake in landfill design, at least at the present time. It is an area in which further research should be undertaken so as to obtain more valid data on the subject. As for now, biological uptake may be considered a safety factor, the value of which is not yet known.

5.2.3   Cation and Anion Exchange

The exchange reactions mainly involve the clay minerals and may be defined as exchange of ions of one type by ions of another type without disturbing the mineral structure (known as isomorphous substitution) (Grim, 1968). The solid phase of a given soil may contain various amounts of crystalline clay and nonclay minerals, noncrystalline clay minerals, organic matter, and salts. Although the amount of nonclay minerals in a given soil is usually considerably greater than the proportion of clay minerals present cation exchange performed by the clay mineral fraction is quite significant. By and large, the nonclay particles are relatively inert.

Anion exchange increases as the pH of the soil system decreases. Since organics are negatively charged ions, attenuation of organic ions in a clayey soil environment will be mostly through anionic exchange. Therefore, a low pH system should attenuate organics well (Mitchell, 1976). Organic anions may be adsorbed by clays and inorganic ions. However, note that the pH of the leachate–soil system converges to a near neutral value (refer to Section 5.2.6). Thus, significant attenuation of organic ions through anion exchange is not expected. Adsorption on particle surfaces in place of previously adsorbed water molecules is a possibility (Van Olphen, 1963).

Time taken to complete anion- and cation-exchange reactions is not well documented. However, it is reported that the exchange reaction in kaolinite is almost instantaneous because of easily accessible sites located at broken bond edges. A longer reaction time is expected in smectites because a majority of the exchange sites are located within layers (Mitchell, 1976). Further research is needed to determine the time required for completion of exchange reactions in clayey soil.

5.2.4 Dilution

This is not a mechanism by which leachate constituents are chemically altered or attenuated by the soil. It reduces the concentration of leachate constituents. To what extent dilution should be used for the design of NA type landfills depends on the policy of the regulatory agency. If a policy of nondegradation of groundwater is to be pursued, then dilution cannot be taken into account, because all natural attenuation type landfills will degrade groundwater. Chloride, nitrate, hardness, and sulfate found in municipal landfill leachate are not attenuated by soil; the only mechanism by which these parameters are attenuated is dilution. The concentration of these and other parameters may be diluted in a groundwater aquifer to such a level that the quality of down-gradient water at a certain distance is degraded only slightly compared to the background water quality and therefore the water quality remains acceptable for specific uses. When locating a NA type landfill the designer should ensure that groundwater quality at a distance [the distance is sometimes fixed by local regulators; in the absence of a regulation a distance of 300–360 m (1000–1200 ft) or the distance to the nearest down-gradient drinking water well, whichever is less, should be used] is safe for drinking purposes.

The major factors that influence dilution are the density difference of leachate and ambient groundwater, leachate entry velocity, groundwater velocity, diffusion–dispersion coefficients of leachate constituents in the aquifer, soil stratigraphy beneath the landfill base, and the area of the landfill base. These factors are all site specific, although for some a range of value may be estimated. Contaminant transport models can be used to estimate concentration of leachate constituents at a down-gradient point from a landfill; however, these models must be field calibrated, which appears to be a fairly costly proposal. Because most potential users of natural attenuation type landfills (e.g., small townships and cities) are incapable of funding such costly studies, realistically contaminant transport modeling is of very little help in designing NA type landfills. An alternative method of estimating the approximate average concentration of leachate constituents at a down-gradient distance from the landfill is available (Bagchi, 1983).

Diffusion and Dispersion. Diffusion and dispersion are two mechanisms by which leachate is diluted by the aquifer. Because the leachate has a chemical concentration that is different from the background water, it tries to equilibrate with the ambient water quality through diffusion.

Diffusion is essentially a physicochemical phenomenon. Dispersion, on the other hand, is more of a mechanical phenomenon. Dispersion can occur in a longitudinal or in a transverse direction. Longitudinal dispersion occurs in the direction of flow and is caused by different macroscopic velocities, as some parts of the invading fluid move through wider and less tortuous pores. For example, leachate entering at point 2 or 3 in Fig. 5.5 will advance more slowly than leachate entering at point 1.

Transverse dispersion occurs normal to the direction of flow and results from the repeated splitting and deflection of the flow by the solid particles in the aquifer. Transverse dispersion is effective only at the edges of a contamination source (Bouwer, 1979). It should be noted that the dispersion theories are applicable to sand and gravel type deposits. (Harleman et al., 1963). For nonfractured clayey soil deposits with a highly oriented bedding plane, transverse dispersion may not be pronounced.

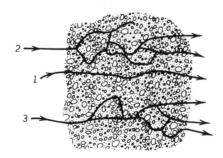

FIG. 5.5. Dispersion of contaminants in soil.

The relative importance of diffusion and dispersion has been extensively studied (Blackwell, 1959; Bruch and Street, 1967; Raudkivi and Callandar, 1976). It was found experimentally that diffusion was important when Reynold's number is less than 1×10^{-3} (Raimondi et al., 1959). Perkins and Johnston (1963) have provided a comprehensive literature survey on the subject of diffusion and dispersion in porous media. Dispersion in fractured rock or fractured clay deposits will seldom produce a uniform predictable concentration distribution within a plume.

5.2.5 Filtration

Filtration is a mechanism by which leachate constituents are physically trapped. The random pore structure of the soil system serves to physically entrap suspended and settleable solids in a leachate, just as gravity filtration through sand and other media will remove solids in water treatment. Filtration efficiency depends on the pore size and hydraulic gradient of the leachate. Finer soil materials and lower leachate hydraulic gradients will improve filtration. It is difficult to estimate the percentage of attenuation through filtration for any parameter, however, this will be an operative mechanism whenever chemical precipitation, biological growth, and other processes produce undissolved solid particles (Farquhar and Rogers, 1976; Fuller, 1977).

Filtration is a "one time" phenomenon and is not highly significant so far as leachate attenuation is concerned. Therefore, it is suggested that this mechanism not be taken into account in landfill design.

5.2.6 Precipitation

Chemical precipitation involves a phase change in which dissolved chemical species are crystallized and deposited from a solution because their total concentration exceeds their solubility limit. Adsorption is prominent in microconcentration, but precipitation is the dominant mechanism in macroconcentrations (Fuller, 1977). The solubility limits depend on factors such as ionic species and their concentration, temperature, pH, redox potential (EH), and concentration of dissolved substances (Gebhard, 1978). Of these, pH and EH are the most important and can be regarded as "master variables."

The pH of a system controls all acid–base reactions and profoundly influences the equilibrium reactions that determine the relative abundance of hydroxide, carbonate, sulfide, and other ions in a system (Stumm and Morgan, 1970; Ponnamperuma, 1973). The pH of a leachate-saturated soil converges to a near neutral value regardless of the initial pH of the soil; this is because the leachates are typically anaerobic solutions making the soil a reduced media. In acidic soils the reducing reactions consume the available hydrogen ions. In calcareous, alkaline soil reducing reactions increase carbon dioxide pressure. So in both cases the system tends to converge to a near neutral value (Ponnamperuma, 1972, 1973; Roberts and Sangrey, 1977). Therefore, shifting of the pH of groundwater down-gradient from a landfill to a near neutral value may serve as an indicator of plume arrival.

The redox potential also influences precipitation of chemicals. Although solubility plots for many metals under various pH and EH values are available, the main difficulty is in measuring redox potential for a soil–leachate system. Representative values of EH can be obtained only by *in situ* testing of soil. Any determination of redox potential even in Shelby tube samples may not give the actual EH of the soil (Stumm and Morgan, 1970). Thus, although the redox potential influences the solubility of a pollutant, no reliable data regarding the range of EH values of leachate-saturated soil are available for use in a solubility plot.

Table 5.1 provides a summary of the net effect of different attenuation mechanisms.

5.3   EFFECTS OF VARIOUS FACTORS ON ATTENUATION MECHANISMS

The effects of factors such as leachate velocity and degree of saturation of soil are discussed in the following sections. The discussions are qualitative in nature, which will help in designing an NA type landfill (Bagchi, 1987b).

TABLE 5.1. Net Effect of Attenuation Mechanisms

S1 Number	Mechanism	Net Effect
1	Adsorption	Immobilization if change of pH is not high
2	Biological uptake	Temporary immobilization; long-term effect needs additional study
3	Cation and anion exchange	Immobilization of the constituents but elution of some other element causing increase in its concentration
4	Dilution	Reduction of concentration
5	Filtration	Immobilization
6	Precipitation	Immobilization in most cases; change of pH and EH has significant effect

After Bagchi (1987b); courtesy of Waste Management and Research, Denmark.

5.3.1 Leachate Velocity

All attenuation mechanisms are somewhat dependent on leachate velocity. Reaction kinetics will govern four of the five mechanisms: adsorption, biological uptake, cation exchange, anion exchange, and precipitation. Because significant data are not available regarding reaction kinetics of leachate in soil, this is a field worth researching. Presently it is assumed that the flow of leachate is slow enough to complete all applicable attenuation mechanisms.

5.3.2 Degree of Saturation of the Unsaturated Zones

The degree of saturation will influence biological uptake significantly and precipitation reactions to some extent. A higher saturation would mean less available oxygen for biological activity, leading to predominantly anaerobic bacterial growth. Since at worst, soil in the unsaturated zone could be fully or nearly saturated, attenuation should be focused on biological uptake under similar conditions.

5.3.3 Organic and Inorganic Matter in Soil, Clay Type, and Soil Fabric

Organic and inorganic matter in clayey soil, usually considerably greater in proportion than clay minerals, influences adsorption, precipitation, and biological uptake. The surfaces of organic matter provide some adsorption sites; in addition, they may serve as an energy source for microorganisms. Inorganic matter such as oxides and hydroxides of iron, aluminum, and manganese participates in the precipitation reaction and influences the pH and EH of the leachate–soil system. Clay types influence cation and anion exchange and precipitation reactions. Soil fabric and texture influence filtration. It must be borne in mind that there are mechanisms other than the cation exchange capacity of soil that control attenuation of leachate in soil. Under appropriate conditions one or more of these factors could become dominant and exert a controlling influence (Fuller, 1977; Griffin et al., 1976a; Roberts and Sangrey, 1977).

5.3.4 Soil Stratigraphy

Soil stratigraphy or the sequence in which different soil type (i.e., sand, silt, and clay) layers exist at a particular site can significantly influence natural attenuation of leachate. The soil stratigraphy best suited for natural attenuation is (Fig. 5.4) (1) an unsaturated zone consisting mainly of silty clay with a fairly high cation exchange capacity (30–40 mEq/100 g) and a permeability of 1×10^{-4} to 1×10^{-5} cm/sec, (2) a groundwater table that is below the base of the unsaturated silty clay and that is within a sandy layer with fairly high permeability (i.e., 1×10^{-3} cm/sec or more) layer. Because it is difficult to find the ideal soil stratigraphy and because in reality numerous soil stratigraphies exist, it is impossible to discuss each situation. Instead, the following comments regarding soil stratigraphy is included to help in decision making:

1. If a clayey stratum is not available beneath the landfill, construction of a retarder base similar in characteristics to the unsaturated zone mentioned above must be undertaken. Guidelines for retarder base construction are included in Chapter 9.

2. Macrofabric features such as fissuring would cause channeling of leachate without proper contact with the soil in the unsaturated zone. This geologic feature must be properly investigated at the time of subsoil investigation (Rowe, 1972).

3. Uneven weathering of bedrock may create a buried mountain-type bedrock profile in residual soils. Thus, while siting a landfill in residual soils, a bedrock profile below a proposed NA type landfill site should be properly investigated. Landfill base should not be constructed directly on bedrock.

4. Low permeability (1×10^{-7} cm/sec or less) of the unsaturated zone will not allow proper percolation of leachate, leading to ponding and subsequent leachate seep; a natural attenuation type landfill should not be sited in such areas.

5. Soil stratigraphy will influence dilution. If the underlying aquifer is not thick enough, then plume development will be restricted, and sufficient dilution may not take place.

6. Soil stratigraphy also influences plume geometry. Paschke and Hoppes (1984) have shown that among other variables the plume shapes depend on the downward velocity of the leachate entering the groundwater and on the velocity of the groundwater. Thus, the permeability of the unsaturated zone, the leachate head within the landfill, and the permeability of the aquifer will influence dilution of leachate parameters in the aquifer.

5.4   ATTENUATION MECHANISMS OF SPECIFIC POLLUTANTS

In the United States 66,000 chemicals are used commercially and worldwide 45,000 substances are traded. It is estimated that 1000 new chemicals are added to the list each year (Richards and Shieh, 1986). No study of the attenuation mechanism(s) for all chemicals in soil is available. The purpose of this section, therefore, is to provide a general overview of the attenuation mechanisms offered by soil and groundwater for the constituents usually found in nonhazardous landfill leachate. Discussions of attenuation mechanisms of pollutants usually found in hazardous waste leachate are not included because hazardous wastes cannot be disposed of in NA type landfills. However, many of the pollutants mentioned in the following sections may not be present in a particular leachate. At the end of the discussion on each pollutant, a qualitative estimate of the attenuation potential for each pollutant in a clayey environment is included, which will be useful for NA type landfill design and land-spreading. Attempts have been made to include as many pollutants as possible to form a broad data base.

5.4.1 Aluminum

The major attenuation mechanism of aluminum is precipitation. In a high (alkaline) or near neutral pH environment, aluminum readily forms insoluble oxides, hydroxides, and silicates. However, in a low pH (acidic) environment, aluminum is quite soluble. The mobility of aluminum in a clayey environment is low (Griffin, 1977).

5.4.2 Ammonium

The major attenuation mechanisms of ammonium are cation exchange and biological uptake. Biological uptake converts ammonium to nitrate and/or organic nitrogen. These transformations may be reconverted at a later time to ammonia or other products. In the long run, however, denitrification and immobilization as organic nitrogen can combine with sorption and fixation to yield a net reduction of both ammonium and total nitrogen in a leachate (Gebhard, 1978).

Nitrification is a two-step process, carried out by autotropic bacteria. Conditions under which this process proceeds at a maximum rate are adequate supply of ammonium, presence of a moderate pH (optimum around 8.5), adequate oxygen, moderate moisture content, and optimum temperature (about 30°C) (Tisdale and Nelson, 1975). These conditions, particularly the presence of adequate oxygen, may not be fulfilled in a landfill. The ammonium ion may also transform to gaseous ammonia under alkaline conditions in the presence of free lime (Carter and Allison, 1961). More transformation apparently occurs from coarser soils (Gebhard, 1978). The mobility of ammonium in clayey soil is moderate.

5.4.3 Arsenic

The major mechanisms of attenuation of arsenic are precipitation and adsorption. In an aerobic environment, arsenic reacts with iron, aluminum, calcium, and many other metals to form arsenate, which is only slightly soluble. If saturation occurs, arsenic appears in the reduced arsenic form, which is more soluble and mobile than the oxidized form (Swaine and Mitchell, 1960). When saturated soils return to oxidic conditions, arsenite is oxidized to arsenate (Quastel and Scholefield, 1953).

Adsorption of arsenite at pH 7 is directly proportional to the amount of lime applied and small amounts of arsenite are irreversibly bound to the soil (Fuller, 1977). Maximum removal of arsenate from leachate occurs in the range of about pH 4–6 and generally removal of arsenite increase from pH 3 to 9. Adsorption of arsenite by montmorillonite clay consistently showed an unexplained discontinuous peak at pH 6–7. Montmorillonite clay was found to adsorb approximately twice as much as kaolinite clay. The mobility of arsenic is moderate in clayey soil (Griffin, et al., 1976a).

5.4.4 Barium

The major attenuation mechanisms of barium in soils are adsorption, ion exchange, and precipitation. Experiments have shown that adsorption of barium increases with the increasing cation exchange capacity (CEC) of the soil and decreases with increasing interference from other leachate constituents (Griffin et al., 1976a; Farquhar and Rovers, 1976). Studies with soil organic matter showed that differences in relative stability and retention value of barium, calcium, magnesium, and copper complexes are small. Retention values were ranked in the order copper > barium = calcium > magnesium (Broadbent and Ott, 1957). If free lime is present in soil, barium will be precipitated as barium carbonate. Barium carbonate is only slightly soluble in water and barium is, therefore, effectively attenuated. Soil factors favoring attenuation of barium will include a high clay percentage and the presence of other colloidal material. An alkaline condition and free lime will also tend to favor attenuation by ion exchange and chemical precipitation (Gebhard, 1978). The mobility of barium in clayey soil is low.

5.4.5 Beryllium

The major attenuation mechanisms of beryllium are precipitation and cation exchange. The chemistry of beryllium is similar to aluminum. Beryllium can be mobile in very low or very high pH soil due to hydrolysis (Griffin, 1977). It is highly attenuated in soils, particularly those containing montmorillonite and illite type clay. Beryllium may displace divalent cations already on common adsorption sites in the exchange complex (Fuller, 1977). In general, the mobility of beryllim is low in clayey soil.

5.4.6 Boron

The major attenuation mechanisms of boron include adsorption as borate on various inorganic surfaces and precipitation or coprecipitation with hydrous iron or aluminum oxides (Gebhard, 1978). Therefore, the activity of boron or borate in soil systems is related to that of aluminum and ferric iron. Leaching experiments using simulated or real landfill leachate have shown a net borate addition to the leachate (Streng, 1976; Griffin et al., 1976a). Its mobility in clayey soil is high (Griffin, 1977).

5.4.7   Cadmium

The major attenuation mechanisms of cadmium are precipitation and adsorption. Adsorption on colloidal surfaces due to coulomb-type forces is said to be primarily responsible for the immobility of cadmium in soils (Fuller, 1977). Cadmium, like zinc, mercury, and lead, undergoes hydrolysis at pH values normally encountered in soil environments. Several studies have shown that pH is the most important factor in controlling the attenuation. Chemical precipitation of cadmium with anions such as phosphate, sulfide, and carbonate can also effectively attenuate cadmium (Santillan-Medrano and Jarinak, 1975, Huang et al., 1977; Griffin et al., 1976b).

Data obtained in leaching experiments using specially prepared cadmium solutions and cadmium-spiked municipal leachate indicate that removal of cadmium by various hydroxides in clayey soils increases from a relatively negligible level to a very significant level as the pH rises from about 6 to 8 and that attenuation is quite stable at pHs greater than 8. Chemical precipitation of cadmium is highly dependent on that available anions, pH, and redox potential (Huang et al., 1977). Precipitation and possible coprecipitation with phosphate, sulfide, and carbonate is likely to occur in a near neutral pH environment (Santillan-Medrano and Jarinak, 1975; Huang et al., 1977). A limestone (calcium carbonate) barrier between landfill materials and soils can provide an additional attenuation capacity for cadmium (Fuller, 1977). However, such use of limestone in an NA type landfill may decrease permeability of the base, which may cause leachate ponding; an NA landfill will not be successful if leachate ponding occurs.

Houle (1976) found that changes in leachate salts and available complexing ions could alter the degree to which cadmium is attenuated. Thus, it appears that cadmium adsorption in leachate–soil systems is not irreversible and remobilization can easily occur. The mobility of cadmium in clayey soil is moderate.

5.4.8   Calcium

The major attenuation mechanisms for calcium are precipitation and cation exchange. It readily forms carbonate precipitation under alkaline pH. Since calcium is the dominant ion in the soil exchange complex, it is not adsorbed but in most cases is eluted (Griffin, 1977). Montmorillonite was found to elute calcium to a significantly greater degree than illite and kaolinite. It has been postulated that calcium elution from soil is due to exchange of calcium by sodium, potassium, ammonium, and magnesium from leachate. This elution adds to the hardness of the groundwater beneath municipal sanitary landfills. Thus, calcium is not only highly mobile in clays, there is a good possibility of this pollutant being eluted.

5.4.9   Chemical Oxygen Demand (COD)

Rather than being an individual ionic species that can be studied in its various states, COD represents a group of compounds that can be oxidized by a boiling solution of potassium dichromate and sulfuric acid. The molecular species falling into this category include many (but not all) organic materials. For example, straight chain aliphatic compounds, aromatic hydrocarbons, and pyridine are not oxidized by a chromic acid–sulfuric acid solution (Gebhard, 1978).

Dissolved organic compounds occur in waters in such low concentrations that they usually have to be concentrated and separated from the inorganic salts before they can be identified chemically. Bioassay methods are frequently of great sensitivity and permit direct determination of very low concentrations of organic nutrients in water (Stumm and Morgan, 1970).

The most important mechanism of COD attenuation is biological uptake, which produces sulfur dioxide and methane. Filtration is a minor mechanism of attenuation for COD. In a municipal landfill, decomposition will usually take place under anaerobic conditions, which results in slower decomposition rates, chemically reducing conditions, and production of acid by-products (Patrick and Mahapatra, 1968).

Vigorous microbial activity (controlled by near neutral pH, adequate supplies of nutrients, and dissolved oxygen where aerobic growth is possible), absorption, and ion exchange (controlled mainly by high soil organic matter, percentage clay, cation-exchange capacity, iron oxide content, etc.) favor attenuation of organic constituents of leachate. It is suggested that for some organic compounds the best overall corelation is with soil organic matter (Greenland, 1970). In a sanitary landfill, maintenance of an aerobic condition is not usual even though experiments were performed to study the feasibility of maintaining such a condition (Stone, 1975). A fine grain soil will probably favor COD attenuation due to increased surface area and improved mixing between solution and solids. COD is relatively mobile in a clayey environment (Gebhard, 1978).

5.4.10   Chloride

Chloride is not attenuated by any soil type and is highly mobile under all conditions (Apgar and Langmuir, 1971; Polkowski and Boyle, 1970; Gerhardt, 1977). Dilution is the only attenuation mechanism for this leachate constituent.

5.4.11   Chromium

Precipitation and cation exchange and/or adsorption are the principal mechanisms of attenuation of chromium in soil (Griffin, 1977; Gebhard, 1978). The redox potential has a marked effect on attenuation. The importance of each mechanism is dependent on the form of chromium. Chromium is found in two valent states: hexavalent and trivalent. Hexavalent chromium is anionic in form and trivalent chromium is cationic in form. The dominant species appears to be trivalent (Fuller, 1977). In a municipal landfill, hexavalent chromium could be of concern. The concentration of chromium in the soil is reduced by adsorption on organic matter, clay minerals and hydrous oxides of iron, manganese, and aluminum and precipitates as an oxide (Basu et al., 1964). Data obtained in leaching experiments indicate that trivalent chromium is attenuated effectively in soil systems (Griffin, 1977). At a pH exceeding 6, migration will be controlled by precipitation as an oxide, carbonate, or sulfide (Griffin et al., 1976b; Fuller, 1977). Below pH 4, trivalent chromium species are effectively attenuated by adsorption on both kaolinite and montmorillonite (Griffin, 1977). Between this pH range, a combination of the two mechanisms is effective with a radical increase in attenuation to a maximum at pH 6 and above. Attenuation of hexavalent chromium was found to be a function of both concentration and soil pH. Results of leaching experiments indicate that montmorillonite is more effective than kaolinite in attenuating chromium (Griffin, 1977). Retention of hexavalent chromium appears to correlate best with amount of iron oxide, manganese, and clay in the soil (Korte et al., 1976). In clayey soil trivalent chromium is immobile whereas hexavalent chromium is highly mobile. Soil materials contributing to attenuation of chromium will include organic matter, clay minerals, and hydrous metal oxides. The impact of pH will depend on the valence state of the chromium. Less attenuation is expected in coarse textured soils than in fine textured soils because of the larger pores, greater permeability, and smaller amounts of clay minerals (Fuller, 1977).

5.4.12 Copper

Most important attenuation mechanisms for copper include adsorption, ion exchange, and chemical precipitation. The majority of the available information concerning the activity of copper in soil–water environments concerns divalent copper (Gebhard, 1978). Attenuation studies indicate that the removal of copper varies somewhat with clay type. The soil pH is the most important factor controlling removal of copper with a given absorbant (Griffin et al., 1976b; Huang et al., 1977). Montmorillonite appears to be more effective in removing copper than kaolinite; however, the amount of copper removed from solutions was not equal to the cation-exchange capacity of the clay mineral. It was suspected that desorbing calcium ions effectively competed with the heavy metals present in solutions (Griffin et al., 1976a). There are some copper compounds that bcome soluble under acidic conditions. In the pH range of 5–6 precipitation of copper compounds can occur when copper concentrations are high (Griffin et al., 1976b).

Copper attenuation by organic matter is extensive and indicates strong complexing with organic matter. Soil column testing indicates that copper is of very low mobility for a wide range of soils (Korte et al., 1976).

Soil materials favoring attenuation of copper include colloidal matter, free lime, hydrous oxides of manganese and iron, a high clay content, and organic matter content. Use of ground limestone resulted in a significant improvement in attenuation of copper (Fuller and Korte, 1976). It appears that a near neutral pH may be the most effective in copper attenuation by clays (Griffin et al., 1976b), although solubility of copper in soil systems may continue to decrease as the pH rises (Lindsay, 1972). The mobility of copper in a clayey environment is low.

5.4.13   Cyanide

The only attenuation mechanism for cyanide is adsorption. Cyanide is an anion and as such is not strongly retained in soils. The adsorption is dependent on the pH of the soil (Griffin, 1977). Based on the limited information available it appears that cyanide is highly mobile in clayey environments.

5.4.14   Fluoride

The major attenuation mechanism of fluoride in soil appears to be anion exchange. Although Bower and Hatcher (1967) found that acidic soils adsorb fluoride more readily than alkaline soils, Larsen and Widowsen (1971) found that the solubility of fluoride increases in both acidic and alkaline soils. In general, the mobility of fluoride in clayey soil is high.

5.4.15   Iron

Precipitation, cation exchange, adsorption, and biological uptake are the important attenuation mechanisms of iron. Divalent and trivalent iron are present in almost all leachate–soil systems (Gebhard, 1978). Below approximately neutral pH conditions, the solubility of divalent iron increases about 100-fold for each unit decrease in pH (Lindsay, 1972). Iron oxides have been found to be among the most significant factors influencing attenuation processes (Fuller and Korte, 1976). In general, iron compounds appeared to be moderately attenuated in soil (Fuller and Korte, 1976). Although Griffin et al. (1976a) found no significant corelation with cation exchange, Farquhar (1977, cited in Gebhard, 1978) found that iron attenuation did increase with the cation-exchange capacity of soil.

Biological activity can influence iron activity in two ways. Anaerobic growth creates reducing conditions and acid by-products that will convert ferric iron to ferrous iron and increase iron mobility.

A zone may exist in a soil system in which the solubility of iron is considerably greater than the drinking water standard. This zone has near neutral or acidic pH and moderately reducing conditions. Such a zone may form the bulk of migrating leachate plumes. High iron levels in groundwater near landfills may only be a consequence of migration of a moderately reduced zone rather than of migration of iron from the landfill (Roberts and Sangrey, 1977).

The mobility of trivalent iron in soil is low and divalent iron is high (Griffin, 1977). Since it is difficult to assess which species is present, it is better to assume that the mobility of iron is high to moderate in clayey soil.

5.4.16 Lead

The major attenuation mechanisms for lead are adsorption, cation exchange, and precipitation. Although lead may be present in two valence states, practically all the common lead compounds correspond to the divalent state (Gebhard, 1978). Lead attenuation in clays increases as the pH rises above 5 (Griffin, 1977). It was found that the lead removal capacity of montmorillonite is higher than that of kaolinite (Griffin et al., 1976a). In a separate study of soil conducted in New York, the major attenuation mechanism for lead has been reported to be precipitation, which depends on the EH–pH state of the soil (Roberts and Sangrey, 1977). In soil systems, lead can be expected to form poorly soluble precipitates with sulfate, carbonate, phosphate, and sulfide anions. Evidence of a relatively insoluble organic matter complex has also been found (Fuller, 1977). Experimental data indicate that lead hydroxide probably regulates lead activity in soils at pH less than 6.6 (Santillan-Medrano and Jarinak, 1975).

In a municipal landfill, where anaerobic conditions are likely to occur, lead should become more mobile (Fuller, 1977). This is consistent with findings of Griffin et al. (1976a), which indicate that competitive effects of other constituents in municipal landfill leachate can also lower the removal of lead (Gebhard, 1978). Soil materials favoring attenuation of lead in leachate will include organic matter, clays, and free lime. Most effective removal will require a pH greater than 5 or 6. It was also reported that there is a great affinity between lead and organic matter that results in immobilization of lead. Lead is generally more strongly attenuated than many other divalent heavy metals (Fuller, 1977). The mobility of lead in clayey soil is low.

5.4.17 Magnesium

Cation exchange and precipitation are the major attenuation mechanisms for magnesium. Under neutral to alkaline pH, magnesium may form a carbonate precipitate under favorable conditions (Griffin, 1977). Magnesium attenuates moderately in clayey soil.

5.4.18 Manganese

Precipitation and cation exchange are the major attenuation mechanisms for manganese. Manganese is a very common element in soils and can have valence states of $2+$, $3+$, $4+$, $6+$, and $7+$. It has been suggested that under normal conditions manganese exists in soils primarily as insoluble oxides, although under reducing conditions Mn^{2+} is formed and can increase in solubility to the point at which it can participate in ion exchange (Ellis and Knezek, 1972). A divalent manganese ion is formed from a tetravalent ion when the redox potential is in the range of $+200$ to $+400$ mV (Lucas and Knezek, 1972). Investigations of the activity of manganese in soil systems under the influence of synthetic or real landfill leachate are contradictory. Using leachate collected from an Illinois landfill, Griffin et al. (1976a) found that at near neutral pH, manganese was actually removed from kaolinite and montmorillonite. In contrast, Farquhar (1977, cited in Gebhard, 1978) found that manganese was moderately attenuated in several soils. He also found very limited desorption of the retained manganese with the addition of water. Therefore, he concluded that manganese has a high selectivity in the exchange sequence. Farquhar also found that manganese removal in soils appeared to increase with the cation-exchange capacity.

Under either alternate wetting and drying conditions or saturated (anaerobic) conditions, the absorption of manganese is highest on bentonite (montmorillonite), intermediate on illite, and lowest on kaolinite.

Soil materials favoring manganese attenuation include clays, organic matter, hydrous metal oxides, and free lime. Alkaline conditions and an abundance of anions such as sulfide and carbonate will improve retention (Gebhard, 1978). The mobility of manganese in clayey soil is high.

5.4.19 Mercury

The dominant attenuation mechanisms of mercury are adsorption, precipitation, and redox reactions resulting in volatilization of this heavy metal (Gebhard, 1978). In aerated water with a neutral pH, the inorganic mercury species distribution is dominated by mercurous hydroxide, but under reducing conditions, elemental mercury is formed readily and can be lost by volatilization. The methyl mercury ion is common in nature above pH 6, and decomposes slowly to methane at a temperature of around 20°C. In aqueous systems such as a landfill leachate, the predominant species is often mercurous chloride (Fuller, 1977, cited in Gebhard, 1978).

The conversion of inorganic mercury compounds to toxic mono- or dimethyl mercury has been shown to be the result of activity by various bacteria. Variation of the anion initially associated with divalent mercury will affect the rate at which methylation takes place (Ridley et al., 1977). Enzymes produced by enteric bacteria have been found to be effective in demethylating methyl mercury. The products of those reactions include dimethylselenide, methane, and reduced mercury (Wood et al., 1975).

In clayey soils, mercury compounds are likely to be attenuated by adsorption with iron oxide, organic matter, and clays. Experimental results indicate that adsorption of mercury, in either specially prepared mercurous chloride solutions or in mercury-spiked municipal landfill leachate, is pH dependent and increases steadily as the pH rises from about 2 to 8 (Griffin, 1977). However, large amounts of mercury were also removed from solution in the absence of clay, suggesting that precipitation and/or volatilization accounted for the removal of 70–80% of the mercury from leachate solutions (Griffin, 1977).

Using extraction by 0.1 N HCl, Korte et al. (1976) found that sorbed mercury was more easily remobilized by acid solution, especially with several clay soils. Extraction with water yielded negligible to low remobilization. In another test, use of a crushed limestone liner had a low to moderate effect on mercury attenuation and improvement was better with longer contact times. Mercury applied to the environment will not remain in the same state and dimethyl mercury or reduced mercury is expected to be expelled as gas (Fuller, 1977, cited in Gebhard, 1978).

Results of investigations indicate that maximum removal of mercury from leachate should be expected under alkaline conditions. It appears that attenuation can be improved by soil colloidal matter, especially clays and iron oxides. In general, it appears that mercury is highly mobile in soils (Gebhard, 1978).

5.4.20 Nickel

The major attenuation mechanism of nickel includes sorption and precipitation. Jenne (1968) found that nickel is removed from solution by hydrous metal oxide precipitates. Nickel appears to have much greater affinity for manganese oxide than ferric oxide.

Korte et al. (1976) found that nickel retention in soil from a nickel-spiked municipal landfill corelates well with surface area, cation exchange capacity, and clay content. Other soil factors favoring retention of nickel include alkaline conditions, high concentrations of hydrous metal oxides, and free lime. The mobility of nickel is moderate in clayey soil (Griffin, 1977).

5.4.21 Nitrate

The major attenuation mechanism for nitrate is biological uptake. The biological denitrification or reduction to gaseous nitrogen or nitrous oxide requires anaerobic conditions and a carbon source. At a pH of 7 and redox potential of about +225 mV denitrification will begin to take place (Patrick and Mahapatra, 1968). The pH of the system (in the range of 5–8.4) does not seem to have a pronounced effect on the rate of denitrification, but does affect the final product. Limited biological loss of nitrate has also been found to occur by way of direct volatilization of nitrate as nitric acid under conditions of low soil pH or with the presence of high exchangeable aluminum in soil (Gebhard, 1978). Immobilization of nitrate nitrogen by adsorption in the bodies of microorganisms has been suggested (Allison and Klein, 1962). As the carbon source is exhausted and the microorganism population begins to decline, nitrogen with the potential for reconversion to nitrate is again released. Nitrate is considered to be highly mobile in soil (Tisdale and Nelson, 1975; Preul, 1964; Griffin, 1977).

5.4.22   Polychlorinated Biphenyls (PCBs)

The major attenuation mechanisms of PCBs in soil are adsorption and biodegradation. Volatilization and plant uptake are important attenuation mechanisms in the context of landspreading. Attenuation of Aroclor 1254 (one variety of PCB) was found to be dependent on soil type (Iwata et al., 1973). The quantities or sorbent required to sorb 50% of applied PCBs increased in the following order: a peaty muck < montmorillonite < sand < peroxide-treated sand. The presence of activated carbon enhanced the attenuation of PCB (Strek, 1980, cited in Girvin and Sklarew, 1986). Less chlorinated PCBs were observed to have a higher degree of biodegradation (Griffin and Chian, 1980). The position of chlorine atoms also affects attenuation of PCBs in soil. Biodegradation is more significant for monochloro-, dichloro-, and trichloro-PCBs than for pentrachloro- and highly chlorinated PCBs (Pal et al., 1980). There is an active debate regarding rates and reversibility of PCB sorption/desorption in soil. In general, the mobility of PCBs in clayey soil is considered to be high to moderate.

5.4.23   Potassium

The major attenuation mechanisms affecting potassium are precipitation and cation exchange. Griffin et al. (1976a) conclude from experiments that potassium is well attenuated in clayey soil In mica-like minerals such as illite, potassium is readily fixed in a nonexchangeable position. The attenuation is maximum under neutral to alkaline conditions. The mobility of potassium in clayey soil is moderate (Griffin, 1977).

5.4.24   Selenium

The major attenuation mechanisms for selenium are adsorption and anion exchange. Selenium is typically present in soils as an inorganic anion associated with iron, calcium, or sodium. Under the action of soil microorganisms and atmospheric agents, selenium can be oxidized and reduced repeatedly (Fuller, 1977).

Griffin et al. (1976a) studied removal of selenium by kaolinite and montmorillonite using specially prepared selenium-deionized water solutions and selenium-spiked municipal landfill leachate. Selenium removal by montmorillonite was found to be two to three times greater than kaolinite. A distinct pH dependency was observed, indicating that selenium removal improved as the pH dropped to values in the range of 2–4, below which selenium removals decreased. The mobility of selenium in clayey soils is moderate.

5.4.25 Silica

The major attenuation mechanism is precipitation. Not much information is available in the literature regarding the attenuation of silica. Silica readily precipitates in silicate mineral phases. It is moderately mobile in soil, however, mobility increases under alkaline condition (Griffin et al., 1976b).

5.4.26 Sodium

Cation exchange is the major attenuation mechanism for sodium. It may be totally attenuated, but since it is a monovalent ion, a low concentration of sodium could pass through soil without being attenuated at all (Griffin et al., 1976a).

5.4.27 Sulfate

The major attenuation mechanism of sulfate is anion exchange. Adsorption on a clay surface, on organic matter, and on hydrous oxides of aluminum and iron has also been reported (Chao et al., 1962). Because sulfate is relatively weakly held, leaching losses are proportional to the amount of water passing through the soil (Tisdale and Nelson, 1975).

Sulfate is also a product and a reactant of soil microorganisms. It is produced from sulfides and free sulfur by the action of aerobic bacteria; optimal conditions for their growth include abundant oxygen, temperatures near 30°C, and low pH (Tisdale and Nelson, 1975). At a pH of 7 sulfate reduction will occur at a redox potential of about 150 mV, which is much lower than the redox potentials necessary for reduction of either nitrate or ferric iron. If iron is not present sulfide loss as hydrogen sulfide gas can occur (Patrick and Mahapatra, 1968). Since sulfate is an anion its mobility is high in soil (Griffin, 1977).

5.4.28 Viruses

Virus survival within soil depends on soil moisture content, temperature, pH, and nutrient availability. Inactivation of viruses near the soil surface is much more rapid than when it penetrates into the soil (Keswich and Gerba, 1980). Transport of viruses for a long distance through sandy soil has been reported (Wellings et al., 1974). Factors that favor virus removal from leachate include clayey soil, low pH, and the presence of cations (Lu et al., 1985). The mobility of viruses in clayey soil is low.

5.4.29 Volatile Organic Compounds (VOC)

These are organic compounds that volatilize at normal temperature and pressure (NTP), with some exceptions. Biological uptake and dilution are the major attenuation mechanisms for these compounds. The existence of

TABLE 5.2. Major Attenuation Mechanism(s) of Landfill Leachate Constituents

Leachate Constituent	Major Attenuation Mechanism	Mobility in Clayey Environment
1. Aluminum	Precipitation	Low
2. Ammonium	Exchange, biological uptake	Moderate
3. Arsenic	Precipitation, adsorption	Moderate
4. Barium	Adsorption, exchange, precipitation	Low
5. Berillium	Precipitation, exchange	Low
6. Boron	Adsorption, precipitation	High
7. Cadmium	Precipitation, adsorption	Moderate
8. Calcium	Precipitation, exchange	High
9. Chemical oxygen demand	Biological uptake, filtration	Moderate
10. Chloride	Dilution	High
11. Chromium	Precipitation, exchange, adsorption	Low (Cr^{3+}); high (Cr^{6+})
12. Copper	Adsorption, exchange, precipitation	Low
13. Cyanide	Adsorption	High
14. Fluoride	Exchange	High
15. Iron	Precipitation, exchange adsorption	Moderate to high
16. Lead	Adsorption, exchange precipitation	Low
17. Magnesium	Exchange, precipitation	Moderate
18. Manganese	Precipitation, exchange	High
19. Mercury	Adsorption, precipitation	High
20. Nickel	Adsorption, precipitation	Moderate
21. Nitrate	Biological uptake, dilution	High
22. PCBs	Biological uptake, adsorption	Moderate to high
23. Potassium	Adsorption, exchange	Moderate
24. Selenium	Adsorption, exchange	Moderate
25. Silica	Precipitation	Moderate
26. Sodium	Exchange	Low to high
27. Sulfate	Exchange, dilution	High
28. Zinc	Exchange, adsorption, precipitation	Low
29. Virus	Unknown	Low
30. Volatile organic compound	Biological uptake, dilution	Moderate

After Bagchi (1987b); courtesy of Waste Management and Research, Denmark.

several VOCs in landfill leachate has been reported (Sridharan and Didier, 1988; McGinley and Kmet, 1984). Although biodegradation of organics has been demonstrated by several researchers (Callahan et al., 1979; Tabak et al., 1981; Petrasek et al., 1983; Alexander, 1981; Barker et al., 1986), the rate of metabolism is the main factor (Richards and Shieh, 1986). In a soil–leachate system attenuation of VOC is not expected to be high (Edil et al., 1992). VOC concentrations of groundwater around landfills with a volume of more than 38,000 m^3 (50,000 yd^3) were found to be high (Friedman, 1988; Battista and Connelly, 1988).

5.4.30 Zinc

The major attenuation mechanisms for zinc are adsorption, cation exchange, and precipitation. Zinc is a common cation in soil systems. As is true with other cations, the pH of the leachate–soil system is a crucial factor in zinc removal, reflecting the influence of dominant hydrolysis species on both the affinity for soil colloids and the sollubility of zinc (Gebhard, 1978).

The attenuation of zinc was found to increase rapidly for a pH change from 2 to 8 with a significant rise around 6 to 8 (Griffin et al., 1976b). Precipitation of zinc with a variety of anions—including sulfide, phosphate, carbonate, and silicate—has also been found to be important in zinc immobilization (Stumm and Morgan, 1970; Fuller, 1977). Experimental results suggest that the removal of zinc is also dependent on clay type and cation-exchange capacity (Griffin, 1977; Farquhar, 1977). Organic matter improves zinc immobilization (Folett and Lindsay, 1971; Huang et al., 1977; Norcell, 1972), and zinc chelates are most stable at pHs between about 5 and 7.5 (Folett and Lindsay, 1971; Huang et al., 1977; Norcell, 1972). Soil material favoring attenuation of zinc includes clays, organic material, hydrous metal oxides, and free lime. Zinc attenuation will be most favored by an alkaline condition. In general, mobility of zinc in a clayey environment is low (Griffin, 1977).

5.4.31 Summary

Attenuation mechanisms for 30 common landfill leachate constituents are discussed. A summary of the attenuation mechanisms of these constituents is included in Table 5.2. From the above discussion, the following general trends can be observed: (1) Most metals attenuate well in clayey soil, (2) nonmetals are not attenuated well in clayey soil, (3) very low attenuation of nitrate, sulfate, VOC, and COD is expected to occur in clayey soil, and (4) chloride is not attenuated in soil at all: dilution is the only mechanism of attenuation for choloride.

5.5 DESIGN APPROACH

Although the methodology discussed in this section is applicable to the design of municipal landfills, the approach may be adapted to design nonhazardous, nonmunicipal landfills. Because of the microorganisms present in municipal landfill leachate it is difficult to duplicate leachate in a laboratory that can represent a field leachate. The exact composition of leachate and its variation with time cannot be established until the landfill has been operated for some time. Because of this problem exact quantitative analysis of the chemical reaction between a landfill leachate and soil, before a landfill siting, is not possible. As an alternative, a semiquantitative design method is used utilizing the information included in Sections 5.2, 5.3, and 5.4. The following issues need to be addressed to develop a design approach for NA type landfills (Bagchi, 1983):

1. A list of leachate constituents with significant concentrations in munici-pal leachate, and the expected total mass of some of these constituents during the design life of the landfill.
2. Attenuation mechanism(s) of the leachate constituents with significant concentrations and the overall impact from the waste specific leachate on groundwater (if available from field study).
3. The volume of soil in the unsaturated zone that is involved in the attenuation reaction.
4. Dilution of leachate constituents in the groundwater aquifer.

Soil stratification shown in Figure 5.4 is assumed in developing this design approach.

5.5.1 Major Leachate Parameters

The pollutants of concern in municipal landfill leachate are copper, lead, zinc, iron, ammonium, potassium, sodium, magnesium, BOD, COD, nitrate, chloride, and sulfate (ASCE, 1976; Chian and Dewalle, 1975; Garland and Mosher, 1975; Ham and Anderson, 1974, 1975a,b). Chemical constituents and their concentrations in leachate vary over a wide range. Refuse composition, landfill age, and climate are the main factors causing variability. The general-ized concentration variation plot, shown in Fig. 5.6, is based on field observa-

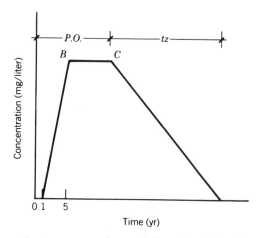

FIG. 5.6. Generalized concentration variation plot for leachate constituents.

tions and laboratory studies (Griffin et al., 1976a; Meyer, 1973). This plot is used to calculate the total mass leached for each of the following: ammonium, potassium, sodium, and magnesium.

In Fig. 5.6, "P.O." represents the period from the start of the landfill to closure. For P.O. < 5 years point C coincides with point B. tz is the time (since closure of the landfill) at which the leachate concentration is low enough so as not to cause any environmental problems. Use of this plot to calculate the total mass of ammonium, potassium, sodium, and magnesium is shown in an example, worked out at the end of Section 5.5.5.

5.5.2 Attenuation Mechanisms of Pollutants and Major Impact of Leachate

Five mechanisms, adsorption, biological uptake, cation-exchange reaction, filtration, and precipitation, are operative in the clayey stratum (first stage). Although all six attenuation mechanisms could be operative in the aquifer (second stage), only one mechanism, dilution, is assumed to be operative in the groundwater aquifer. This is because the aquifer sand is assumed to have negligible amounts of fines. This assumption leads to a conservative design.

The first stage of attenuation, which takes place in the initially unsaturated zone, will cause the pH of the leachate–soil system to merge to near neutral value (Roberts and Sangrey, 1977). Attenuation of copper, lead, zinc, iron (in part), ammonium, magnesium, potassium, and sodium will occur in the first stage under the near neutral pH environment. Ammonium, magnesium, and potassium will be exchanged to elute calcium and thus will increase the hardness of the groundwater (Griffin et al., 1976a). BOD, COD, iron (in part), nitrate, chloride, and sulfate are attenuated through dilution.

Field observations on the impact of leachate on groundwater are supportive of the laboratory observations indicated in Section 5.4. Many have observed a predominant increase in hardness of the groundwater in the vicinity of municipal landfills (Andersen and Dornbush, 1967; Walker, 1969; Ziezel et al., 1962).

5.5.3 Soil Volume Involved in Attenuation

Seepage of water from a channel to the groundwater table, the channel bottom being covered with a thin layer of sediment, is discussed by Bear (1969). This model fits the municipal landfill situation. Neglecting the length effect and leachate head buildup in a municipal landfill (which, in a well-designed municipal landfill, is normally low) the volume of soil involved in attenuation reactions could be approximated by

$$V_s = R \times A \times H \tag{5.1}$$

in which V_s = the soil volume available for attenuation reaction in cubic meters and R = the reduction factor to account for the soil fabric effect (varies between 1/1.2 and 1/1.3). The flow of pore fluid through the soil is mostly via large pores, therefore the total cation exchange capacity of the attenuating soil mass must be reduced to account for the soil that does not come into contact with the leachate. A = the base area of the landfill in square meters and H = the average depth of the unsaturated zone beneath the landfill in meters. The value of V_s is somewhat conservative but is reasonable for practical purposes (Mundell, 1984; Bagchi, 1984).

5.5.4 Dillution of Pollutants

Dillution does not reduce the amount of contaminant in the flow system but does reduce the concentration of contaminant. When the exfiltrate (the leachate after flowing through the unsaturated zone) reaches the groundwater table, the concentration of each pollutant is reduced further either due to density differences or diffusion and dispersion (Bouwer, 1979; Bruch and Street, 1967) or a combination of both. The plume geometry suggested in Fig. 5.7 is developed from laboratory and field data (Bouwer, 1979; Bruch and Street, 1967; Fattah, 1974; Freeze and Cherry, 1979; Kimmela and Braids, 1974; Nicholson et al., 1980; Sykes et al., 1969). The plume geometry is applicable for diluting only nonreactive pollutants in a nonreactive aquifer. The dotted line in Fig. 5.7a represents the more probable plume configuration; however, so far as steady-state dilution is concerned, the discrepancy between the probable plume and suggested plume geometry will not significantly affect results.

The area that will be intercepted by the proposed plume at some horizontal distance is given by

$$A_i = [(L_1 + X_1)\tan \theta_1 + X_2 \tan \theta_2] \times [L_2 + 2(X_1 + L_1) \tan \theta_3 + 2X_2 \tan \theta_4] \tag{5.2}$$

in which A_i = the area intercepted by groundwater for effecting pollutant dilution in square meters. X_1 and X_2 = the horizontal distance as shown in Fig. 5.7 in meters. L_1 = the dimension of the landfill parallel to groundwater flow in meters and L_2 = the dimension of the landfill perpendicular to groundwater flow in meters; θ_1 through θ_4 = divergence angles in degrees.

Equation (5.3) is obtained by equating the mass per unit time of a parameter passing through area A_i at section S2–S2 and the sum of pass per unit time of the parameter in the ambient water passing through the same area A_i at section S1–S1 and leachate exfiltrate, which enters the proposed plume configuration through the landfill bottom.

$$C_b A_i U + C_I V_e = C_{dx}(A_i U + V_e) \tag{5.3}$$

in which C_b = the average concentration of a parameter in the background water in mg/m^3, C_I = the peak concentration of a parameter in the leachate in mg/m^3, U = the average groundwater velocity in m/day, V_e = the average

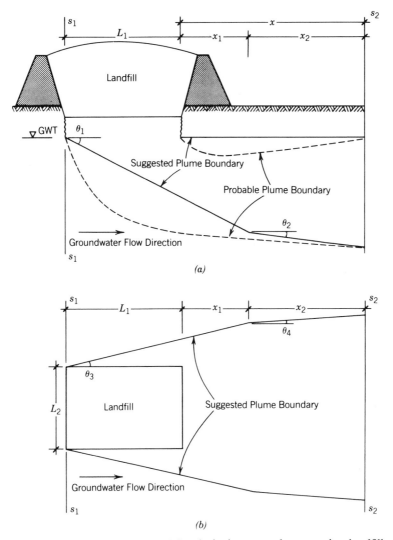

FIG. 5.7. Plume geometry used for designing natural attenuation landfills.

volume of leachate exfiltration per unit time in m³/day, C_{dx} = the average concentration of a parameter in the down-gradient water at distance X ($= X_1 + X_2$) in mg/m³. Rearranging the terms

$$\frac{C_b}{1 + (V_e/A_iU)} + \frac{C_1}{1 + D(A_iU/V_e)} = C_{dx} \qquad (5.4)$$

The plume geometry is applicable close to the landfill, which would mean that X_1 is less than or equal to 90 m (300 ft) and ($X_1 + X_2$) is less than or equal to 300 m (1000 ft). The suggested values of the divergence angles are $\theta_1 = 4$–$7°$, $\theta_2 = 2$–$5°$, $\theta_3 = 2$–$3°$, and $\theta_4 = 1$–$2°$. These values are based on limited field data; if available, additional field data should be taken into consideration.

5.5.5 Design Procedure

The following procedure is applicable to a typical NA type landfill site described earlier (Fig. 5.4):

1. Copper, lead, and zinc: Because the concentrations of these heavy metals are low in leachate they will precipitate out in a clayey environment.

2. Iron: Attenuation of iron is not high in clayey soil. Assume that 50% of peak concentration will be diluted in the sandy aquifer.

3. Ammonium, magnesium, potassium, and sodium: These will undergo cation exchange and in turn will increase the hardness of the exfiltrate from the first stage. First, soil volume should be checked to see whether enough CEC is available for performing the exchange reaction with these four pollutants. If adequate CEC is available then the hardness of the exfiltrate (the hardness of the leachate plus the increase in hardness due to the cation-exchange reaction of these pollutants) should be checked for dilution.

4. BOD, COD, nitrate, chloride, and sulfate: Dilution is the only long-term attenuation mechanism for these constituents. Therefore their peak concentration should be checked for dilution:

The landfill plan dimensions should be increased, if necessary, until acceptable concentrations for all leachate parameters are met within 300 m

TABLE 5.3. Peak Concentration of Selected Parameters in Leachate

S1 Number	Constituent	Leachate (mg/liter)	Background Water (mg/liter)
1	Ammonium	200.00	0.00
2	BOD (5 day)	5,000.00	0.00
3	Chloride	500.00	100.00
4	COD	12,000.00	0.00
5	Copper	0.5	0.01
6	Hardness ($CaCO_3$)	4,000.00	50.00
7	Iron	200.00	0.03
8	Lead	0.2	0.005
9	Magnesium	150.00	0.00
10	Potassium	200.00	0.00
11	Sodium	300.00	0.00
12	Sulfate	300.00	50.00

(1000 ft) of the landfill or the property boundary, whichever is less. The following example will enumerate the design steps.

Example 5.1

The proposed disposal site has a dimension of 100 m (parallel to the groundwater flow $= L_1$) by 150 m (perpendicular to the groundwater flow $= L_2$). The unsaturated zone is 6 m thick and has clayey soil (CEC $= 30$ mEq/100 g of soil). The proposed waste volume is 60,000 m³ with a site life of 10 years. The expected peak concentration of parameters in the leachate and in the background water is given in Table 5.3. Groundwater velocity in the sandy aquifer is $1 = 10^{-1}$ m/day. Given $R = 1/1.2$, tz $= 15$ years, unit weight of soil $= 1.9$ g/cm³, $\theta_1 = 5°$, $\theta_2 = 3°$, $\theta_3 = 2.5$dg, $\theta_4 = 1°$, $X_1 = 60$ m, $X_2 = 100$ m, and rainfall $= 75$ cm/year.

Calculation for Exchange Reaction. The available cation exchange capacity (CEC) of soil: substituting relevant values in Eq. (5.1):

$$V_s = (1/1.2) \times 100 \times 150 \times 6 = 7.5 \times 10^4 \text{ m}^3$$

$$\text{Weight of soil} = (7.5 \times 10^4) \times (1.9 \times 10^6) = 1.425 \times 10^{11}$$

$$\text{Available CEC in the soil} = (1.425 \times 10^{11}) \times (30/100)$$

$$= 42.75 \times 10^9 \text{ mEq}$$

Assuming a uniform filling over the 10-year period, the following formula can be obtained from Fig. 5.6 for the total mass for ammonium, potassium, magnesium, and sodium leached from the refuse.

$$\text{Total mass leached} = \frac{V_f C_1}{4}(1 \times 2 + 2 \times 3 + 3 \times 4 + 4 \times 5)$$

$$+ V_f C_1(6 + 7 + \cdots + 10) \qquad (5.5)$$

$$+ \frac{10 \ V_f C_1}{15}(1 + 2 + 3 + \cdots + 14)$$

$$\text{where } V_f = \text{landfill filling rate (m}^3/\text{year)}$$
$$= 60,000/10 = 6000 \text{ m}^3/\text{year.}$$

Sample Calculation

1. First term (year 3): Mass leached in third year $= C_3 \times 3V_f \times 1$. From Fig. 5.6: $C_3/2 = C_1/4$. Therefore mass leached in the third year $= (V_f \times 2C_1 \times 3 \times 1)/4$.

2. Second term (year 7): Mass leached in the seventh year $= 7V_f \times C_1 \times 1 = V_f \times C_1 \times 7$.
3. Third term (year 23): Mass leached in the twenty-third $= 10V_f \times C_{14} \times 1$. From Fig. 5.6: $C_{23}/2 = C_1/15$. Mass leached in the twenty-third year $= (10V_f \times C_1 \times 2 \times 1)/15$.

Note that the first term of Eq. (5.5) will remain the same so far as P.O. $>$ 5 but the second and third terms will depend on the period of operation (for P.O. > 5).

From Table 5.4 the total milliequivalents available in soil is higher than the total leached. (Note: mEq/liter $=$ concentration (mg/liter $\times$ valence/atomic weight.) Hence, hardness causing constituents in the leachate will be exchanged totally by the clayey stratum. To calculate the increase of hardness in the groundwater, the peak concentration of each of these pollutants shall be used to study the worst case. Peak increase in hardness due to exchange reaction $= 41.6 \times 40.08/2$ (assuming exchange takes place with the calcium ion only) $= 833$ mg/liter. Total hardness of the exfiltrate $= 4833$ mg/liter.

Dilution Calculation. Substituting proper values in Eq. 5.2: $A_i = 3222$ m^2. Assuming 20% infiltration of precipitation, the average exfiltration of leachate into the groundwater system per day will be

$$V_e = (0.72 \times 0.2 \times 100 \times 150)/365 = 6.16 \text{ m}^3/\text{day}$$

Groundwater quality change due to the landfill siting is summarized in Table 5.5.

TABLE 5.4. Total and Peak Milliequivalents of Hardness Causing Pollutants Leached

S1 Number	Pollutants	Peak Concentration (mg/liter)	Peak (mEq/liter)	Total Leached (mEq $\times$ 10^9)
1	Ammonium	200	11.11	8
2	Magnesium	150	12.34	8.9
3	Potassium	200	5.11	3.7
4	Sodium	300	13.04	9.4
		Total	41.60	30

After Bagchi (1983).

TABLE 5.5. Change of Groundwater Quality due to the Landfill Siting at 160 m Down-Gradient

S1 Number	Pollutants	Background Concentration (mg/liter)	Average Concentration 160 m Down-Gradient (mg/liter)
1	Ammonium	0.00	0.00
2	BOD (5 day)	0.00	93.8
3	Chloride	100.00	107.5
4	COD	0.00	225.1
5	Copper	0.01	0.01
6	Hardness	50.00	139.7
7	Iron	0.03	1.9
8	Lead	0.005	0.005
9	Magnesium	0.00	0.00
10	Potassium	0.00	0.00
11	Sodium	0.00	0.00
12	Sulfate	50.00	54.7

After Bagchi (1983).

Sample Calculation. (S1 number 3, Table 5.5):

$$C_{dx} = \frac{100.0}{1 + (6.16/3222 \times 10^{-1})} + \frac{500.0}{1 + (3222 \times 10^{-1}/6.16)}$$

$$= 107.5 \text{ mg/liter}$$

5.6 SUMMARY AND COMMENTS

As is apparent from the previous sections a total quantitative analysis of a NA type landfill is not possible. All of the following items are site specific: quantity and quality of leachate, chemical composition of the soil and therefore the attenuative capacity of the soil in the unsaturated zone, and plume geometry and therefore the dilutive capacity of the groundwater aquifer. So it is difficult to say how much waste can be deposited safely in a NA type landfill at a particular site. In general, approximately 35,000–40,000 m³ of municipal garbage is the maximum volume that can be allowed in a natural attenuation landfill. A higher volume may be allowed for other nonputrescible waste type. The problem is compounded by applicable groundwater rules at each site. Some areas follow a qualitative approach (e.g., a policy of no degradation of groundwater or allowing degradation up to a certain percentage above background water quality) and others follow a quantitative approach (i.e., set concentration limits at a set distance from the landfill).

TABLE 5.6. Chemicals Usually Found in Municipal Landfill Leachate and Their Potential Contributors

Parameters	Products
Organics	
Acetone	Carburetor and fuel injection cleaners, paint thinners, paint strippers and removers, adhesives, fingernail polish removers
Xylene	Oil and fuel additives, carburetor and fuel injection cleaners, adhesives, paints, transmission additives
Methylene chloride	Oven cleaners, tar removers, wax, degreasers, spray deodorants, brush cleaners
Toluene	Contact cement, degreasers, paint brush cleaners, perfume, dandruff shampoo, carburetor and fuel injection cleaners, paint thinners, paint strippers and removers, adhesives, paints
1,1,1-Trichloroethane	Drain and pipe cleaners, oven cleaners, shoe polish, household degreasers, deodorizers, leather dyes, photographic supplies
cis/*trans*-1,2-Dichloroethylene	Contact cement, perfumes, make-up (perfume), upholstery and rug cleaners
Benzene	Adhesives, antiperspirants, deodorants, oven cleaners, tar removers, medicines, solvents and thinners
1,1-Dichloroethane	Degreasers, adhesives
Metals	
Lead	Batteries, electrical solder, paints
Cadmium	Paint, pigment, plastics
Chromium	Cleaners, paint pigments, linoleum, batteries
Nickel	Batteries, spark plugs, electrodes
Zinc	Batteries, solder, TV screens

(Based on a study conducted by Minnesota Pollution Control Agency, Minneapolis, Minnesota, U.S.A.)

Theoretically a NA type landfill will always alter groundwater quality downgradient of the landfill irrespective of the volume of waste deposited at the site. So strictly speaking, a NA type landfill will invariably degrade groundwater. As previously, discussed, the thickness of the unsaturated zone does not dictate the overall impact on groundwater. The hardness of the exfiltrate will always be higher than that of leachate; some of the leachate constituents (e.g., chloride, nitrate, and organics) are not attenuated by the unsaturated zone. So, these constituents will degrade the downgradient water quality compared to background water quality. For some remote sites a minimal impact may be tolerated. However, if the design shows that a severe impact is probable, than a NA type landfill should not be designed for the site.

Because of the above problems, a parametric study may be used to develop the range of allowable waste volume for a site. The designer may then compare the site against an ideal site and arrive at a conclusion after giving due consideration to the local groundwater rules.

A list of organic chemicals usually found in landfill leachate and their potential contributors is included in Table 5.6. Exclusion of discarded containers of these products from the solid waste stream disposed in a landfill will minimize the risk of groundwater contamination from the chemicals listed in Table 5.6.

LIST OF SYMBOLS

V_s	= Soil volume available for attenuation reaction
R	= reduction factor to account for soil fabric effect (varies between $1/1.2$ to $1/1.3$)
A	= base area of landfill
H	= average depth of unsaturated zone beneath the landfill
A_i	= area intercepted by groundwater for effecting pollutant dilution
X_1, X_2	= horizontal distances
L_2	= dimension of landfill perpendicular to groundwater flow
L_1	= dimension of landfill parallel to groundwater flow
θ_1 to θ_4	= divergence angles
C_b	= average concentration of a parameter in the background water
C_1	= peak concentration of a parameter in the leachate
U	= average groundwater velocity
V_e	= average volume of leachate exfiltration per unit time
C_{dx}	= average concentration of a parameter in the down-gradient water at distance X

6 Containment Landfills

The design concept for a containment type landfill consists of restricting leachate seepage into the aquifer so as to minimize groundwater degradation. To satisfy these design criteria, landfills are lined with clay or synthetic membrane or both and a leachate collection system is installed (Fig. 6.1a and 6.1b). For a few waste types and insensitive environments, total containment may be needed. How the regulatory agency defines groundwater must be considered before designing a containment site. In general, groundwater is loosely defined as any water in the ground that can be withdrawn for use; however, it may also be defined as any water that occurs in a saturated subsurface geological formation of rock or soil. Thus, based on the first definition only groundwater aquifers need to be protected whereas the second definition includes saturated clay deposits and perched groundwater (which are usually not "tapped") in addition to groundwater aquifers. Detailed design and groundwater monitoring will depend on how groundwater is defined, which should be clarified by consulting the regulatory agency.

Theoretically leakage through the base of a containment landfill is unavoidable, however, it can be reduced to practically zero. In general, leakage through the liner of a single lined landfill is higher than through the liner of a properly designed and constructed double lined landfill. It may be mentioned that if the primary mode of transport of the leachate constituent(s) is diffusion, low leakage cannot be achieved using the conventional containment landfill design discussed in this book.

In regions with a shallow groundwater table (GWT), the landfill base may have to be constructed below the groundwater table. Therefore, landfills can also be divided into two types, based on whether the GWT is below or above the landfill base.

Usually clay or synthetic material is used in lining a landfill. A detailed discussion regarding the suitability of liner material can be found in Chapter 7. Discussions on how to determine the spacing for the leachate collection pipe and the thickness of the drainage blanket are included in Section 6.6 and a discussion on liner thickness is included in Section 6.4. There are several other elements (e.g., collection pipe) in a containment landfill that need to be designed. Detailed designs of each of these elements are included in Chapter 8.

6.1   SINGLE LINED LANDFILLS

Figure 6.1a and 6.1b shows cross section and plan of a typical single lined landfill. As mentioned, either clay or a synthetic membrane may be used for lining a site. Synthetic materials allow less leakage but are difficult to protect from damage, whereas clay liners are not easily damaged. The chances of damaging the liner in nonsludge landfills are higher. Hence clay is preferred as a liner in such landfills. For sludge landfills a synthetic membrane may be used provided care is taken to avoid running the compaction equipment directly on the liner. Sometimes drag lines are used to dispose of sludge in landfills, which may damage the top portion of the liner. In such cases a combination of clay and synthetic lining may be used; one such scheme is shown in Fig. 6.2.

Leakage through a properly constructed clay liner is not high. Gordon et al. (1989) published some data on clay liner leakage that indicate that leakage through a clay liner reduces over time (probably due to a decrease in perme-

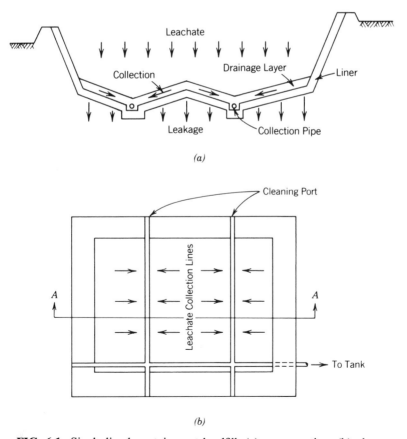

FIG. 6.1. Single lined containment landfill: (a) cross section; (b) plan.

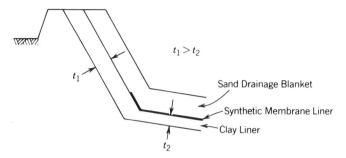

FIG. 6.2. Scheme showing combined clay and synthetic membrane lining.

ability of the linear). Leakage through a synthetic membrane liner could be higher than anticipated if damage (during installation and landfill operation) goes undetected. Thus, if synthetic membrane is used as liner extreme caution must be exercised to protect the liner during both the construction and operation of a landfill.

6.2 DOUBLE OR MULTIPLE LINED LANDFILLS

A double or multiple lined landfill may have one or two leachate collection systems. A landfill constructed with two liners, a synthetic membrane liner placed over a clay liner (Fig. 6.3), is called a composite lined landfill. Figures 6.3 to 6.8 show conceptual drawings for different options for double or multiple lined landfill designs. A discussion of each option is included in the following section.

6.2.1 Comments on Design Issues

To design a landfill with two collection systems it is necessary to understand the mechanics of leachate apportionment in a lined landfill. Current leachate apportionment models (see Section 6.6) show that leachate flow toward a collection pipe is due to the gradient of the phreatic surface; the slope of the liner (usually 2–4%) has very little effect, if any, in driving the leachate

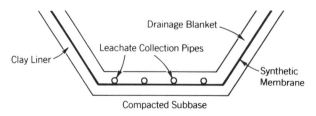

FIG. 6.3. Double lined landfill with a single collection system: scheme 1.

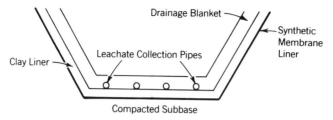

FIG. 6.4. Double lined landfill with a single collection system: scheme 2.

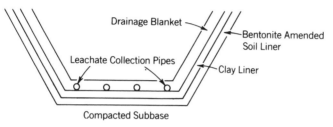

FIG. 6.5. Double lined landfill with a single collection system: scheme 3.

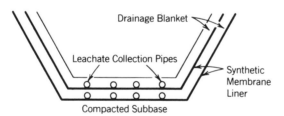

FIG. 6.6. Multiple lined landfill with two collection systems: scheme 1.

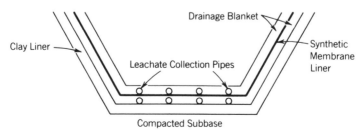

FIG. 6.7. Multiple lined landfill with two collection systems: scheme 2.

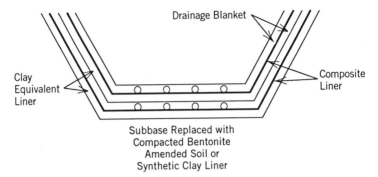

FIG. 6.8. Multiple lined landfill with two collection systems: scheme 3.

toward the collection pipe. For a properly designed and constructed liner, leakage through the first liner will be low. If there is a second collection system below the first liner then the permeability of the second liner installed for the second collection system is critical in effectively collecting leachate. The rate of input on the second liner is the rate of leakage through the first liner. Leachate apportionment models show that leakage through a liner with permeability of 1×10^{-7} cm/sec) is extremely high (80–100% of input rate) if the input rate is low (2–3 cm per year). However, if the second liner has a permeability of 1×10^{-9} cm/sec or less, then the leakage through the second liner will be practically negligible provided advection is the primary mode of transport of the leachate constituents through a porous media.

The next issue that needs to be understood is the thickness of the drainage blanket. All the models discussed in Section 6.6 use a single permeability of the drainage blanket, which means that the thickness of the drainage blanket is more than the maximum height of leachate mound that occurs near the crest of the liner at all times during the active site life of the landfill. The maximum height of the leachate mound fluctuates depending on the amount of leachate volume generated each day. If the permeability of the waste is lower than the permeability of the drainage blanket (e.g., municipal waste, papermill sludge) and if the drainage blanket is not thick enough to accommodate the leachate mound, then part of the leachate mound will penetrate the waste, which will increase the depth of the leachate mound from the point of intersection of the mound with the waste. This increase in depth of leachate mound will increase leakage through the liner (see Section 6.6 for a detailed discussion of this topic).

Leakage through the composite liner shown in Fig. 6.3 will be very low. Chances of damaging the synthetic liner during placement of the sand drainage blanket can be minimized by adopting a proper construction technique (see Section 9.4). Because of construction-related issues the option shown in Fig. 6.4 may not provide as much protection against leakage as the option shown in Fig. 6.3. Theoretically, a synthetic liner will allow significantly lower leakage compared to a clay liner. So the synthetic liner should be able to hold leakage coming from the clay liner. However, damage to the synthetic liner during the construction of the clay liner above it is unavoidable. If the subbase is sandy then even a small damaged area will allow the majority of the leachate to escape. The chance of damaging the first synthetic liner shown in Fig. 6.3 during construction is low; even if there is a damaged area the clay liner below will allow far less leakage compared to a compacted sandy or clayey subbase.

The option shown in Fig. 6.5 is comparable to the option shown in Fig. 6.3. In general, the use of a thin layer (30–45 cm) of bentonite-amended soil liner alone is not recommended, mainly because of the relatively high probability of deterioration due to desiccation cracks and chemical incompatibility with the leachate. Construction of the clay liner above the bentonite liner can provide protection against both of these. However, the clay liner should be constructed immediately after laying the bentonite-amended soil layer to minimize desiccation cracks.

The option shown in Fig. 6.6 will not provide reliable protection against leakage, although there are two layers of synthetic membrane liners and two collection systems. Damage to any of the two synthetic liners during construction will cause significant leaking because the highly permeable drainage blanket will not sufficiently impede the flow through the liners. In the option shown in Fig. 6.7 the leakage through a damaged synthetic liner will be high because of the sand drainage blanket below it. Leakage through the clay liner will also be high because the relatively low volume of leachate leaked through the first liner will not create a significant leachate mound over the second liner; hence most of the leachate leaked through the first liner will escape through this clay liner.

The option shown in Fig. 6.8 has a multiple liner and two collection systems. Of the options shown in Figs. 6.3 to 6.8 this will provide maximum protection against leakage. While a clay liner is recommended for the first composite liner, either bentonite-amended soil or a synthetic clay liner may be substituted for the clay liner in the second composite liner. The subbase need not be replaced with bentonite-amended soil or synthetic clay liner if the subbase soil is clayey. Scarification and recompaction of the subbase will provide a high degree of protection.

The following points should be taken into consideration in designing a double or multiple lined landfill:

1. A clay, synthetic clay liner, or bentonite liner alone should not be used below the second collection system because, even when constructed properly, a clay or bentonite liner will collect only a small percentage of leachate leaked from the first liner.

2. A synthetic liner should not be used alone; it should always be used in combination with a clay, synthetic clay liner, or bentonite-amended soil. If such composite liner is used then placement of the synthetic liner over the clay or bentonite-amended soil is recommended.

3. The drainage blanket over the first liner should be thick enough to accommodate the leachate mound at all times during the active site life. Use of a thin synthetic blanket (e.g., geogrid) is not recommended over the first liner.

4. The drainage blanket over the second liner need not be as thick as the first drainage blanket. A thin synthetic blanket may be used as a drainage blanket above the second liner.

5. Leachate apportionment models included in Section 6.6 are not applicable if a thin synthetic blanket is used as a drainage blanket. The leachate apportionment models are applicable only when a drainage blanket of sufficient thickness to accommodate the entire depth of the leachate mound is used.

6.3 LINER MATERIAL SELECTION CRITERIA

Selection of liner material depends on the waste type and landfill operation. The liner material must be compatible with leachate. In other words, the leachate generated from the waste must not degrade the liner material. A detailed discussion of this issue can be found in Chapter 7. For a double or multiple lined landfill compatibility of the primary liner with the leachate is important; compatibility of the secondary liner with the leachate may not be as important an issue except for some acutely hazardous wastes. Landfill operation also influences design. The thickness of the drainage blanket should be increased if a synthetic membrane is used as the primary liner where heavy compaction equipment is expected to be used in the landfill (e.g., municipal refuse). If the primary liner is a synthetic membrane then extreme care should be taken during the construction of the drainage blanket, and during disposal of the first 1.2 m (4 ft) of waste so that the liner is not damaged. A thicker drainage blanket will provide better protection for a synthetic membrane liner.

6.4 COMMENTS ON LINER THICKNESS

The liner should be thick enough to provide a low permeability layer. Although theoretically, a thin low permeability layer is sufficient to reduce leakage, several other factors need to be considered in determining liner thickness.

For synthetic liners, degradation due to ultraviolet rays and puncture resistance are determining factors in choosing liner thickness. In general synthetic membranes 1.5–2 mm (60–80 mil) thick are used for lining landfills. Additional testing regarding loss of strength due to exposure to leachate must also be considered in choosing the thickness of a synthetic membrane liner. Refer to Section 7.2 for a more detailed discussion of these issues.

Clay liner thickness is more dependent on construction-related issues and degradation due to freeze–thaw and desiccation than on theoretical design.

Desiccation crack and freeze–thaw degradation increase the permeability of a compacted clay liner (refer to Section 7.1). Therefore although a 15- to 30-cm (6-in. to 1-ft)-thick clay liner is theoretically acceptable for providing a low permeability layer, a thicker liner should be constructed for the reasons mentioned above.

Benson (1990) reported the use of a stochastic model to evaluate the minimum thickness of a clay liner considering only the hydraulic behavior of soil. The hydraulic conductivity of a soil liner can vary significantly within a few meters (Rogowski, 1985). Benson (1990) indicated that a 1500 cm^2 area of soil liner is sufficient to incorporate the variability in hydraulic property. The following factors influence the overall hydraulic conductivity of clay liners: flow through macropores (both vertical and horizontal), variation in soil properties, variation in leachate production rate, degradation by chemicals present in leachate, cracks due to desiccation, and freeze–thaw cycles. Benson's stochastic model incorporated the variability and uncertainty associated with hydraulic properties of soil due to improper construction practices. The schematic of macropores used in the model is shown in Fig. 6.9. The variation of equivalent hydraulic conductivity with liner thickness predicted by using the models is included in Fig. 6.10, which shows that the sensitivity of uncertainty associated with *in situ* hydraulic conductivity, for certain typical hydraulic conductivities, is negligible for liner thicknesses of 90 cm or more. In other words, an overall hydraulic conductivity of 1 × 10^{-7} cm/sec can be easily achieved in a 90 cm or more clay liner even when quality control during liner construction is poor. Based on *in situ* hydraulic conductivity measurement of several landfill liners, Benson (1990) concluded

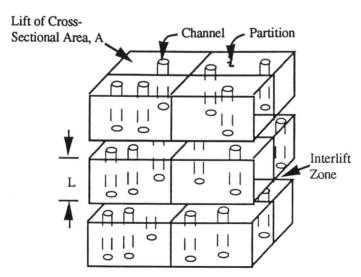

FIG. 6.9. Schematic of soil liner described by the Stochastic model. After Benson, (1990.)

that *in situ* hydraulic conductivity of liners is less sensitive to thickness if excellent construction methods are used and that clay liners as thin as 60 cm may perform adequately when excellent construction methods are used.

A liner may be exposed to natural elements for 1 year whereby the permeability of the top 30–60 cm (1–2 ft) may increase by an order of magnitude. A clay liner 1.2–1.5 m (4–5 ft) thick will provide at least a 60- to 90-cm low permeability layer even after degradation of the top 30- to 60-cm layer. In addition, construction of a thick liner (1.2–1.5 m) will ensure an overall low permeability barrier because chances of developing macrofabric features such as stratification, fissuring, and large-scale heterogeneity are low if proper construction and quality control measures are undertaken.

6.5 COMMENTS ON FINAL COVER DESIGN

The design and construction of the final cover on a landfill are as important as the liner at the base of the landfill. The final cover should be such that it will allow infiltration that is less than or equal to the leakage through the base of the landfill. If the infiltration is more than the leakage, then the collection system has to be maintained in perpetuity to avoid continuous leachate head build-up within the landfill. Such head build-up will cause more leakage and will increase the possibility of seep(s) through the berm. The barrier layer (see Section 8.8) in the final cover should be the same as or better than the primary liner at the base of the landfill. Refer to Section 8.8 for a detailed discussion on final cover design.

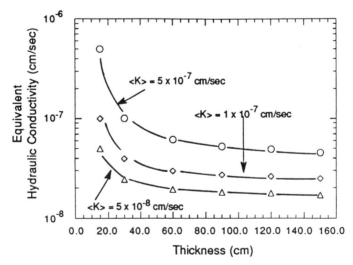

FIG. 6.10. Equivalent Hydraulic conductivity as a function of thickness. After Benson, (1990.)

There is a debate as to whether a final cover should be constructed over a landfill as soon as a reasonable area (from a construction standpoint) reaches the designed final grade (Note: This debate is applicable to a containment type landfill only. Final cover on a NA type landfill should be constructed as soon as a reasonable area reaches the designed final grade) or be deferred to a later date. This debate pertains to municipal waste landfills only; however, the concept is applicable to any putrescible waste that can be composted. The current approach for disposing of municipal waste is to encapsulate it to minimize the impact on groundwater. In following sub sections the opponents' and supporters' views on the issue of whether landfill final cover should be constructed as soon as a reasonable portion reaches final grade are summarized.

Opponents' Views. In theory the encapsulation concept can keep the waste dry, which will not allow waste to decompose leading to prevention of leachate formation. It is postulated that such a design will minimize pollution of groundwater. This approach has significant deficiencies that will ultimately threaten public health and the environment. First, leakage through the liner is practically unavoidable (for both synthetic membrane and clay liners) because of manufacturing and construction-related defects. For a synthetic membrane liner the typical warranty is for about 20 years. It is reasonable to assume that leachate formation and subsequent leakage from the landfill are unavoidable in the long run: encapsulation merely delays the process. Thus the encapsulated waste represents a perpetual threat to groundwater (Lee and Jones, 1990; Harper and Pohland, 1988).

To avoid this threat municipal waste should be composted (or fermented) first in a leach pad whereby all leachable chemicals can be extracted and the waste stabilized. This stabilized waste should then be disposed and encapsulated. Threats to groundwater and public health are negligible from such stabilized waste. It is claimed that recirculation of leachate will enhance fermentation and removal of leachate chemical constituents. It is conceded though that little is known about conditions under which removal of leachables will be maximized.

Supporter's View. The concept of "stabilizing" municipal waste appears to be good in theory but not so in practice because (1) the possibility of leachate generation is high even after stabilizing the waste; (2) for a large volume of waste the cost of handling is estimated to be high, the area of leaching pad needed is quite large, the cost of construction and maintenance of such leaching pad is also high, the odor and bird hazard around the landfill will increase significantly; and (3) in a municipal landfill, precipitation water percolates through preferred channels and not through the entire waste mass. The total amount of chemicals leached out is expected to be significantly low compared to the fermentation/compost operation proposed for stabilizing the waste.

Summary Comments. From an overall management standpoint it appears that "stabilizing" municipal waste prior to permanently landfilling is not realistic for very large landfills. However, the concept of "stabilizing" waste may be considered for landfills of small communities where the yearly waste generation rate is rather low. The following additional issues, either in favor of or against the timing of landfill final cover construction, should be considered prior to making a decision for a particular landfill:

1. An early release of gas and leachable constituents from a waste mass will certainly minimize future risks of groundwater contamination. However, the leachate generated during "stabilization" is expected to be strong during the early phase. A pretreatment or a dedicated treatment system for treating the leachate should be considered if a "stabilizing" operation is proposed.

2. An estimate regarding the time to "flush out" all or the majority of contaminants is needed for estimating the leaching pad area required and for planning purposes.

3. Monitoring requirements and remedial action possibility may be significantly less for landfilling a "stabilized" waste but should not be completely neglected. Necessary funds should be set aside for these purposes.

4. For both "stabilized" and unstabilized waste landfills sufficient funds for monitoring and remedial action should be escrowed during the period when revenue is being generated.

5. Leachate generation from an encapsulated landfill may continue for many years after the final cover construction. Ten years ago it was thought that leachate should be pumped for 20 to 30 years after closure. However, currently the estimate for such long-term maintenance is a minimum of 40 years. This uncertainty in long-term maintenance requirement poses financial and legal problems, especially for privately owned landfills. In view of the existing laws, unless a clear provision is made in the laws governing landfills, "inheritance of unknown future liability" cannot be forced on an individual or a company. For publicly owned landfills appropriation of necessary funds may be difficult politically. Setting up "long-term maintenance funds" for all landfills within the jurisdiction of a regulatory agency is an option that should be investigated. Such funds may be accumulated by requiring a disposal fee from all landfills at the time of active disposal. However, disbursing of such funds for long-term maintenance of a landfill may be difficult administratively.

6. From a regulatory standpoint the encapsulation concept is easily enforceable, especially on privately owned landfills, because the action is taken relatively quickly. If the final cover construction is delayed by several years after revenue generation, the owner's action is difficult to control unless the owner has continued interest in the matter (e.g., owns another landfill controlled by the same regulatory agency). A requirement to escrow the funds necessary for cover construction during active site life may help resolve this problem.

6.6 LEACHATE APPORTIONMENT MODELS

Leachate percolates vertically through the waste until it is influenced by the low permeability liner that retards vertical movement. Leachate then starts ponding on the liner. A portion of the ponded liquid flows laterally to the leachate collection pipes and the other portion flows vertically through the liner. Thus, the leachate produced in a landfill is apportioned into a drainage fraction (flowing laterally) and a leakage fraction (flowing vertically). Several models have been developed to predict the apportionment of leachate. Both steady-state and non-steady-state models are available (Wong, 1977; McBean et al., 1982; Korfiatis and Demetracopoulos, 1986; Demetracopoulos et al., 1984; Demetracopoulos and Korfiatis, 1984; Lentz, 1981; Schroeder et al., 1984; Bagchi and Ganguly, 1990). All these models assume a saturated liner. Bureau (1981) proposed a model in which the liner is assumed to be unsaturated. This model, although applicable to short-term waste impoundments, is not discussed here because landfill liners are permanent and are expected to be saturated.

The first apportionment model was proposed by Wong (1977). He assumed that leachate ponded on a liner develops a phreatic surface parallel to the liner and that the phreatic surface is formed instantaneously in response to leachate input. The rate of collection depends on the velocity with which the rectilinear slab of leachate moves to the collection point (Fig. 6.11). At any time (t), the saturated volume (v) above the linear is given by

$$v = sh \cos \theta \qquad (6.1)$$

for low values of θ, $\cos \theta \approx 1.0$, in which s = length of the rectilinear slab at time t, h = thickness of the rectilinear slab at time t, and θ = slope of the liner. The rate of volume change will be

$$dv/dt = h(ds/dt) + s(dh/dt) \qquad (6.2)$$

The first term represents the collection rate and the second term represents the leakage rate. Each term is integrated separately to find drainage and leakage volumes.

The following equations were derived to estimate drainage and leakage volumes:

$$\frac{v_1}{v_0} = \left[\frac{1}{K}\right][e^{-Kt/t_1} - 1][1 + (d/h_0)] + (d/h_0)(t/t_1) \qquad (6.3)$$

$$\frac{v_2}{v_0} = \left[\frac{1}{K}\right][e^{-Kt/t_1} - 1 + K][1 + d/h_0] \qquad (6.4)$$

$$K = \frac{P}{2d} \frac{K_2}{K_1} \cot \theta \qquad (6.5)$$

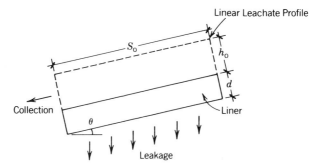

FIG. 6.11. Geometry for Wong's model.

in which v_1 = the total volume of leachate collection, v_2 = the total volume of leachate leakage, v_0 = the total leachate volume, t_1 = the time for the entire length of the leachate slab to drain into the collection line, d = the thickness of the liner, h_0 = the initial thickness of the saturated volume, K_1 = the permeability of the drainage layer, and K_2 = the permeability of the liner, Wong's model was revised by Kmet et al. (1981); Scharch et al. (1985) and Demetracopoulos et al. (1984) extended the model to provide drainage and leakage estimates for quasi-steady-state conditions.

McBean et al. (1982) proposed a steady-state model that used a nonlinear phreatic surface and a nonleaking liner. An implicit nonlinear equation for leachate mound was derived that transforms to the following simple form for horizontal liners:

$$y = \sqrt{y_0^2 + 2\left(Lx - \frac{x^2}{2}\right)} \tag{6.6}$$

in which $y = y_0$ at $x = 0$. The procedure is helpful in obtaining the leachate mound where leachate collection lines are placed on a liner sloped in one direction, but the approach cannot be used for estimating leakage through the liner.

A non-steady state model was proposed by Korfiatis and Demetracopoulos (1986). Principles of mass conservation are used to develop the following equation summing all inflow and outflow to an element:

$$\frac{\partial Q_c}{\partial_x} + I - Q_1 = n\frac{\partial H}{\partial t} \tag{6.7}$$

in which Q_c = the leachate collection rate, I = the leachate input rate, Q_1 = the leakage rate, H = the leachate head over the liner at a distance x from the crest of liner, t = time, and n = porosity of the drainage layer.

The following dimensionless nonlinear partial differential equation is developed, which was solved by explicit finite difference approximation using central differences:

$$\frac{\partial}{\partial x^*}\left(h^*\frac{\partial h^*}{\partial x^*} - h^* \sin \theta\right) + I^* - K^*(h^* + 1) = n\frac{\partial h^*}{\partial t^*} \qquad (6.8)$$

in which $x^* = x/d$; d = the thickness of the liner, $h^* = h/d$, $I^* = I/Kd$, K_1 = the saturated permeability of the drainage layer, $K^* = K_2/K_1$, K_2 = the saturated permeability of the liner, and $t^* = tK_2/d$.

Lentz (1981) proposed a leachate apportionment model for landfill cover that can predict leakage for water surplus on a daily basis. The model can be used for the bottom liner. Principles of mass balance are used in developing a partial differential equation that was linearized to obtain a solution. Schroeder et al. (1984) proposed a computer-based model known as HELP (Hydrologic Evaluation of Landfill Performance) for estimating leachate apportionment on a daily basis. The model couples both water infiltration into a landfill through cover (also known as water balance) and leachate apportionment at the bottom liner level.

The following linearized semiempirical equation was developed for a steady-state condition for estimating the leachate collection rate:

$$Q_c = \frac{2C_1K_1\overline{Y}(Y_0 + \theta p)}{p^2} \qquad (6.9)$$

$$C_1 = 0.51 + 0.00205\theta p \qquad (6.10)$$

in which $\overline{Y}$ = the average thickness of the water profile above the liner between the leachate collection pipe and the crest of a module, θ = the liner slope, $2p$ = pipe spacing. Y_0 is estimated as

$$Y_0 = \overline{Y}\left(\frac{\overline{Y}}{\theta p}\right)^{0.16} \qquad (6.11)$$

The vertical leakage is essentially estimated by the following equation:

$$Q_1 = K_2\frac{\overline{Y} + d}{d} \qquad (6.12)$$

An iterative procedure with four equal time steps per day is used to estimate the collection and leakage rate for a degree of accuracy of ±5%. The model was subsequently updated, which includes a few modifications. Required data for many U.S. cities are included in the program files. Files containing daily precipitation, temperature, and solar radiation data need to be added for use in other countries.

The theory for leachate apportionment developed by the author is discussed below. Error of prediction using this model was found to be low (Bagchi and Ganguly, 1990). A general equation of the leachate mound is developed first, which is then used to predict the daily collection and leakage rate. The liner and collection pipe configuration used in developing the general steady-state equation are shown in Fig. 6.12. Equating the leachate generation for collection and leakage,

$$(Q_x)_c + (Q_x)_l = (p - x)I \tag{6.13}$$

in which $(Q_x)_c$ = the leachate flow rate toward the collection point, through a section located at a distance of x (i.e., through GH as shown in Fig. 6.12) $(Q_x)_l$ = the leachate leakage rate through the liner between the apex and a point x (i.e., through the liner GF area as shown in Fig. 6.12, p = the distance of the liner apex from the origin, and I = the leachate generation rate (i.e., recharge rate on liner) per unit length of the liner.

$$Z = y \sec \theta + x \sin \theta \tag{6.14}$$

or

$$dZ/dx = \frac{dy}{dx} \sec \theta + \sin \theta \tag{6.15}$$

$\sin \theta$ can be neglected for small values of θ. Hence,

$$dZ/dx = (dy/dx) \sec \theta \tag{6.16}$$

From Darcy's law,

$$(Q_x)_c = K_1 y(dZ/dx) \tag{6.17}$$

or

$$(Q_x)_c = K_1 y \sec \theta \, (dy/dx) \tag{6.18}$$

in which K_1 = the permeability of the drainage layer. From Eqs. (6.13) and (6.18)

$$(1 + R)K_1 y \sec \theta \, (dy/dx) = (p - x)I \tag{6.19}$$

in which $R = (Q_x)_l/(Q_x)_c$.

From symmetry, the phreatic surface is assumed to be passing through the center line of the collection trench. From Fig. 6.12 the coordinates of M are $x_c = -1.5d \sec \theta$ and $y_c = 0$, in which d = diameter of the leachate

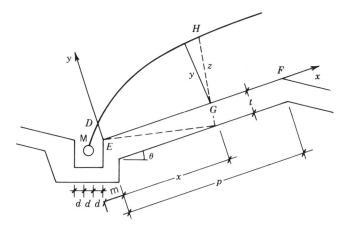

FIG. 6.12. Geometry for the unsteady-state leachate apportionment model.

collection pipe. The general equation of the phreatic surface [Eq. (6.20)] is obtained by integrating Eq. (6.19) and using the coordinates of M. A detailed derivation of Eq. (6.20) can be found elsewhere (Bagchi and Ganguly, 1990). Leachate mound curves (phreatic surface) predicted by this model were found to be parallel to the experimental curves obtained by Schroeder (1985) for a base slope of up to 4%.

$$y^2 = -Ax^2 + Bx + C \qquad (6.20)$$

The general equation of the phreatic surface is used to estimate the daily leakage and collection rate. The following assumptions are used to estimate the daily events:

1. The phreatic surface passes through point M (Fig. 6.12).
2. The gradient of the phreatic surface is zero at $x = p$; i.e. $(dy/dx)_{x=p} = 0$.
3. The slope of the liner is not greater than 4%.

Using the formula for the area of an ellipse and assumption 1

$$V_d = \frac{n\pi(p + m)h_{max}}{4} \qquad (6.21)$$

or

$$h_{max} = 4V/[n\pi(p + m)] \qquad (6.22)$$

in which V_d = the remaining volume of leachate in a day, n = the porosity of the drainage blanket, p = the width of the module (= $1/2$ pipe spacing approximately), m = the distance shown in Fig. 6.12 (= $d \sec \theta$), θ = the slope of the liner, and h_{max} = the height of the leachate mound at $x = p$. Put $y = 0$ and $x = -m$ in Eq. (6.20):

$$0 = -Am^2 - Bm + C \qquad (6.23)$$

Using assumption 2 and Eq. (6.20)

$$-2Ap + B = 0 \qquad (6.24)$$

Put $y = h_{max}$ and $x = p$ in Eq. (6.20)

$$h_{max}^2 = -Ap^2 + Bp + C \qquad (6.25)$$

From Eqs. (6.22) to (6.25)

$$A = \frac{h_{max}^2}{m^2 + 2pm + p^2}$$

$$B = \frac{2h_{max}^2\, p}{m^2 + 2pm + p^2}$$

$$C = h_{max}^2 \left(\frac{m^2 + 2pm}{m^2 + 2pm + p^2} \right)$$

For a precipitation event V_i is given by

$$V_i = I(p + m) \qquad (6.26)$$

in which I = the depth of precipitation and V_i = the volume of precipitation. For any day V_d is the remaining volume, which is the carryover volume from the previous day and the leachate generated due to any precipitation that falls in that day (i.e., V_d = the remaining volume + V_i). The leachate volume generated by the precipitation is added to the remaining volume before apportionment is started for the day. Figure 6.13 shows the routing scheme. V_c for a day is calculated first using the remaining volume, assuming no leakage during that process. The collected volume is subtracted from the remaining volume V_d, which is then used to calculate the leakage for that day. The process is reversed to calculate leakage first and collection next. The averages of the two collection and leakage volumes are taken, which are the respective volumes for the day.

The volume of collection (V_c) and leakage (V_l) per day are calculated as follows:

$$V_c = K_1 \times \text{area } DE \times (dy/dx)_{x=0} \qquad (6.27)$$

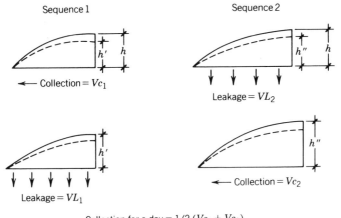

FIG. 6.13. Apportionment schemes.

$$\text{area } DE = \sqrt{(-Ax^2 + Bx + C)}_{x=0} = \sqrt{C} \tag{6.28}$$

$$(dy/dx)_{x=0} = \frac{B}{2\sqrt{C}} \tag{6.29}$$

so

$$\tag{6.30}$$

$$V_c = (K_1 B)/2$$

$$V_1 = \int_0^p K_2 \frac{y \sec \theta + t}{t} dx \tag{6.31}$$

or

$$V_1 = K_2 p + \int_0^p \frac{K_2 Y \sec \theta}{t} dx \tag{6.32}$$

Y can be expressed as

$$Y = \sqrt{A}(\sqrt{V^2 - U^2}) \tag{6.33}$$

in which

$$U = x - B/2A; \quad V = \sqrt{\frac{C}{A} + B^2/4A^2}$$

Since $(dy/dx)_{x=p} = 0$, $B = 2AP$ when $x = p$, $U = 0$, and when $x = 0$, $U = -p$, and $dU = dx$. Equation (6.32) can be written as

$$V_1 = K_2P + \frac{K_2 \sec \theta \sqrt{A}}{t} \int_{-p}^{0} \sqrt{V^2 - U^2} \, dU \qquad (6.34)$$

Integrating Eq. (6.34)

$$V_1 = \left({}^{1/2}(U\sqrt{V^2 - U^2} + V^2 \sin^{-1} \frac{U}{V} \right)_{-p}^{0} + K_2p \qquad (6.35)$$

or

$$V_1 = K_2p + \frac{K_2 \sec \theta \sqrt{A}}{t} \left[\frac{p}{2} \sqrt{C/A} + {}^{1/2}\left(\frac{C}{A} + p^2 \right) \sin^{-1} \frac{p}{\sqrt{C/A + p^2}} \right]$$
$$(6.36)$$

The leachate volume in a landfill is less than the precipitation that falls on a landfill. Part of the precipitation is lost by evaporation from the surface of the landfill and part is absorbed by the waste (nonsludge type waste only). In landfills located in cold regions, snow accumulated during the winter months melts in spring. Thus, the precipitation needs to be adjusted to reflect the effects of these factors. Leachate data obtained from an operating landfill need to be used to determine the daily values of the three factors (namely evaporation factor, absorption factor, and snow accumulation factor). Once these factors are obtained using field data for at least 1 year, they can be used to predict future apportionment. The thickness of the drainage blanket is dictated by the maximum depth of the leachate mound at the crest (h) in a year. The drainage blanket should be thick enough to accommodate the leachate mound throughout the year or most of the year. The model can be used to predict the daily leachate collection and leakage rate, the variation of h over a year for a set of design parameters (namely: pipe spacing, permeabilities of the drainage blanket and liner, and thickness and slope of the liner). The observed leachate head at the mid-point of a module for a landfill in Wisconsin was reported to be in good agreement with the value predicted by the model (Bagchi and Ganguly, 1990). Variation of h in a year for a typical landfill in midwest United States is expected to be between 30 and 58 cm. Variation of h is dependent primarily on precipitation and waste type. If the drainage blanket is not thick enough to accommodate the leachate mound in a day then the mound will rise (Fig. 6.14), leading to an increase in leakage; this rise in leachate mound will occur only if the permeability of the waste is lower than the permeability of the drainage blanket. Leakage from part of the module between point a and point b will increase because of the increased height of the phreatic surface between these two points. The increase in leakage rate will depend on the permeability of the waste

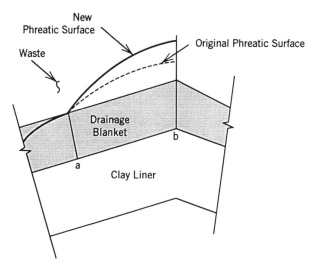

FIG. 6.14. Change in phreatic surface due to inadequate drainage blanket thickness.

(which will dictate the increase in the height of the phreatic surface) and the distance between point *a* and *b* (which is dependent on the thickness of the drainage blanket). A designer should attempt to minimize the occurrence of the situation (e.g., not more than 30 days) depicted in Fig. 6.14 at a predetermined point on the module (e.g., *ab* = 1/4th of module width). In other words the drainage blanket should be thick enough such that the leachate mound does not cross the drainage blanket line at point *a* for (say) more than 30 days a year.

The leachate apportionment models indicate that leakage through the landfill base depends heavily on the liner permeability and pipe spacing. Leakage can be reduced by lowering the liner permeability and pipe spacing. However, one should note that the required drainage blanket thickness increases with a decrease in pipe spacing. A leachate apportionment model should be used to estimate collection and leakage. Both peak and average collection volumes are important for treatment plant design and may become critical for a treatment plant with marginal available capacity. In the absence of access to an apportionment model the following guidelines may be used for landfill liner and collection system design:

1. *Liner*
   Thickness: 1–1.5 m (3–5 ft)
   Permeability: 1×10^{-7} cm/sec or less
2. *Pipe spacing*
   15–30 m (50–100 ft)

3. *Diameter of collection pipes*
   15 cm (6 in) minimum
4. *Slope of liner*
   2–4%
5. *Thickness of the drainage blanket* (Note: The drainage blanket thickness increases with decrease in pipe spacing)
   30–120 cm (1–4 ft)

6.7   COMMENTS ON DESIGNING THE LANDFILL BASE BELOW THE GROUNDWATER TABLE

The landfill base should be above the seasonally high GWT. However, it is difficult to follow this guideline at all times. The design approach is different if the subbase is entirely clayey or sandy. A liner should be constructed even in a clayey environment because in most cases clays have sand seams, vertical and horizontal fractures that are difficult to ascertain during subsoil investigation. Even field permeability tests may not detect these problems if testing is not performed carefully. Excessive leakage may take place through the cracks and so on. Some regulatory agencies may require a minimum thickness of the saturated clay stratum, which should be verified with the agency before designing a landfill in a saturated clayey environment. The chances of leakage through the liner of such landfills is not higher than a landfill sited above the groundwater table.

In a sandy environment the groundwater table may be lowered temporarily by dewatering during construction or permanently by installing a groundwater dewatering system (Fig. 6.15). A groundwater dewatering system, if installed, must be maintained properly by cleaning the pipes at least once a year. Raising of the base grade by constructing a retarder subbase may be considered in some cases to avoid the construction of a dewatering system. Since maintenance of the dewatering system is likely to become a perpetual activity, its installation should be avoided as far as possible. The following approach may be used to estimate pipe spacing.

The Donnan formula [Eq. (6.37)] is used to calculate pipe spacing in draining agricultural fields; a steady-state condition is assumed.

$$L^2 = \frac{4K(b^2 - a^2)}{Q_d} \qquad (6.37)$$

in which L = drain spacing (ft); K = hydraulic conductivity (ft/day), a = the distance between the pipe and the barrier (ft), b = the maximum allowable water table height measured from the barrier (ft), Q_d = recharge rate (ft^3/ft^2/day). Note: Donnan's formula is valid in SI units also.)

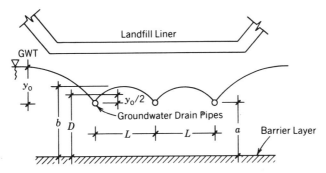

FIG. 6.15. Geometry for the groundwater dewatering system design—pipes above the barrier layer.

For groundwater dewatering systems, neglecting landfill leakage, which is small compared to groundwater flow,

$$Q_d = ki \tag{6.38}$$

Equation (6.37) changes to

$$L^2 = \frac{4(b^2 - a^2)}{i} \tag{6.39}$$

Equation (6.37) is used for the case in which the pipes are installed above a barrier (Fig. 6.15). For pipes installed on a barrier, $a = 0$ (Fig. 6.16). In a sensitivity analysis of several formulas for drain tile design, Slane (1987) found that (1) the pipe spacing and mound height depend on the permeability of the medium and recharge rate, and (2) the Donnan formula predicts the lowest mound height for a pipe spacing compared to other formulas. However, the Donnan formula is widely used for its simplicity. The pipe spacing obtained from it is within ±20 ft of the spacing obtained using unsteady-state models (Bureau of Reclamation, 1978); hence a reduction factor of 0.8–0.9 is suggested to obtain a conservative estimate.

For pipes above a barrier the discharge from the pipes is given by (Bureau of Reclamation, 1978)

$$q_p = \frac{2\pi K y_0 D}{86,400L} \tag{6.40}$$

For pipes on the barrier it is given by

$$q_0 = \frac{4KH^2}{86,400L} \tag{6.41}$$

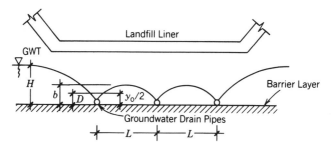

FIG. 6.16. Geometry for the groundwater dewatering system—pipes on the barrier layer.

in which q_p or q_0 = the discharge from two sides per unit length of drain (ft³/sec/ft), y_0 or H = the maximum height of the water table above the pipe invert (ft), K = the weighted average hydraulic conductivity of the soil between the maximum water table and the barrier or drain (ft/day), D = the average flow depth (= $d + y_0/2$) (ft), and L = the pipe spacing (ft).

The total discharge from a pipe is obtained by multiplying q_p or q_0 by the total length of the pipe. The pipe size can be calculated from the known discharge rate and from the velocity of water in the pipe using the Manning equation [Eq. (6.42)]

$$V = \frac{1.486}{n_r} r^{2/3} s^{1/2} \tag{6.42}$$

in which V = velocity in ft/sec, r = the hydraulic radius in ft, s = the slope of the groundwater collection pipe in ft/ft, and n_r = the coefficient of roughness. Example 6.1 enumerates the design steps. The minimum suggested pipe size is 10 cm (4 in.).

Example 6.1 (in F.P.S. units)

Design a groundwater collection system for a landfill in which the soil has a K of 2.83 ft/day (1×10^{-3} cm/sec) and a gradient of 0.1. The maximum permissible height of the water table is 10 ft above the barrier and the pipe invert is 5 ft above the barrier. The length of each pipe is proposed to be 1000 ft laid at a slope of 1%. n = 0.015 for plastic pipes. From Eq. (6.39)

$$L^2 = \frac{4(b^2 - a^2)}{i}$$

$$= \frac{4(10^2 - 5^2)}{0.1}$$

$$= 55 \text{ ft}$$

Using a reduction factor of 0.8, $L = 44$. From Eq. (6.40)

$$q_p = \frac{2\pi K y_0 D}{86,400L}$$

$$D = 5 + 5/2 = 7.5 \text{ ft}$$

$$q_p = \frac{2\pi \times 2.83 \times 5 \times 7.5}{86,400 \times 4}$$

$$= 0.00017 \text{ ft}^3/\text{sec ft}$$

The total discharge is $(0.00017/2) \times 1000 = 0.09$ cfs. From Eq. (6.41) assuming a 4-in. pipe flowing full,

$$V = \frac{1.486}{0.015} \times \left(\frac{4}{2 \times 12}\right)^{2/3} \times (0.01)^{1/2}$$

$$= 3 \text{ ft/sec}$$

The capacity of this pipe flowing full is

$$3 \times \pi(2/12)^2 = 0.26 \text{ ft}^3/\text{sec}$$

which is more than the total discharge.

6.8 CHECK FOR LINER BLOWOUT

Blowout of the landfill liner must be checked if the landfill base is constructed below the water table, especially in a sandy environment. Dewatering wells can be used temporarily to lower the groundwater table during construction to avoid blowout. The analysis and example given below will help in determining whether there is a possibility of blowout of clay liners and when to turn off a dewatering system if one is used.

Land disposal facilities constructed under the water table encounter water pressure at the base and side of the site. An empty landfill, which is the worst condition (as far as blowout is concerned), should be assumed. Figure 6.17 shows a situation in which the clay liner may blowout because of a high water table. Analyses of both base and side liners for different situations are available.

When a composite liner is used, uplift of the synthetic membrane liner at the base and its slippage at the side wall due to seepage of water through the clay liner should be considered. Failure of the synthetic membrane liner due to such water seepage can be avoided if sufficient volume of waste is disposed of relatively quickly (e.g., within a year). This operational issue should be considered while designing composite liners below GWT. In gen-

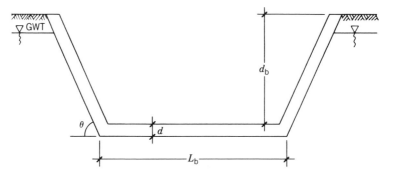

FIG. 6.17. Possible liner blowout situation.

eral, for big landfills uplift of the synthetic membrane liner at the base is not likely because usually sufficient waste is disposed within a year. However, slippage of the synthetic membrane liner on a side slope due to reduced interfacial friction should be carefully evaluated. Failure can be avoided if the interfacial friction between the synthetic membrane and clay is high enough even under saturated condition. If necessary a layer of sand (30–60 cm thick) or geo composite below the clay liner on the side slope should be included.

6.8.1 Blowout of Base Liner

The following possible causes of blowout failure for the base liner were analyzed: (1) shear failure, (2) bending failure, and (3) punching shear failure (Bagchi, 1986b). The water head at which the stress due to water pressure equals the strength of the liner was defined as the critical head. It was observed that critical head expressions could be divided into two terms: one containing the shear or bending strength of clay and another containing the ratio of unit weights of soil and water. It was shown that high bending stress is the most probable cause of failure. Equation (6.43) gives the critical head (h_b):

$$h_b = \frac{4}{3} \frac{K_b C_u}{\gamma_w} \left(d/L_b \right)^2 + \frac{\gamma_s}{\gamma_w} d \tag{6.43}$$

in which C_u = the unit shear strength of clay, K_b = constant (≈ 0.25), γ_w = the unit weight of water, γ_s = the unit weight of soil, d = the thickness of the liner, and L_b = the length of the landfill base. Based on a parametric study it was concluded that the numerical contribution of the first term is negligible for practical design purposes (Bagchi, 1986b).

6.8.2 Blowout of Side Liner

Failure of a side liner below the water table or due to artesian pressure in a sand seam is discussed in this section. Failure at section $Y-Y$ due to a high water table is discussed first. The possible causes of failure at section

Y–Y include shear failure and bending failure. It was observed that in this case also, the critical depth of the water table above the liner can be subdivided into two terms: one containing the shear or bending strength of the clay and another containing the ratio of unit weights of soil and water. It was shown that high shear stress is the most probable cause of failure (Bagchi, 1986b). Figure 6.18 shows the force diagram for this case. Equation (6.44) gives the critical depth of water for shear failure at section Y–Y.

$$C_{DW} = \sqrt{\left(\frac{2C_u d \sin^2 \theta}{\gamma_w} + \frac{\gamma_s}{\gamma_w} d \sin^2 \theta \, d_b\right)} \qquad (6.44)$$

in which C_{DW} = the critical depth of the water table above the liner, D_b = the depth of the top of the base liner from the land surface, and other terms are as previously defined.

However, unlike base liner blowout the numerical value of the first term is not negligible compared to the numerical value of the second term for the range of values of different parameters usually encountered in the field.

Analysis of the blowout of a side liner due to a narrow water-bearing stratum (Fig. 6.19) is given by Oakley (1987). The factor of safety (FS) against failure is given by

$$FS = \frac{\gamma_T d \cos \alpha}{Z\gamma_w} + \frac{2d^2 f}{Z\gamma_w(T/\sin \alpha)^2} \qquad (6.45)$$

A factor of safety of 2–3 should be assumed. The following range of values for C_u is reasonable: 7425–9900 kg/m² (1500–2000 psf).

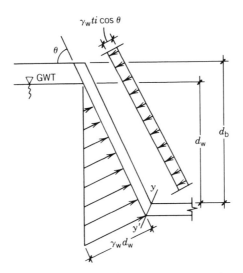

FIG. 6.18. Force scheme for the analysis of side liner blowout.

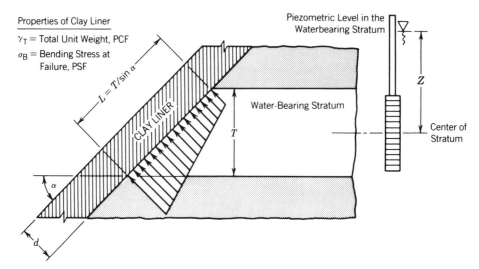

FIG. 6.19. Possible side liner blowout situation due to artesian pressure in sand seam. After Oakley (1987).

Example 6.2

Following are the dimensions of a square landfill, constructed using a dewatering system for which the possibility of liner blowout is to be checked: Assume C_u = 7500 kg/m², L_b = 50 m, d_b = 5 m, d_w = 3 m, d = 1.5 m, θ = 18.4°; unit weight of solid waste = 816 kg/m³, and γ_s = 1650 kg/m³.

Check for the Base Liner. From Eq. (6.43) and the discussion, the critical head for the base liner is

$$(1650/1000) \times 1.5 = 2.5 \text{ m}$$

Since the height of the GWT is (= $d + d_w$ = 1.5 + 3) 4.5 m above the bottom of the liner, blowout will occur if the operation of the dewatering system is stopped immediately after construction of the liner. One way to avoid blowout is to place a certain height of solid waste in the landfill before turning off the dewatering system.

The head of water to be balanced by the solid waste is

$$4.5 - 2.5 = 2 \text{ m}$$

The equivalent height of solid waste that must be placed to resist blowout is

$$(2 \times 1000)/816 = 2.45 \text{ m}$$

Therefore, to prevent liner blowout a minimum of 2.45 m of solid waste must be disposed of on the entire area of the lined landfill base, before the dewatering system can be shut off, or a thicker liner should be designed.

Check for the Side Liner. Failure near the base at section $Y-Y$. From Eq. (6.44)

$$C_{DW} = \sqrt{\left(\frac{2 \times 7500 \times 1.5 \times \sin^2 18.4}{10^3} + \frac{816 \times 1.5 \times 5 \times \sin^2 18.4}{10^3}\right)}$$

$$= 1.69\,m$$

Therefore, a side liner failure will occur at section $Y-Y$ if the operation of the dewatering system is stopped immediately after construction of the liner. To prevent blowout either the liner should be thickened along the corner (section $Y-Y$) or a certain height of solid waste should be disposed in the landfill before turning off the dewatering system. Thickening of the corner is preferred because it will increase the shear resistance of the liner. Reliance on the shear resistance of the waste is not suggested.

Faulty construction may also lead to blowout of the base as well as the side liner. Liner blowout due to bad quality control cannot be analyzed. The best way to ensure proper quality control is to arrange on-site engineering supervision.

6.9   NATURAL ATTENUATION VERSUS CONTAINMENT LANDFILL

Because both of NA type and containment type landfills have advantages and disadvantages, a designer has to consider various factors in choosing the correct type. A designer must verify whether any regulatory restrictions exist regarding siting of a natural attenuation type landfill in a region. Some of the points discussed below are rather broad based and are sometimes beyond the control of an individual designer; however, they are discussed to help in developing a proper perspective and attitude toward each landfill type.

A NA type landfill can be designed for a small waste disposal rate (non-hazardous wastes only) but a containment landfill cannot be designed and operated economically for a small waste volume because the cost of constructing and operating a containment landfill is high. The leachate has to be treated both during the active site life and after closure of a landfill, the cost of which must be obtained from the landfill customers. A small segment of liner cannot be constructed economically, and a larger liner may remain unused for several years causing severe deterioration due to exposure to the natural elements. If waste is disposed in a small area of a containment landfill, all precipitation water will be collected by the leachate collection system and must be treated, although the leachate strength is likely to be low. For these technical reasons a containment landfill for a small waste disposal rate is not the preferred option; in addition, operation of such a landfill could be very costly.

NA type landfills cannot be used for a large waste volume because of unavoidable groundwater impact. A containment landfill is the only choice for large waste volumes. NA type landfills put very little demand on the environment as a whole (only the landfill site is disturbed), whereas containment landfills disturb not only the areas in which they are sited but also the areas from which clay and sand is borrowed (especially clay lined landfills). The total energy consumption for construction, operation, and maintenance of a containment landfill is much higher than for an NA type landfill. However, this overall impact can be reduced only if the waste type is such that it can be easily disposed of in an NA type landfill without the risk of groundwater contamination. But the individual designer has no control over waste type. Control on waste type is a waste management issue dictated largely by local and state regulations.

Thus, in summary, a designer should use judgment as to which type of landfill may be used in a particular situation. In general, if regulations allow, properly designed natural attenuation landfills are suitable for a small waste volume in some cases and are cheaper to operate, whereas containment landfills are suitable for larger waste volumes and may be the most economical option for certain communities. If the waste generation rate in a community is small but regulations do not allow the use of NA type landfills, then collecting the waste in a container and shipping that container periodically to a nearby large landfill may be cheaper than constructing and operating a small containment landfill. This option should be investigated for small community landfills.

LIST OF SYMBOLS

v	= saturated volume above liner at time t
s	= length of the rectilinear saturated slab at time t
h	= thickness of the rectilinear saturated slab at time t
θ	= slope of the liner
v_1	= total volume of leachate collection
v_2	= total volume of leachate leakage
v_0	= total leachate volume
t	= time for the entire length of the leachate slab to drain into the collection line
d	= thickness of liner
h_0	= initial thickness of the rectilinear saturated slab
K_1	= permeability of drainage layer

K_2	=	permeability of the liner
S_0	=	initial length of the saturated liner
Q_c	=	leachate collection rate
I	=	leachate input rate
Q_1	=	leakage rate
H	=	leachate head over the liner at any distance x from the crest of the liner
n	=	porosity of the drainage layer
$\overline{Y}$	=	average thickness of water profile above the liner between the leachate collection pipe and the crest of a module
$2p$	=	pipe spacing
$(Q_x)_c$	=	leachate flow rate toward the collection point, through a section located at a distance x
$(Q_x)_1$	=	leachate leakage rate through the liner between the liner crest and a point x
V_d	=	remaining volume of leachate in a day
h_{max}	=	height of the leachate mound at $x = p$
V_i	=	volume of precipitation
K_b	=	constant
C_u	=	unit shear strength of clay
γ_w	=	unit weight of water
d_b	=	depth of the top of base liner from land surface
γ_s	=	unit weight of soil
h	=	critical head of waste to cause failure of base liner
L_b	=	length of landfill base
C_{DW}	=	critical depth of water table to cause shear failure of side liner
L	=	drain spacing
K	=	hydraulic conductivity
Q_d	=	recharge rate
a	=	distance between the pipe and the barrier
b	=	maximum allowable water table height measured from the barrier
q_p or q_0	=	discharge from two sides per unit length of drain
y_0 or H	=	maximum height of water table above the pipe invert
D	=	average flow depth
V	=	velocity
r	=	hydraulic radius
n_r	=	coefficient of roughness
s	=	slope of collection pipes

7 Liner Materials

The different types of materials used to construct landfill liners and final covers fall into three categories: (1) clayey soil, (2) synthetic membranes or other artificially manufactured materials, and (3) amended soil or other admixtures. As discussed in Section 6.2, either only clay or a combination of two or all three types of materials is used for base liner and final cover construction. Each type has advantages and disadvantages that must be considered when choosing a particular liner material. A cost comparison must also be done prior to selecting an option. Material specifications, quality control tests and specifications, and minimum allowable thickness of the liner may vary from one state to another or in different countries. It is therefore essential that the local regulatory agency be contacted for acceptance criteria. The following discussions on each material type are provided as general guidelines and to provide a better understanding so that proper judgment regarding selection criteria can be developed. Inyang (1994) proposed a model to judge the long-term performance of waste containment systems.

7.1 CLAY

Clayey soil is widely used for lining nonhazardous waste landfills. However, there is some debate regarding the definition of clay. Clay can be defined as the soil fraction that has particles equal to or finer than 0.002 mm (or 2 μm) or as the soil fraction that has particles equal to or finer than 0.005 mm (or 5 μm). Three commonly used grain size classifications are the unified soil classification system, the international classification system, and the MIT classification system (Means and Parcher, 1963). Both the international (proposed in 1927 and adopted by most countries except the United States) and MIT classification systems define clay by the 2-μm criterion. The MIT classification, which is simple, widely used in the United States, and easy to remember, uses the following particle size range:

Sand: 2–0.06 mm
Silt: 0.06–0.002 mm
Clay: 0.002 mm and less

Unless otherwise specified in this book clay is defined by the 2-μm criterion.

The shape of the grain size curve is an important issue in obtaining a low permeability compacted soil liner. A soil with a grain size curve closely following the classical inverted "S" shape is expected to develop lower permeability.

7.1.1 Effect of Double layer

An understanding of the role of the electrical double layer in controlling soil behavior is helpful. Clay surfaces usually have an excess negative charge. The space between two clay particles contains water with cations (positively charged ions) and anions (negatively charged ions). The density of cations near the negatively charged clay surface is high and becomes lower with increasing distance. The negatively charged clay surface together with the liquid phase adjacent to it is called the diffuse double layer. Based on Gouy–Chapman theory the thickness of the double layer is expressed as

$$Th = \left(\frac{DKT}{8\pi n_0 \varepsilon^2 v^2} \right)^{1/2} \tag{7.1}$$

in which Th = the thickness of the double layer, D = the dielectric constant, K = Boltzmann's constant, T = absolute temperature (in Kelvin), n_0 = ion concentration, ε = the unit electronic charge, and v = the valence of the ion.

Equation (7.1) indicates that the thickness is inversely proportional to the valence of the ion and square root of the ion concentration. The thickness of the double layer increases with the dielectric constant and temperature. The following conclusions were drawn from a parametric study of Eq. (7.1) (Mitchell, 1976).

1. The swelling behavior of clay depends partially on the electrolyte concentration within the double layer.
2. The addition of small amounts of di- or trivalent cations to a monovalent double layer system can influence the physical properties significantly.
3. Double layer thickness controls the tendency to flocculate and influences the swelling pressure of clays.
4. A decrease in double layer thickness may cause shrinkage of clay. This is an important issue for waste disposal sites because the dielectric constant of leachate is expected to be different from water.
5. The effect of temperature change on the double layer is difficult to predict because the dielectric constant is also temperature dependent. For water, the change in the value of $D \times T$ is not significant for a change of temperature from 0° to 60°C, which indicates that the double layer thickness will not be influenced significantly due to a change in temperature in that range.

It should be noted that Gouy–Chapman theory does not take into account the following important effects on the double layer thickness: secondary energy terms (Bolt, 1955), superimposing electric field and structure of water on the double layer (Low, 1961; Mitchell, 1976), ion size in the double layer (Van Olphen, 1963), and pH (Mitchell, 1976). The Gouy–Chapman theory cannot explain the observed behavior of natural clay completely because natural clayey soils are mostly a mixture of a number of different clay minerals, each of which has a different double layer thickness. Change in the electrolyte concentration in the diffused double layer causes a change in its thickness leading to a change in permeability and shear strength properties of clay.

7.1.2   Effects of Various Parameters on Clay Properties

The mechanical properties of clay depend on several interacting factors such as mineral composition, percentage of amorphous material, absorbed cation, distribution and shape of particles, pore fluid chemistry, soil fabric, and degree of saturation. The effect on the mechanical properties of soil due to a change in any of these factors can be predicted qualitatively using physicochemical theories. However, quantitative prediction of soil behavior based on the above factors and any improved double layer theory is almost impossible, mainly because of the inadequacy of the available physicochemical theories and the difficulties in taking into account the effect of soil fabric and other *in situ* environmental factors. Several studies have been reported on the effect of change in pore fluid chemistry on the strength of clayey soil (Torrance, 1974; Yong, 1986; Yong et al., 1979), and on the permeability of clayey soil (Acar and Seals, 1984; Acar and Ghosh, 1986; Brown and Anderson, 1980; Brown et al., 1983, 1984; Fernandez and Quigley, 1985; Green et al., 1981; Reeve and Tamaddoni, 1965; Foreman and Daniel, 1984; Daniel et al., 1984). Some of the above studies used pure solvents and high concentrations of chemical compounds and the rest used actual landfill leachate or diluted solvent.

Mitchell and Madsen (1987) summarized the literature on the effect of inorganic and organic chemicals on clay permeability. The summary indicates that although dilute solutions of inorganic chemicals may change clay permeability, dilute solutions or organic chemicals have virtually no effect. Significant changes in permeability of the clay samples were not observed when leached with diluted solvent or salt solutions (Carpenter and Stephenson, 1986; Acar and Seals, 1984; Brown and Thomas, 1985). Fang and Evans (1988) observed practically no difference in the permeability of clay samples permeated with tap water and landfill leachate. Whether a chemical species in the leachate will trigger a change in the permeability of a clay liner or not, it may leak through the liner. The rate of leakage will depend on the diffusion coefficient and permeability of the chemical through the liner (Daniel and Shackelford, 1988).

Most of the studies cited above are rather short term and so the effect on permeability due to long-term exposure is not readily known. Bowders et al. (1984) suggested a decision tree (Fig. 7.1), which can be used to decide whether short-term or long-term permeability tests should be used. At present the general trend in the United States is not to use clayey soil as the primary liner in hazardous waste sites. Since pore fluid chemistry plays a significant role in changing the permeability of a soil, the effect of leachate from a proposed hazardous waste landfill on the chosen soil should be studied even if it is not required by the regulatory agency.

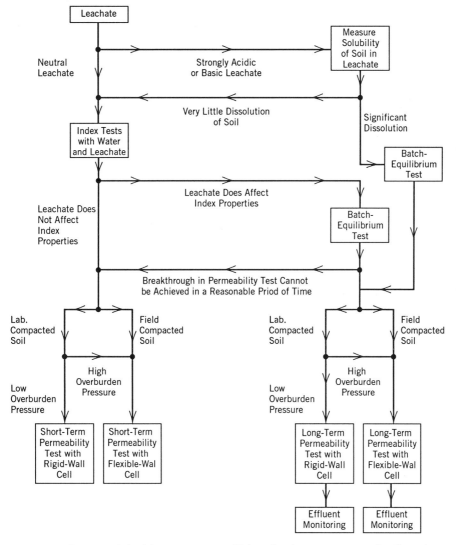

FIG. 7.1. Suggested decision tree as an aid in selecting a program of soil testing. After Bowders et al. (1984). Copyright ASTM. Reprinted with permission.

Because of the problem of running a long-term permeability test one may use the decision tree shown in Fig. 7.1 to decide whether a long-term permeability test should be undertaken. As previously discussed, changes in pore fluid chemistry will influence all the mechanical properties of soil. It is postulated that the Atterberg limits reflect the strength and permeability characteristics of soil (Terzaghi, 1936).

The liquid limit of clay corresponds approximately to the water content at which the shear strength is between 2 and 2.5 kN/m² (Norman, 1958). The plastic limit indicates the lower boundary of the water content range below which the soil no longer behaves as a plastic. In other words a soil can be deformed without volume change or cracking above the plastic limit and will retain its deformed shape (Mitchell, 1976). The plastic limit is a reflection of the structure of water within the pore and the nature of interparticle forces (Terzaghi, 1936; Yong and Warkentin, 1966). If the moisture content at which compaction is proposed (compaction moisture is usually 2–3% above optimum moisture) is less than the plastic limit then the tendency to develop microcracks during liner construction is expected to increase. Therefore, it is essential to ensure that the plastic limit of soil is lower than the proposed compaction moisture content. Seed et al. (1962, 1964a,b) developed a theoretical relationship between liquid limit, plasticity index, and clay content (Fig. 7.2) that is similar to the plasticity chart used for

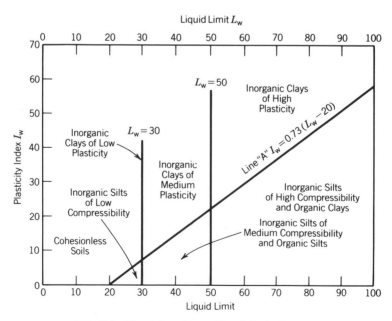

FIG. 7.2. Plasticity chart. After Mitchell (1976).

classifying clay using the USCS method. Inorganic clays of medium plasticity are preferred and caution must be exercised in using low plasticity inorganic clays for liner construction purposes. The ratio of the plasticity index and the percentage clay fraction is termed activity (Skempton, 1953). Studies of artificially prepared sand–clay mineral mixtures indicate that the relationship between the plasticity index and percentage clay fraction does not necessarily pass through the origin but may intersect the percentage clay fraction axis between 0 and 10. Based on two studies Mitchell (1976) defined activity as

$$A_c = \frac{PI}{P_c - n} \tag{7.2}$$

in which A_c = activity, PI = plasticity index, P_c = percentage clay fraction, and n = constant ($=5$ for natural soil and 10 for artificial mixtures).

The activity of smectites is highest and that of kaolinites is lowest. The higher the activity of a soil the higher its susceptibility to property changes due to a change in factors such as pore fluid chemistry and water content.

7.1.3   Effect of Compaction on Permeability of Clay

Compaction of soil during construction causes a change in soil fabric that influences mechanical properties, especially the permeability of soil. A dispersed soil fabric is expected to have lower pore diameter and higher tortuosity, which will influence permeability. A study of the Kozeny–Carman equation (Carman, 1956; Mitchell, 1976) is helpful in understanding the effect of pore size on soil permeability. The equation, although usable for sand, is considered inadequate in predicting permeability for clayey soil (Lambe, 1954; Michaels and Lin, 1954; Olsen, 1962). However, the equation is useful in understanding the effect of various *in situ* parameters on soil permeability. According to the theory the permeability of saturated soil can be expressed as

$$K = \frac{1}{K_0 T_f^2 S_0^2}\left(\frac{e^3}{1 + e}\right)\left(\frac{\gamma}{\mu}\right) \tag{7.3}$$

in which K = the permeability of soil, K_0 = the pore shape factor, T_f = the tortuosity factor, S_0 = the specific surface per unit volume of particle, e = the void ratio, γ = the unit weight of the permeant liquid, and μ = the viscosity of the permeant liquid.

The equation shows the following:

1. Permeability will be reduced due to a reduction in the void ratio (e).
2. Permeability is dependent on the ratio of the unit weight and viscosity of the permeant liquid.

3. Permeability will decrease due to an increase in tortuosity (i.e., if a more zigzag path is followed by the liquid).
4. Permeability will decrease due to an increase in the specific surface area.

The permeability of soil is controlled mostly by large pores (Mitchell and Madsen, 1987), the distribution and size of which depend on the fabric of the soil. The following is a simplistic description of compacted clay soil fabric observable under the electron microscope: the clay particles aggregate together to form "miniclods" (or assemblages) that are packed tightly to form a larger clod. The sizes of the pores within the miniclods are smaller than the pores formed by the space between the miniclods. In addition to these two types of pores, a third type may exist in naturally occurring clayey soil due to cracks and fissures (Collins and McGown, 1974). The pores caused by cracks and fissures are the largest and transport the maximum amount of fluid. So, when compacting clay, measures should be taken to avoid the formation of cracks and fissures. The factors that influence the fabric of compacted clay are the water content during compaction, the method and effort involved in compaction, the clod size of clay, and the interlocking of layers.

Since dispersed fabric will increase the tortuosity and reduce pore size, the aim of field compaction should be to create a dispersed fabric. Kneading compaction induces high shearing strain, which breaks down flocculated fabric and helps create a dispersed fabric (Mitchell, 1976). Water content during compaction influences the rearrangement of clay particles, which greatly influences permeability. The method of compaction also influences permeability (Mitchell et al., 1965; Mitchell, 1976). The effects of water content and the method of compaction on permeability are shown in Figs. 7.3 and 7.4, respectively. Because of these reasons, a sheep's foot roller should be used, which will provide kneading compaction.

In summary, the liner should be compacted wet of optimum moisture using a sheep's foot roller. More construction-related comments for clay compaction are included in Section 9.2

7.1.4 Effect of Clod Size on Clay Permeability

Precompaction clod size may influence the structure of compacted clay and thereby the permeability of clay (Daniel, 1981; Barden and Sides, 1970). Influence of clod size on the hydraulic conductivity of a highly plastic soil (note: the soil was classified as CH per the USCS and had a clay fraction of 42%) was reported by Benson and Daniel (1990). For the clay compacted dry of optimum using standard proctor procedure, the hydraulic conductivity was 1×10^6 times higher than when the soil was prepared from large (19 mm) rather than small (4.6 mm) clods. Clod size did not influence the hydraulic

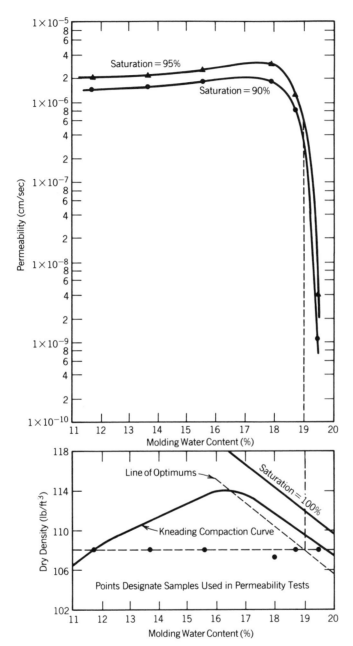

FIG. 7.3. Effect of molding water content on permeability of a silty clay (kneading compaction was used in preparing the samples). After Mitchell (1976).

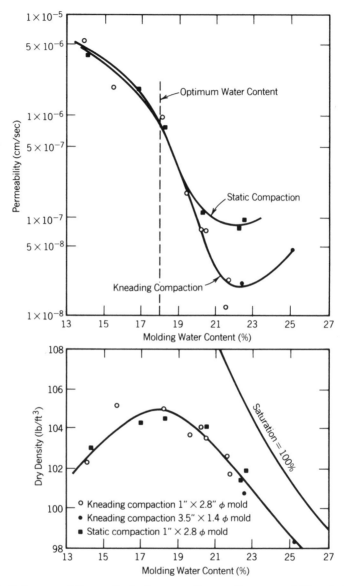

FIG. 7.4. Influence of the method of compaction on the permeability of silty clay. After Mitchell (1976).

conductivity when the soil was compacted wet of optimum using a modified proctor procedure.

It has been observed that the liquid limit of clay may be altered due to drying and rewetting, indicating a lack of rehydration to the original state (Sangrey et al., 1976). Therefore, an attempt should be made to maintain a natural moisture content whenever possible and sufficient time should be allowed to moisten the soil in the field.

7.1.5   Effects of Natural Elements on Clay Permeability

Changes of permeability due to exposure to the natural elements could be due to desiccation cracks and freeze–thaw degradation.

Desiccation Cracks. Desiccation cracks in a liner may develop because compacted clayey soil liners are subjected to periods of drying usually immediately after construction. In a laboratory study the permeability of desiccated samples increased by one order of magnitude or more (Dunn, 1986). The depth of desiccation cracks can be as high as 1 ft, which is of concern for liner construction. Desiccation cracks are essentially due to shrinkage of the soil. Soils with a high liquid limit are expected to shrink more. The degree of shrinkage (S_r), defined by Eq. (7.4), can be used to classify the soil qualitatively (Jumikis, 1962).

$$S_r = \frac{V_i - V_f}{V_i} \tag{7.4}$$

in which V_i = the initial volume of the sample before drying and V_f = the final volume of the sample after drying. Standard shrinkage limit tests are used to find V_i and V_f.

Table 7.1 provides a soil grouping based on shrinkage. According to these criteria good soils shrink less.

It is essential that a clayey soil liner be protected from excessive drying after construction by spraying water if necessary. Disposal of waste will reduce the chance of development of desiccation cracks.

Freeze–Thaw Degradation. Freeze–thaw degradation is due to alternate freezing and thawing of the clay liner exposed to subzero temperatures. Cracks in natural soils and an increase in vertical permeability due to the freeze–thaw effect have been observed by many (Chamberlain and Gow, 1978; Vallejo and Edil, 1981; Chamberlain and Blouin, 1977). Many researchers reported a two orders of magnitude increase in hydraulic conductivity

TABLE 7.1. Soil Grouping Based on Shrinkage

Degree of Shrinkage (S_r) (%)	Quality
Less than 5	Good
10–15	Poor
More than 15	Very poor

of clay due to freeze–thaw in laboratory studies (Chamberlain and Bouin, 1976; Chamberlain and Gow, 1979; Chamberlain et al., 1990; Zimmie and La Plante, 1990; Otheman and Benson, 1992; Kim and Daniel, 1992). Paruvakat (1993) raised doubts regarding validity of laboratory data when three dimensional freezing technique is used in the laboratory. Although all laboratory tests showed a significant increase in hydraulic conductivity of frozen clay specimens, field studies reported by researchers indicated an increase in hydraulic conductivity only up to three-fourths of an order of magnitude (Starke, 1989; Paruvakat et al., 1990; Bagchi, 1993).

A layer of frozen clay develops at the surface of the liner when the ambient temperature falls below 0°C. Frozen soil develops a reticulate ice vein network that subdivides the soil into irregular blocks (Vallejo, 1980). The water present in the pores of the clay layer forms ice crystals. Since the volume of ice is 9% more than the volume of water, the ice crystals exert pressure on the surrounding soil causing a change in fabric and possibly a localized consolidation of the clay (Taylor and Luthin, 1978; Andersland and Anderson, 1978; Chamberlain and Bouin, 1976; Alkire and Morrison, 1982). This layer of frozen soil (known as frozen fringe) creates a suction causing water from lower portions of the layer to move upward. As a result, a localized hydraulic gradient develops within the soil that causes development of a nonuniform water content profile within the layer (Williams, 1966; Konrad and Morgenstern, 1980; Dirksen and Miller, 1966; Hoekstra, 1966). Migration of water to the frozen clay layer from the lower portion of a layer may result in vertical shrinkage cracks depending on the plasticity index of the clay and the water content during compaction. If sufficient moisture is available during freezing, then a reticulate ice vein structure may develop. An equilibrium develops within the compacted clay liner whereby the penetration of the frozen fringe stops at a depth that is related to ambient temperature and the temperature of the compacted clay liner. The clay liner temperature depends on solar radiation, soil moisture, evaporation rate, heat storage, and insulation on the liner (e.g., snow cover). The depth of the frozen fringe depends primarily on the duration and degree of the subzero temperature. When the ambient temperature rises above 0°C the ice lenses thaw. This thawing may increase the size of pores and its interconnectedness. Thus a layer of degraded clay forms that could be as deep as the frozen fringe penetration. When the ambient temperature falls below freezing again, the upper layer of the liner provides less resistance to frost penetration.

The presence of desiccation cracks at the liner surface, an almost unavoidable phenomenon, probably triggers the degradation due to freeze–thaw. The presence of interconnected pores that is dictated by compaction during construction is another key issue in freeze–thaw degradation.

Thus, the depth of liner degradation due to freeze–thaw depends primarily on the duration and degree of subzero temperature, available solar radiation, and thickness of cover on the liner. However, the degree of liner degradation or, in other words, increase in hydraulic conductivity depends primarily on the number of freeze-thaw cycles and the size and distribution of interconnected pores. The size and distribution of interconnected pores are a reflection of compactness of the liner, which depends on soil type, compaction equipment and effort used, and the quality control program used during liner construction.

The necessary conditions for the formation and growth of ice veins are a low overburden pressure, a slow freezing rate, a sufficient supply of moisture to keep the soil saturated at the beginning and during the freezing process, and soil with 3-10% clay (Martin, 1958; Vallejo, 1980; Linell and Kaplar, 1959; Anderson et al., 1978). Several field studies indicated that the depth of liner degradation due to exposure in one winter in Wisconsin (note: in Wisconsin the winter temperature ranges from $-9.5°$ to $-1°C$ and there are usually two or three freeze-thaw cycles) in one winter is approximately 30 cm (Starke, 1989; Paruvakat et al., 1990; Bagchi, 1993). The focus of laboratory studies was to gain an insight into the effect on intrinsic property of permeability of compacted clay samples due to freeze-thaw cycles. A direct use of the data for predicting field behavior of compacted clay liner due to winter exposure is bound to lead to an extremely conservative design. Discrepancy between laboratory and field data is often reported in technical literature. Caution must be excercised in extrapolating laboratory data to predict field behavior. Field studies have clearly shown that the effect of freeze-thaw on clay liner permeability is not as severe as predicted by laboratory studies and, most importantly, the depth of degradation is not expected to exceed 30 cm (for winter conditions similar to Wisconsin, USA.) if proper care is taken during and after liner construction.

The following recommendations are made for protecting clay liners from freeze-thaw degradation:

1. Clay liners should not be exposed to freeze-thaw for two or more consecutive winters. Sufficient frost protection material should be placed over the liner before the onset of the second winter.
2. Good quality clay (CL or CL-ML type soil per USCS) and a proper QA/QC program for liner compaction should be used to reduce the effect of freeze-thaw on clay liners (note: see Chapter 9 for a QA/QC program).
3. Appropriate care (e.g. sufficiently thick layer of sand or waste) for protecting clay liners from freeze-thaw degradation should be taken from the first winter if the average infield hydraulic conductivity of the liner is 1×10^{-7} cm/sec or more.
4. Appropriate care should be used in protecting clay liners from freeze-thaw degradation if the thickness of the liner is 60 cm or less. A minimum liner thickness of 120 cm is recommended in all cases.

7.1.6 Permeability Test

Proper permeability tests should be used to assess the permeability of liners. The terms permeability and hydraulic conductivity are used interchangeably in the literature. When the term hydraulic conductivity is used it specifically refers to the conduction property related to water, whereas the term permeability is used to refer to a conduction property related to any liquid, which also includes water. Although the use of laboratory permeability tests is met with skepticism because of macrofabric features (Dunn, 1986; Day and Daniel, 1985), current practice continues to be to obtain several undisturbed samples and to test them in the laboratory.

7.1.6.1 Laboratory Permeability Test Peirce et al. (1986) advocated the use of a consolidation test to determine hydraulic conductivity of clay in the laboratory, because in their opinion the test will simulate field conditions. However, others prefer a flexible wall permeameter, with back pressure, over fixed wall permeameter, or a consolidometer (Bagchi, 1987d; Daniel et al., 1984). Standard test method for measuring hydraulic conductivity of fine grained soil using a flexible wall permeameter (ASTM D5084-90) has since been adopted by ASTM. The gradient during permeation, the chemistry of the permeating fluid, the degree of saturation, and confining pressure influence the permeability of a soil sample (Olsen, 1962; Mitchell, 1976). All these factors must be carefully monitored to simulate field conditions as far as practicable. Trainor (1986) reported a dedicated flexible wall permeameter that is equipped with back pressure to saturate soil samples before performing a permeability test using leachate (Fig. 7.5). It is essential that 90–100% saturation is achieved during a permeability test. A low confining pressure and low gradient (maximum 10) are preferred during these permeability tests to simulate field conditions.

There are several sources of error in a laboratory permeability test (Olson and Daniel, 1979; Zimmie et al., 1981). Sources of error also exist in estimating field permeability using laboratory test results (Daniel, 1981). Most of the error sources can be eliminated if proper quality control is exercised during construction and laboratory testing. Use of a triaxial permeameter and landfill leachate as the permeant is preferred. It may be difficult and time consuming to run a permeability test using landfill leachate as permeant, especially if the bacterial population is high in the leachate (i.e., leachate obtained from putrescible waste). Such a special test procedure would mean special laboratory arrangements and therefore higher costs.

The following approach may be used to minimize the total cost yet maintain a good quality control on tests. A small number of samples (10–15%) is tested under stricter test conditions (i.e., testing in triaxial cells with low confining pressure and hydraulic gradient) and then the results are compared with those obtained by the standard falling head permeability test. The degree of saturation of each of the samples of the second set must be determined at the beginning and end of the test. If the degree of saturation is 90% or

FIG. 7.5. Apparatus for the clay permeability test using triaxial pressure. (Courtesy of RMT Inc., Madison, WI.)

more and the results of the two sets are close then the results of the falling head permeability test may be acceptable. The degree of saturation should be obtained carefully because it greatly influences the permeability of soil samples (Johnson, 1954).

7.1.6.2* In situ *Permeability Test Several *in situ* permeability test methods are available (Daniel, 1989). Each method has its advantages and disadvantages, which are included in Table 7.2. The methods can be subdivided into four major categories: borehole tests, porous probe, underdrain, and infiltrometer. The following is a brief description of each of the methods.

7.1.6.2.1 Borehole Tests. Two different methods are available in this category: Boutwell permeameter and constant head permeameter.

BOUTWELL PERMEAMETER. This is a two-stage borehole test. The test schematic is shown in Fig. 7.6. A hole is drilled in the compacted liner in which the casing is placed; the annular space is grouted. Falling head tests are performed. Next the hole is deepened by augering or by pushing a thin walled sampling tube into the compacted liner; the smeared soil is removed from the hole. The test is performed again after assembling the device. Both horizontal and vertical hydraulic conductivity of the compacted liner can be

TABLE 7.2. Advantages and Disadvantages of In situ Permeability Test Methods

Type of test	Device	Advantages	Disadvantages
Borehole	Boutwell permeameter	Low equipment cost (<$200 per unit) Easy to install Hydraulic conductivity is measured in vertical and horizontal direction Can measure low hydraulic conductivity (down to about 10^{-9} cm/sec) Can be used at great depths and on slopes	Volume of soil tested is small Unsaturated nature of soil not properly taken into account Testing times are somewhat long (typically several days to several weeks for hydraulic conductivities <10^{-7} cm/sec)
	Constant head permeameter	Low equipment cost (<$1,000 per unit) Easy to install Unsaturated nature of soil taken into account relatively rigorously Relatively short testing times (a few hours to several days) The hydraulic conductivity that is measured is primarily the horizontal value (which is an advantage if this is the deisred value) Can be used at great depths	Volume of soil tested is small The hydraulic conductivity that is measured is primarily the horizontal value (in some applications, the value in the vertical direction is desired) The device is not well suited to measuring very low hydraulic conductivities (less than 10^{-7} cm/sec)
Porous probe	BAT® permeameter	Easy to install Short testing times (usually a few minutes to a few hours) Probe can also be used to measure pore-water pressures Can measure low hydraulic conductivity (down to about 10^{-10} cm/sec) The hydraulic conductivity that is measured is primarily the horizontal value (which is an advantage if this is the desired value) Can be used at large depths	High equipment cost (>$6,000) Volume of soil tested is very small Soil smeared across probe during installation may lead to underestimation of hydraulic conductivity The hydraulic conductivity that is measured is primarily the horizontal value (in some applications the value in the vertical direction is desired) The unsaturated nature of the soil is not properly taken into account

Infiltrometer		
Open, single-ring infiltrometer	Low cost (<$1,000) Easy to install Very large infiltrometer can be used to test a large volume of soil Hydraulic conductivity in the vertical direction is determined	Low hydraulic conductivity ($<10^{-7}$ cm/sec) is difficult to measure accurately Must eliminate, or make a correction for, evaporation May need to correct for lateral spreading of water beneath infiltrometer Testing times are relatively long (usually several weeks to several months for hydraulic conductivities $<10^{-7}$ cm/sec) Must estimate wetting-front suction head Cannot be used on steep slopes unless a flat bench is cut
Open, double-ring infiltrometer	Low equipment cost (<$1,000) Hydraulic conductivity in the vertical direction is determined Minimal lateral spreading of water that infiltrates from inner ring	Low hydraulic conductivity (10^{-7} cm/sec) is difficult to measure accurately Must eliminate or make a correction for evaporation Testing times are somewhat long (usually several days to several months for hydraulic conductivites $< 10^{-7}$ cm/sec) Must estimate wetting-front suction head Cannot be used on steep slopes unless a flat bench is cut
Closed, single-ring infiltrometer	Low equipment cost (<$1,000) Hydraulic conductivity in the vertical direction is measured Can measure low hydraulic conductivity (down to 10^{-8}–10^{-9} cm/sec)	Volume of soil tested is somewhat small because diameter of ring is < 1 m Need to correct for lateral spreading of water if wetting front penetrates below the base of the ring Testing times are long (usually several weeks to several months) Must estimate wetting-front suction head Very difficult to use on steeply sloping ground

TABLE 7.2. (*Continued*)

Type of test	Device	Advantages	Disadvantages
	Sealed, double-ring infiltrometer	Moderate equipment cost (<$2,500) Hydraulic conductivity in the vertical direction is determined Can measure low hydraulic conductivity (down to about 10^{-8} cm/sec) Minimal lateral spreading of water that infiltrates from inner ring Relatively large volume of soil is permeated	Testing times are relatively long (usually several weeks to several months) Must estimate wetting-front suction head Cannot be used on slopes unless a flat bench is cut
	Air-entry permeameter	Modest equipment cost (<$3,000) Relatively short testing times (a few hours to a few days) Hydraulic conductivity in the vertical direction is measured Can measure low hydraulic conductivity (down to 10^{-8}–10^{-9} cm/sec) Wetting-front suction head is estimated in second stage of test	A relatively small volume of soil is permeated because the wetting front usually does not penetrate more than a few centimeters into compacted clay Cannot be used on slopes unless flat bench is cut Several important assumptions are required
Underdrain	Lysimeter pan	Low cost The hydraulic conductivity in the vertical direction is measured Large volumes of soil can be tested Few experimental ambiguities No disturbance of soil	Must install underdrain before the liner is constructed Relatively long testing times (usually several weeks to several months for hydraulic conductivities less than 10^{-7} cm/sec) Must collect and measure seepage from underdrain, which usually necessitates a sump and a pump.

After Daniel (1989). Reprinted by permission from ASCE.

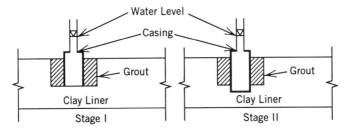

FIG. 7.6. Boutwell permeater.

calculated using this test. It is reported that the field hydraulic conductivity measured using this method compared well with laboratory test; this test also indicated that horizontal hydraulic conductivity is typically 5 to 10 times vertical hydraulic conductivity for compacted clay soils (Daniel, 1989). It takes several days to several weeks to complete this test.

CONSTANT HEAD BOREHOLE PERMEAMETER. In this method (Fig. 7.7) a constant head of water is maintained using a Mariotte system (Olson and Daniel, 1981; Reynolds and Elrick, 1985). The rate of flow needed to maintain two constant water levels is measured, which is then used to calculate the hydraulic conductivity. *In situ* hydraulic conductivity obtained using a bore-hole permeameter was reported to be an order of magnitude higher than laboratory values (Stephens et al., 1988). The test can be completed in a relatively short time—a few hours to a few days.

7.1.6.2.2 Porous Probes. In this method either a constant or falling head test is performed using a porous probe pushed into the soil (Fig. 7.8). BAT® permeameter is similar to porous probe (Torstensson, 1984). In a BAT® permeameter a chamber is lowered into a bore hole created by the probe and brought into contact with the porous probe using a hypodermic needle and a septum. The chamber contains both air and water. The mixture is either pressurized or evacuated causing water to flow out or in of the chamber.

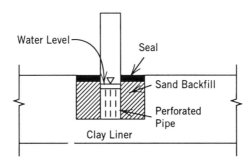

FIG. 7.7. Constant head borehole permeater.

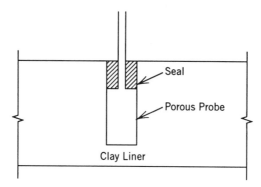

FIG. 7.8. Porous probe.

A pressure transducer records the change of air pressure within the chamber due to the flow of water. The hydraulic conductivity is measured from the time required for a change in air pressure. *In situ* hydraulic conductivity of compacted clay soil compared well with laboratory values (Chen and Yamamoto, 1987). The time taken to perform the test is only a few minutes to few hours.

7.1.6.2.3   Air Entry Permeameter. The air entry permeameter (AEP) consists of a sealed ring, about 60 cm in diameter, embedded approximately 10 cm into the soil (Fig. 7.9). The test is done in two stages. In the first stage the rate of infiltration is determined from falling or constant head test until the wetting front penetrates to the base of the ring (note: this may take several weeks for compacted soil). In the second stage the valve that allows

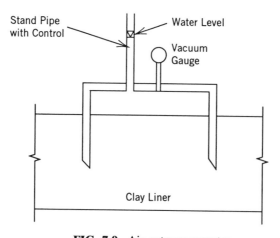

FIG. 7.9. Air entry permeater.

flow of water is closed. This closing of the valve creates a negative pressure, which is measured with a gauge. The negative pressure develops because of the soil suction, which tries to suck water out of the AEP. When the vacuum gauge reaches a peak, the AEP is disassembled and the depth to the wetting front is measured by finding out the variation of water content with depth. The hydraulic conductivity is calculated from the water entry suction pressure and air entry suction pressure; these values are obtained from a plot of the suction pressure with volumeric water content. *In situ* hydraulic conductivity obtained by using AEP was about one-half order of magnitude higher than laboratory values (Daniel, 1989). The test takes several weeks to complete.

7.1.6.2.4 Lysimeter. A lysimeter (Fig. 10.5) is essentially a pan constructed with material of very low permeability (e.g., synthetic membrane) and back-filled with highly permeable material (e.g., sand); the water collected in the pan is drained to a stand pipe or similar device (see Section 10.3.1 for details). Hydraulic conductivity is calculated by Darcy's law and the measured flow rate in the pan. These are installed below the compacted clay liner and are used mainly for landfill monitoring purposes. However, use of the lysimeter to calculate the *in situ* hydraulic conductivity of compacted clay liners has been reported, which indicated good agreement with other *in situ* tests (Day and Daniel, 1985; Rogowski et al; 1985). Lahti et al. (1987) reported good agreement with laboratory values. Time taken to complete the test is several weeks to several months.

7.1.6.2.5 Infiltrometers. Four different types of infiltrometers are available. Of these the sealed double ring infiltrometer (SDRI) appears to provide representative *in situ* values of compacted soil liners (Sai and Anderson, 1991).

OPEN, SINGLE-RING INFILTROMETER. In this method a ring is embedded in the compacted soil (Fig. 7.10). The ring is filled with water. Hydraulic conductivity of the soil is calculated by measuring the rate of change of the water head within the ring. *In situ* hydraulic conductivity of compacted clay liners was reported to be in good agreement with values obtained by using

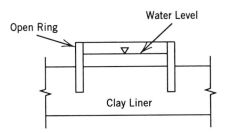

FIG. 7.10. Open single ring Infiltometer.

other methods (Daniel, 1984; Day and Daniel, 1985). Time taken to complete the test is several weeks to several months.

SEALED, SINGLE-RING INFILTROMETER. The apparatus is similar to the open single-ring infiltrometer (Fig. 7.11) except that the ring is sealed, which minimizes loss due to evaporation from an open ring. Hydraulic conductivity is calculated from the rate of change of level in the stand pipe. Testing time using this method varies from several weeks to several months.

OPEN, DOUBLE-RING INFILTROMETER. Two rings or boxes are embedded in soil, which are filled with water; the rings are covered to minimize evaporation (Fig. 7.12). Hydraulic conductivity is calculated using the rate of change of water head in the inner ring. The purpose of the outer ring is to limit lateral spreading of water originating from the inner ring. *In situ* hydraulic conductivity of compacted soil compared well with the values obtained using other methods (Daniel, 1984; Day and Daniel, 1985). Testing time using this method varies from several weeks to several months.

SEALED, DOUBLE-RING INFILTROMETER. In this method two circular or square rings are embedded in the compacted soil (Fig. 7.13). A small flexible bag is attached to the inner ring, which is sealed. Water is siphoned into the inner ring up to a depth of 25 mm. The outer ring is filled with water until the water level is 10 cm above the top of the inner ring. Flow of water from the inner ring is replaced automatically by the flexible bag. Volume of water through the inner ring is calculated by finding the loss of weight of the flexible bag. The infiltration rate is calculated from the weight loss and the area of the inner ring. The test is continued until the infiltration rate becomes steady or reaches a specified value. While Daniel and Trantwein (1987) and Chen and Yamamoto (1987) reported that the *in situ* hydraulic conductivity obtained using this method was up to one order of magnitude higher than laboratory values, Elsbury et al. (1988) (cited in Daniel, 1989) reported exel-

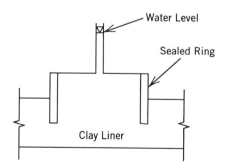

FIG. 7.11. Sealed single ring infiltometer.

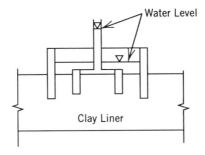

FIG. 7.12. Open double ring infiltometer.

lent agreement between values obtained using SDRI and lysimeter. Testing time using this method varies from several weeks to several months. A standard test method for SDRI is available (ASTM; D5093-90).

7.1.7 Density Tests

Knowledge about two types of density testing is required for clay liner construction purposes: (1) modified Proctor's density test (ASTM D1557-78) performed in the laboratory to find the maximum density and optimum moisture content relationship and (2) field density and moisture content test for quality control purposes.

Laboratory Test. The dependency of soil density on the moisture content of soil was first investigated by Proctor (1933). The compaction procedure was later standardized to develop the commonly known standard Proctor's method. The weight of the hammer in the standard Proctor's test is 5.5 lb (2.49 kg), having a free fall of 12 in. (30.48 cm); the sample is compacted

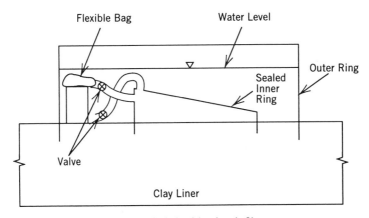

FIG. 7.13. Sealed double ring infiltometer.

in a 4-in. (10.16-cm) mold in three layers, each layer receiving 25 blows. In a modified Proctor's test the soil is compacted in five layers in a 4-in. (10.15-cm) mold using a 10-lb hammer (4.54 kg) with a 18-in. (45.7-cm) free fall. Each layer is compacted with 25 blows. Bulk density and moisture content of the compacted soil obtained from several samples compacted at different moisture contents is used to develop a moisture–density relationship (Fig. 7.14). The optimum moisture is the moisture content at which the soil exhibits maximum density. The modified Proctor's test is expected to simulate compaction offered by a sheep's foot roller.

Field Test. Two types of field tests are available: (1) direct measurement using a sand cone, drive cylinder, or rubber balloon and (2) indirect measurement using a nuclear gauge.

Direct Methods. The sand cone test (ASTM D1556) consists of digging out a sample of the soil, measuring the volume of soil by filling it with sand, and then finding the dry weight of the removed sample. The dry unit density is obtained by dividing the weight by the volume of sand. The rubber balloon method (ASTM D2167) is the same as the sand cone method except that the volume of the hole is measured by a water-filled balloon under constant pressure. The drive cylinder test (ASTM D-2937) consists of driving a standard tube into soil, trimming the soil flush with the ends of the tube, and finding the dry weight of the soil in the tube. Dry unit weight is obtained from the weight of the soil and the known volume of the cylinder. The moisture content of the soil is obtained by oven drying a field soil sample (ASTM D2216-80). A microwave oven may also be used for drying the soil samples (Gee and Dodson, 1981).

Indirect Method. A nuclear gauge is used to determine both density and moisture content (ASTM D2922). The method is much faster and is widely used currently for quality control purposes. The nuclear gauge essentially

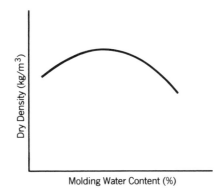

FIG. 7.14. Typical dry density versus water content relationship for clayey soils.

consists of a gamma ray source and a detector. The reflected emmision sensed by the detector is inversely proportional to the density of the soil. The density can be measured by any of two operation modes: back scattering and direct transmission. In the back scattering mode the gauge is kept at the top of the compacted surface. Any air void between the gauge and the liner will lead to a lower density reading. In the transmission mode the source of the gamma rays is lowered into a hole (5–30 cm deep). The error in density measurement is minimal in this mode. It is essential to calibrate the gauge for each project and on a regular basis thereafter against a direct method of density measurement (e.g., sand cone method) because the composition of the soil and handling may affect results.

A nuclear moisture gauge consists of a fast neutron source and a neutron detector. The water content is proportional to the hydrogen atom concentration in the medium. It is essential to calibrate the gauge against oven-dried samples on a regular basis. Proper precautions must be taken while handling nuclear meters because the devices use radioactive material.

7.1.8 Specifications

The criteria for choosing a clay is primarily based on the recompacted permeability achievable under field conditions. A clay that can be compacted to obtain a low permeability (1×10^{-7} cm/sec or less) sample when compacted to 90–95% of the maximum Proctor's dry density at wet of optimum moisture is chosen for landfill liner construction. Note that clay with a high liquid limit (LL) tends to develop more desiccation cracks, clay with a very low plasticity index (PI) or plastic limit (PL) is less workable, and a well-graded soil is expected to develop low permeability when compacted properly. Therefore, the PI, LL, and some minimum requirements regarding grain size distribution should be specified. Inorganic clays of medium plasticity (Fig. 7.2) are best suited for liner construction. Usually a soil with the following specifications would prove suitable for liner construction.

LL greater than or equal to 30%
PI greater than or equal to 15%
0.074 mm and less fraction (P200) greater than or equal to 50%
Clay fraction greater than or equal to 25%

The minimum percentage compaction (usually 90–95% of the maximum modified Proctor's density) should be specified. To obtain better kneading action and lower permeability, all compaction must be done at wet of optimum moisture. Note that the shape of the grain size distribution curves should be studied to confirm that they are close to an "inverted S" shape. The specifications regarding Atterbergs limits and grain size percentages are also helpful in quality control during construction (refer to Section 9.2 for more details).

In reality it may be difficult to obtain clay that will satisfy all the specifications described above. Field experience indicates that clayey soil with the following values can also be compacted to obtain a low permeability liner:

PI between 10 and 15%
LL between 25 and 30%
0.074 mm or less fraction between 40 and 50%
Clay content between 18 and 25%

However, it is prudent to perform field trials for such marginal clayey soils.

7.1.8.1 Field Trial for Marginal Clays. Necessary laboratory test data described in Section 7.1.6 through 7.1.8 must be obtained prior to the field trial. Then the soil should be spreaded in thin lifts [20–25 cm (8–10 in.)] before compaction in two or more plots and compacted at two or more different percentages using the equipment proposed for use for the actual liner construction (for more guidance on construction refer to Sections 9.1 and 9.2). Undisturbed soil samples should be collected per the recommendations provided in Section 9.2 and tested for permeability. The actual moisture content and percentage compaction that will provide the desired permeability are decided based on the field test data. Field trials should be undertaken for all major landfill projects.

7.2 SYNTHETIC MEMBRANE

There are several polymers that are combined with different additives to form a thermoplastic widely known as a geosynthetic membrane (or geomembrane). The molecules that make up plastics are very large and are called polymers. The factors that influence the physical properties of polymers include the size of the molecules, the distribution of different molecular sizes within a polymer, and the shape and structure of individual molecules. Different additives are added to the polymers to improve manufacturing and the usefulness of the final product.

From the numerous polymers and additives available, thousands of possible formulations can be developed. However, in practice only a few are used and these are named after the major polymers. The following polymers are in common use: butyl rubber, chlorinated polyethylene (CPE), chlorosulfonated polyethylene (CSPE), ethylene-propylene rubber (EPDM), high-density polyethylene (HDPE), medium, low- and very low-density polyethylene (MDPE, LDPE, and VLDPE), linear low-density polyethylene (LLDPE), and polyvinyl chloride (PVC). Each has certain advantages and disadvantages, which are summarized in Table 7.3. It may be noted that all membranes in the same category do not have the same composition. Thus, testing of each lot of synthetic membrane is essential to determine whether the one proposed for a project conforms to certain standard physical properties.

TABLE 7.3. Advantages and Disadvantages of Commonly Used Synthetic Membranes

S1 Number	Synthetic Membrane	Advantages/Disadvantages
1	Butyl rubber	Good resistance to ultraviolet (UV) ray ozone, and weathering elements Good performance at high and low temperatures Low swelling in water Low strength characteristics Low resistance to hydrocarbons Difficult to seam
2	Chlorinated polyethylene (CPE)	Good resistance to UV, ozone, and weather elements Good performance at low temperatures Good strength characteristics Easy to seam but poor seam reliability Moderate resistance to chemicals, acids, and oils
3	Chlorosulfonated polyethylene (CSPE)	Good resistance to UV, ozone, and weather elements Good performance at low temperatures Good resistance to chemicals, acids, and oils Good resistance to bacteria Low strength characteristics Minor problem during seaming
4	Ethylene-propylene rubber (EPDM)	Good resistance to UV, ozone, and weather elements High strength characteristics Good performance at low temperatures Low water absorbance Poor resistance to oils, hydrocarbons, and solvents Poor seam quality
5	Low-density and high-density polyethylene (LDPE and HDPE)	Good resistance to most chemicals Good strength and seam characteristics Good performance at low temperatures Poor puncture resistance

TABLE 7.3. (*Continued*)

S1 Number	Synthetic Membrane	Advantages/Disadvantages
6	Medium, very low, and linear low-density polyethylene (MDPE, VLDPE, LLDPE)	Good to excellent chemical resistance Good seam quality (Note: These products are newly introduced in the market; properties should be reviewed before use)
7	Polyvinyl chloride (PVC)	Good workability High strength characteristics Easy to seam Poor resistance to UV, ozone, sulfide, and weather elements Poor performance at high and low temperatures

Discussions on factors considered for choosing a membrane are included in Sections 7.2.1 to 7.2.4 and a discussion on polyethylene is included in Section 7.2.5.

7.2.1 Damage by Soil Microbes, Rodents, and Vegetation

In general, the resins in membranes are resistant to microbiological attack, but the additives are not (Connolly, 1972; Kuster and Azadi-Bukhsh, 1973; Potts et al., 1973). Since microbial attack is expected for some waste types, care should be taken to choose a proper additive. The two following tests may be used to determine microbial resistance: ASTM G-21 and ASTM G-22 (ASTM). Both are short-term tests and at present no long term-test is available.

Insects and burrowing rodents may severely damage plastics (Connolly, 1972). PVC are more susceptible to attack by rats than polyethelene. Hoofed animals such as deer may puncture a synthetic membrane.

Certain grass species may germinate and penetrate through synthetic membrane. To prevent such damage, use of an herbicide prior to synthetic membrane installation is recommended (Schult and McKias, 1980). However, study of canals lined with synthetic membranes indicate that roots of vegetation are not likely to penetrate synthetic membranes (Hickey, 1969).

7.2.2 Workability

The cost of installation will depend on the ease with which the membrane can be handled. Thicker membranes [1.5 mm (60 mils) and above] have the advantage of having more tolerance to handling abuse. A thicker membrane is less likely to be weakened by the seaming process. The main disadvantage of a thicker liner is that it is heavier and may require special equipment to handle it. Synthetic membrane liners will be exposed to temperature variation prior to and during installation. Therefore whether the membranes stick together at anticipated field temperatures should be investigated. Hypalon may be susceptible to this problem of sticking or "blocking" (Lee, 1974; Woodley, 1978). ASTM D-1893 and D-3354 (ASTM) can be used to test blocking properties of synthetic membranes.

Ease of seaming is an important consideration in the installation of a synthetic membrane. In general, it is more difficult to seam a membrane that is more chemically resistant (Forseth and Kmet, 1983). The long-term durability of the seam must be considered. The chemical compatibility of the solvent may vary considerably (Haxo, 1982). The problem of seaming a membrane after long-term exposure to waste should also be considered. Membrane aging may affect the ability to repair a damaged area or seam a new membrane to an old portion when expanding a landfill. Hypalon is known to lose seaming ability within a year even when protected by a layer of soil (Forseth and Kmet, 1983).

7.2.3 Compatibility with Waste

The chemical compatibility of the synthetic membrane with the waste leachate must be determined. Koerner (1986) reported a chemical resistance chart indicating chemical resistance of many common synthetic membranes against several generic chemicals. Waste/membrane compatibility tests typically involve immersing a membrane coupon in a waste leachate or waste slurry. Membrane samples are exposed to actual or synthetic leachate either on both side (immersion) or on one side only (tub or pouch encapsulation). The samples are removed after a period of time (up to 120 days) and tested for several physical properties (Koerner, 1986). Because of the magnitude of possible error, attempts are being made to develop a test based on diffusion parameters (Lord and Koerner, 1984). At present, however, the 9090 test, which is an immersion type test proposed by the United States EPA, is commonly used in landfill projects in the United States. A designer should check with the regulatory authority as to whether a compatibility test should be undertaken and, if so, an acceptable test procedure for the type of membrane selected. In addition to exposure to waste, synthetic membranes will be exposed to soil. Therefore compatibility with the on-site soil should also be evaluated. Naturally occurring oxides of metals, chlorides, and sulfur compounds and some organic compounds may react with the membrane; high or low pH soil can also degrade the membrane (Connolly, 1972). Short-term burial tests such as ASTM D-3083 (ASTM) can be used to determine short-term compatibility with the on-site soil. Table 7.4 includes suggested values of property change that may be allowed after exposure to the chemical or waste (Little, 1985).

TABLE 7.4. Suggested Limits of Different Test Values for Incubated Synthetic Membranes Used as Liner

Property	Resistant
Permeation rate	<0.9 g/m^2-hr
Change in weight (%)	<10
Change in volume (%)	<10
Change in tensile strength (%)	<20
Change in elongation at break (%)	<30
Change in 100 or 200% modulus (%)	<30
Change in hardness	10 points

After Koerner (1990). Reprinted with permission from the Industrial Fabrics Association International.

7.2.4 Mechanical Properties

The mechanical properties of concern and corresponding test methods are listed in Table 7.5. The details of test procedures can be found in the relevant ASTM book or in the cited references. Resistance to degradation due to leachate exposure is also a physical property needing assessment. It is not included in Table 7.5 because it has already been discussed in Section 7.2.3. The name of each test clearly indicates the property under investigation. Most of the tests are related to installation; a detailed discussion regarding their need can be found in Section 9.2. It should be mentioned that the permeability of synthetic membranes is extremely low and doubt exists regarding the use of Darcy's equation in predicting membrane permeability (Giroud, 1984).

TABLE 7.5. Standard Tests Used in the United States for Testing Physical Properties of Synthetic Membranes

S1 Number	Physical Properties	Standard Test Method
1	Tensile strength	ASTM D638
2	Tear resistance	ASTM D1004
3	Puncture resistance	ASTM D4833
4	Low-temperature brittleness	ASTM D746 procedure B
5	Environmental stress crack resistance	ASTM D1693 condition C
6	Permeability	Refer to Section 7.2.5.1
7	Carbon black percent	ASTM D1603
	Carbon black dispersion	ASTM D2663
	Accelerated heat aging	ASTM D573, D1349
8	Density	ASTM D1505
9	Melt flow index	ASTM D1238

7.2.5 Discussion on Polyethylene

Although many polymers have been used in the past for landfill liner construction it appears that HDPE and LDPE are the preferred choices in many instances. A comparison of physical properties of some synthetic membranes and their performance as liners can be found elsewhere (Fong and Haxo, 1981; Cadwalladar, 1986). It should be mentioned that new polymers are also being developed (Brookman et al., 1984; Wollak, 1984) that may provide alternatives to HDPE and LDPE (e.g., VLDPE, LLDPE). Polyethylene is the general name for many polymer resins. Proper selection of polyethylene resin is important. Resins used for pipe manufacturing are ideal for manufacturing quality liner material. Polyethylene is a crystalline polymer and hence tends to have a regular crystal lattice. The molecules crystallize by folding of the polymer chains forming lamellae or plate-like polymer crystals (Fig. 7.15). These lamellae are arranged to form larger aggregates known as spherulites (Fig. 7.16). The physical properties of polyethylene are greatly influenced by the size, shape, and arrangement of the spherulites. The smaller the diameter of the spherulites the greater is the resistance to stress cracking. Brittleness in polyethylene increases due to the increase in spherulite dimension. The molecular weight distribution also plays an important role in stress cracking. Low-molecular-weight species occupy the space between high-molecular-weight species thereby resisting crack propagation. Both crystalline and amorphous materials are present in a polymer. Cracks tend to propagate through amorphous material. There are essentially three amorphous materials or zones that link the crystalline zones (Fig. 7.17). (1) Tie molecules: these are chains that tie two lamellae; (2) loose loops: these are closed chains that loosely hang from a lamellae but do not interconnect lamellae; (3) celia: these are open chains loosely hanging from lamellae. Thus, tie molecules form a bridge between lamellae and spherulites. Polyethylenes containing few tie molecules are brittle in nature. However, with a high number of tie molecules, a polyethylene will be highly ductile but not very stiff (Lustiger and Corneliussen, 1988). Thus, a balance must be obtained in the number of tie molecules. The three parameters that control tie molecule numbers are molecular weight, comonomer content, and density.

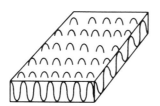

FIG. 7.15. Polymer crystal with folded chain that forms lamellar crystal plates.

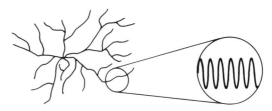

FIG. 7.16. Pattern and interconnection between lamellae and spherulite.

Molecular Weight. A low melt flow index indicates longer average polymer chains (or high molecular weight). The distribution of molecular weight is also important to obtain better control over tie molecule.

Comonomer Content. Comonomers are long-chain olefins (1-butene or 1-hexene) that inhibit crystallinity and cause short branches. Short-chain branching increases entanglements of tie molecule.

Density. This is an indirect measure of crystallinity. The more crystalline the polymer, the fewer the number of tie molecule. A typically good value for density is 0.95 g/cm^3 and for melt flow index is 0.22 g/10 min (Cadwalladar, 1986).

7.2.5.1 Permeability of Polyethyene (PE). In general, the hydraulic conductivity of PE liners is extremely low (of the order of $1 \times 10^{-10} \text{ cm/sec}$) and hence the leakage of water is negligible when used as landfill liners. Water vapor transmission (*WVT*) through a synthetic membrane is calculated using the following equation (Koerner, 1990):

$$WVT = \frac{m \times 24}{t \times a} \text{ g/m}^2/\text{day} \qquad (7.5)$$

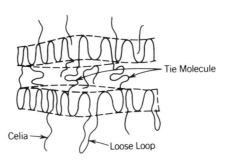

FIG. 7.17. Typical structure of polyethylene lamellae.

in which m = weight loss (gram), t = time interval (hours), and a = area of the specimen (m). In the water vapor transmission test, a specimen is sealed over an aluminum cup with water in it. A controlled relative humidity is maintained on either side of the cup. The entire assembly is placed in a chamber in which the relative humidity is controlled. The weight change of the cup is monitored over a length of time. The test time varies between 3 and 30 days. Permeance is defined as change of water vapor transmission due to the difference in relative humidity, which is mathematically expressed as follows:

$$\frac{WVT}{\delta_p} = \frac{WVT}{S(R_1 - R_2)} \tag{7.6}$$

in which δ_p = the vapor pressure difference across the membrane (in millimeters of mercury), S = the saturation vapor pressure at test temperature (in millimeters of mercury), R_1 = the relative humidity within a cup and R_2 = the relative humidity outside the cup. The permeability is calculated by the following equation:

$$\text{Permeability} = \text{permeance} \times \text{thickness} \tag{7.7}$$

As mentioned earlier in this section the hydraulic conductivity of synthetic membranes is extremely low. However, the permeation of other chemicals, especially organic species, is not as low. Synthetic membrane liners exhibit permselective properties, i.e., rate of permeation through the synthetic liner depends on the chemical specie. Table 7.6 shows permeation rates of a mixture of hydrocarbons through a 1-mm (40-mil)-thick HDPE. The transport of nonelectrolytic contaminants through synthetic liners should be determined experimentally.

ASTM E96 test method may be employed by replacing the water in the cup with the solvent of interest (Koerner, 1990). Additional relevent standard tests are (1) ASTM D1434 for gas vapor transmission and (2) ASTM D814

TABLE 7.6. Permeation Rates of a Mixture of Hydrocarbons through HDPE

Hydrocarbon	Permeation Rate (g/m²/day)
Trichloroethylene	9.4
Tetrachloroethylene	8.1
Xylene	3.0
iso-Octane	0.8
Acetone	1.4
Methanol	0.7

After Telles et al. (1988).

for organic vapor transmission. Telles et al. (1988) suggested a method called the "Permachor Method," in which the permeability coefficient for a given polymer/permeant system is calculated using apparent permeation energy (E_p) at temperature T using the following equation:

$$\ln P = \ln P_0 - [E_p/(R_0 - T)] \tag{7.8}$$

in which P_0 and R_0 are universal gas constant and preexponential factor, respectively. Several researchers have reported their findings on permeation of liquid through various polymers (Salame, 1961; Nasim et al., 1972; Haxo et al., 1984). The concentration of organic species in leachates from most nonhazardous waste landfills (e.g., municipal waste) is not high and, hence, special test(s) to find the permeation rate of such chemicals need not be undertaken. However, if the concentration of organic species is expected to be relatively high for a particular landfill then permeation test(s) should be performed. Small reductions in densities below 0.93 g/cm^3 result in large increases in permeability of polyethylene (Telles et al., 1988).

Leakage through holes in synthetic membrane, though not an intrinsic property like permeability, need to be considered for estimating leakage through a liner. The number of holes or flaws per square meter of a liner, arial distribution of such holes, and their sizes are difficult, if not impossible, to predict. However, even under strict quality control few holes are expected to be present either at pre- or postconstruction stage. Leakage rate through a synthetic liner depends on the permeability of the material immediately below the membrane and the permeability of the drainage blanket above the membrane. Leakage rate in three possible situations are discussed:

1. Flow of liquid through a hole in the synthetic membrane may be considered as free flow if it is overlain and underlain by high permeability material (e.g., gravel).

2. Flow of liquid through a hole in the synthetic material is lesser than case 1 if it is underlain by material of moderate permeability (e.g., silty soil).

3. Flow of liquid through a hole in the synthetic material is significantly low if it is overlain by a high to moderate permeability material (e.g., sand) and underlain by a low permeability material (e.g., compacted clay). This scenario models a composite liner. The following empirical equations may be used for this case (Bonaparte et al., 1989):

For good contact:
$$Q = 0.21 \, a^{0.1} \, h^{0.9} \, k_s^{0.74} \tag{7.9}$$

For poor contact:
$$Q = 1.15 \, a^{0.1} \, h^{0.9} \, k_s^{0.74} \tag{7.10}$$

in which Q = steady-state leakage rate through a hole in the synthetic liner (m^3/sec), a = area of the hole (m^3), h = head of liquid on the synthetic membrane at the hole location (m), and k_s = hydraulic conductivity of the underlying low permeability layer (m/sec).

The good and poor contact conditions are defined as follows (Bonaparte et al., 1989):

Good contact: A synthetic membrane installed with very few wrinkles on a smooth well compacted clay surface.

Bad contact: A synthetic membrane with several wrinkles; the underlying surface is not well compacted and appears to be rough.

7.2.6 Types of Commonly Available Synthetic Membranes

Three types of membranes are usually available.

Nonreinforced. These are manufactured in plants with an extrusion or callendering process. The ranges of available width and thickness when the extrusion process is used are width, 4.85–10 m (16–33 ft) and thickness, 0.25–4 mm (10–160 mils). The ranges of available width and thickness when the callendering process is used are width, 1.5–2.4 m (5–8 ft) and thickness, 0.25–2 mm (10–80 mils).

Reinforced. Woven or nonwoven geofabrics are coated with polymeric compounds. The typical thickness of reinforced synthetic membranes ranges between 0.75 and 1.5 mm (30 and 60 mils). The width of the membrane depends on the width of the geofabric.

Laminated. A nonwoven geofabric is callendered to a synthetic membrane (usually nonreinforced). The thickness and width depend on the thickness and width of the geofabric and synthetic membrane.

7.3 SYNTHETIC CLAY LINER

Synthetic clay liner is a new line of product that is gaining popularity. Synthetic clay liners are manufactured by sandwiching a uniform layer of dry bentonite between two geotextiles. The bentonite is kept in place by using a water-soluble adhesive. When a synthetic clay liner comes in contact with water it swells, forming a continuous layer of bentonite 12 to 25 mm in thickness.

Table 7.7 includes a comparison between synthetic clay liners and compacted clay liners. The hydraulic conductivity of synthetic clay liners varies between 1×10^{-7} and 1×10^{-9} cm/sec, which may decrease by almost an order of magnitude due to a 10-fold increase in effective stress (James Clem

TABLE 7.7. Comparison between Synthetic Clay Liners and Compacted Clay Liners

Synthetic Clay Liner	Compacted Clay Liner
Easy to install	Difficult to construct correctly
Light equipment can be used for construction	Heavy equipment is necessary for construction
Choice of equipment is not critical for construction	Choice of equipment used for compaction is critical
Majority of the quality control tests are done in factory; very little quality control test is necessary during installation	Some of the quality control tests are done at the borrow source; substantial quality control tests are necessary during construction
Construction time is short	Construction time is long
Construction is interupted due to rain	Construction is interupted due to rain
Installation at low temperature (up to −5°C) is permissible	Construction at or below freezing temperature not permissible
Appears to have little effect due to desiccation and freeze–thaw (Note: based on limited information)	Effect of desiccation and freeze–thaw is pronounced
Has relatively high tensile and shear strength properties	Has low tensile and shear strength properties
Can be installed easily at relatively steeper slopes (up to 2H:1V)	Cannot be constructed easily in steeper slopes
Adjusts to differential settlement	Performance is poor in case of a differential settlement
Does not consume significant airspace	Consumers relatively higher air space
Cannot be used in direct contact with most leachate	Can be used in direct contact with most leachate
Overall cost is low in most cases	Overall cost is high in most cases (note: cost depends primarily on haul distance and liner thickness)

Corporation, 1992). The interface shear values of synthetic clay liners depend on the contact material on which it is laid. It is prefered that an appropriate test be undertaken for determining the shear strength parameters. However, the following values may be assumed in the absence of actual test data:

Smooth HDPE	$\phi = 11$,	$C = 0$
Textured HDPE	$\phi = 27$,	$C = 1.92$ kPa
Wet clay	$\phi = 24$,	$C = 1.44$ kPa
Sand	$\phi = 24$,	$C = 3.84$ kPa

(Source: James Clem Corporation, 1992.)

It may be noted that the shear strength parameters will depend on the type of geotextile used in the manufacture of the synthetic clay liner. As discussed in Section 7.4.2, the permeability of bentonite may increase due to a change in chemistry of the permeating liquid. Since synthetic clay liners contain bentonite, it is preferable not to expose them to leachate directly. Thus they should not be used as the primary liner in a double lined landfill. It is also a prudent decision not to replace conventional compacted clay liners with synthetic clay liners but to use them in conjunction with compacted clay liners to reduce the thickness of the clay liner. It appears that when properly installed a synthetic clay liner can replace about 60 cm in clay thickness in a conventional compacted clay liner.

7.4   AMENDED SOIL AND OTHER ADMIXTURES

Results of studies on three types of amended soil [the amenders used are bentonite, asphalt, and cement and two admixes (which are asphaltic concrete)] are available. Bentonite amended soil has proved to be more useful in landfill projects; hence, the discussion is divided into two categories: nonbentonite mixes and bentonite amended soil.

7.4.1   Nonbentonite Mixes

Fong and Haxo (1981) reported findings in which the following four admix materials were exposed to municipal solid waste leachate for a fairly long time (56 months).

Paving Asphalt Concrete. Concrete made of a hot mix of 7% asphalt (60–70 penetration grade) and granite proportioned to meet a 0.25-in. maximum gradation for dense graded asphalt concrete.

Hydraulic Asphalt Concrete. A concrete made of a hot mix of 9% asphalt (60–70 penetration grade) and granite proportioned to meet an 0.25-in. maximum gradation for dense graded asphalt concrete.

Soil Asphalt. This was a mixture of 100 parts of soil with 7 parts of liquid asphalt, grade SC-800.

Soil Cement. This was a mixture of 9.5 parts of soil, 5 parts of kaolinite, 10 parts of portland cement, and 8.5 parts of water.
     Although the permeability and other physical properties of the asphaltic mixtures did not deteriorate much, a significant loss in compressive strength due to absorption of water or leachate was observed. The soil cement probably gained strength and decreased permeability; however, a caveat was issued regarding the cracking of soil cement as observed in highway projects. The study did not favor the use of any of these mixtures in lining a landfill. Additional studies (Haxo et al., 1985) indicate that except for soil cement

all other mixes performed poorly when exposed to different industrial waste leachate.

No study is available regarding the use of these admixtures in landfill final cover construction. Use of the admixtures as a layer in a multiple layered cap (i.e., as a layer below a synthetic membrane) may be studied for chemical compatibility and retention of structural integrity under differential settlement conditions. Since asphaltic concrete is not as flexible as soil, it is highly likely that it will not be able to withstand stresses induced by differential settlement. A detailed report on the subject is available (Haxo et al., 1985).

7.4.2 Bentonite-Amended Soil

Bentonite-amended soil is used in different civil engineering projects. Bentonites are essentially clay minerals of the smectite group. The basic structure of a smectite consists of repeated stacking of layers; each layer consists of an octahedral sheet sandwiched between two silica sheet, as schematically shown in Fig. 7.18. The bond between the layers is weak and has an unbalanced charge. Water is easily absorbed between the layers whereby the basal spacing increases causing swelling of the clay. The basal spacing can vary from 9.64 to complete separation (Mitchell, 1976). Because of the unbalanced charge, smectite exhibits a high cation-exchange capacity whereby ions are absorbed in the interlayer spaces.

Montmorillonite, the most common mineral of the smectite group, is formed essentially from the substitution of aluminum by magnesium in the octahedral sheet. Naturally occurring bentonite is a variety of montmorillonite. The interlayer swelling of bentonite and its fabric is influenced by the composition of the pore liquid. A change in the pore liquid composition may trigger isomorphous substitution whereby the swelling may increase or decrease, which may cause a change in the fabric. Bentonite swells heavily due to addition of water and forms a low-permeability fabric. However, if

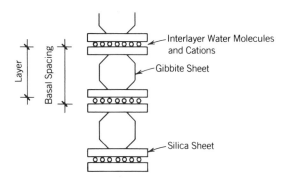

FIG. 7.18. Schematic diagram showing structure of bentonite.

the pore fluid is changed the fabric as well as the basal spacing may be reduced causing an increase in permeability. The effect of pore fluid on the permeability of bentonite can be reduced by treating the bentonite with special polymer(s). Since the porefluid chemistry has significant effect on permeability of bentonite the compatibility of leachate (expected from the landfill) with the bentonite amended soil must be studied. Changes in the permeability of the bentonite mix due to a change in the permeating fluid is well documented (D'Appolonia, 1980; Alther, 1982).

7.4.2.1 Mixing Process. Central plant mixing (Fig. 7.19) is reported to be more effective than in place mixing using agricultural equipment (Bagchi, 1986a; Goldman et al., 1986; Lundgren, 1981). For constructing final cover over a papermill landfill in Wisconsin a well-graded sand with 15% of 0.074 or less fraction (P200) content was mixed with a 4% commercially available powered bentonite (a polymer-enriched natural bentonite) to develop a low-permeability $(1 \times 10^{-7}$ cm/sec or less) mix (Bagchi, 1986a). The use of small truck-mounted concrete batch plants for mixing bentonite is not documented and is worth studying.

The quality of the mix must be checked to ensure uniformity and correctness of the bentonite percentage. The methylene blue test (ASTM C-837) and the sand equivalent test (ASTM D-2419) were found to be useful in estimating the bentonite percentage in the mix (Alther, 1983; Bagchi, 1986a;

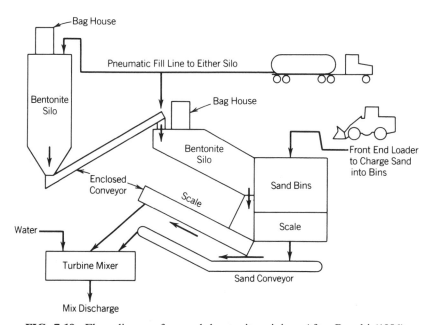

FIG. 7.19. Flow diagram for sand–bentonite mixing. After Bagchi (1986).

Barvenic et al., 1985). A minimum of five trial runs should be made to check the quality of the mix visually and using grain size analysis. The bentonite percentage should be checked using either of the tests mentioned above during the trial runs. The permeability should also be checked using the field mix compacted in the laboratory. Quality control during construction of the liner is discussed in Section 9.2 The following steps may be followed to determine the suitability of the bentonite amendment for a project.

1. Find the percentage of bentonite to be added to well-graded sand (use the sand borrow proposed for the project) that will develop a low permeability mix. A proper permeant must be used for the permeability tests. It is recommended that the sand contain a minimum of 10–15% of 0.074 mm or less fraction (P200) soil. The percentage of bentonite may vary between 3 and 15%, which depends on the sand and the permeating liquid. Only powdered bentonite should be used for mixing because complete hydration of pellets cannot occur in the field.

2. Perform trial runs using the mixing equipment proposed for the project to check quality and permeability of the field mix.

3. Run a sand equivalent or methylene blue test on a regular basis during actual construction to check the percentage bentonite added.

LIST OF SYMBOLS

Th	= thickness of the double layer
D	= dielectric constant
K	= Boltzmann's constant
T	= absolute temperature
n_o	= ion concentration
ε	= unit electronic charge
υ	= valence of ion
A_c	= activity of clay
PI	= plasticity index.
P_c	= percentage clay fraction
n	= constant ($=5$ for natural soil and 10 for artificial mixtures)
K	= permeability of soil
K_0	= pore shape factor
T_f	= Tortuosity factor
S_0	= specific surface per unit volume of particle
e	= void ratio
γ	= unit weight of the permeant liquid
μ	= viscosity of the permeant liquid
S_r	= degree of shrinkage
V_i	= initial volume of sample before drying

V_f	= final volume of sample after drying
WVT	= water vapor transmission
m	= weight loss
t	= time
a	= area of specimen
δ_p	= the vapor pressure difference
S	= saturation vapor pressure at test temperature
R_1, R_2	= relative humidity
E_p	= apparent permeation energy
P_0	= universal gas constant
R_0	= preexponential factor
Q	= steady-state leakage
a	= area of hole
h	= head
K_s	= hydraulic conductivity of the underlying layer
ϕ	= angle of internal friction
C	= cohesion

8 Design of Landfill Elements

A detailed design of several landfill elements, in addition to liners, is necessary. The proper functioning of each of these elements is essential to construct and maintain a landfill in an environmentally sound manner. All the elements mentioned in this chapter may not be necessary for all landfills. Engineering designs for most of the elements are already available; however, in some cases changes need to be made to adapt them to a landfill situation.

8.1 LEACHATE COLLECTION SYSTEM

The leachate collection system consists of a leachate trench and pipe, leachate line clean out ports, a leachate collection pump and lift station, and a leachate storage tank. The leachate storage tank may not be necessary for a site in which the leachate is discharged directly to a sewer. Permission from the proper authority is necessary before landfill leachate can be discharged into a sewer.

8.1.1 Leachate Collection System Failure

A knowledge about causes of leachate collection system failure is essential to appreciate the design of various elements of the system. A leachate collection system can fail due to the malfunctioning of one or more of the system elements. Kmet et al. (1988) and Bass (1985) discussed several causes of failure. The pipe may fail because of clogging, crushing, or faulty design.

Clogging. The pipe may clog because of the buildup of fines, the growth of biological organisms, or the precipitation of chemicals (Bass, 1985). Buildup of fines can result from sedimentation from the leachate or migration of fines from the trench (if the liner material is clay). To minimize the possibility of soil buildup, it is a good idea to use geotextile or filter fabric in the trench, as discussed in Section 8.1.2. Migration of surrounding soil into the trench will not occur if the pore space of the filter layer is small enough to hold the 85% size of soil (Cedergren, 1977). Biological clogging occurs because of the presence of microorganisms in the leachate. Factors that may contribute to biological clogging include the carbon–nitrogen ratio in the leachate, nutrient availability, concentration of polyuromides, temperature, and soil moisture (Kristiansen, 1981). Clogging resulting from chemical precipitation could be caused by chemical or biochemical processes. Factors that control chemical precipitation are change in pH, change in partial pressure of CO_2, or evaporation (Bass et al., 1983). The precipitates produced due to biochemical processes are generally mixed with slime consisting of bacterial colonies that adhere to the pipe wall. Polymers and additives are susceptible to biodeterioration from microorganisms (Klausmeier and Andrews, 1981). Cracking of polyurethanes caused by fungal growth has also been reported (Hamilton, 1988). Incorporation of an antimicrobial additive to the plastic composition can delay or reduce the extent of the formation of bioslime (Hamilton, 1988).

Crushing of Pipe. Crushing of a pipe may occur if the strength of the pipe chosen for the landfill is insufficient. Plastic pipes are considered flexible. The structural design of these pipes are discussed in Section 8.1.2.

Faulty Design. A leachate collection pipe may also fail because of faulty design. In general, the leachate flow rate is very low ($\sim$0.5–1.0 cm^3/min); however, in some landfills the flow can be significantly higher for an accidental run-on due to failure of the diversion structure. The size of the leachate collection pipe may be insufficient to effectively manage such situations, although these situations are not common for most landfills. Sizing the leachate collection pipe to drain such an exceptionally high water volume is not standard practice. The pipe may also fail due to uneven settlement, especially near exit points from the landfill and at manhole entry points. In designing a leachate collection system, all these causes of failure must be taken into consideration. A leachate collection system may also fail because of the malfunctioning of a joint. Hence each joint must be designed carefully.

8.1.2 Design of Leachate Trench and Pipe

Leachate pipes are generally installed in trenches that are filled with gravel. The trenches are lined with geotextile to minimize entry of fines from the liner into the trench and eventually into the leachate collection pipe. Typical trench details are shown in Figs. 8.1 and 8.2. Usually the design shown in Fig. 8.1 is used in landfills in which liner material is clay and the design shown in Fig. 8.2 is used in landfills in which the primary liner material is synthetic membrane. It is essential to have a deeper excavation below the collection trench so that the liner has the same minimum design thickness even below the trench.

Leachate Trench. The gravel used in the trench should be mounded as shown in Figs. 8.1 and 8.2 to distribute the load of compaction machinery and thereby provide more protection for the pipe against crushing. The geotextile, which acts as a filter, should be folded over the gravel. Alternatively, a graded sand filter may be designed to minimize the infiltration of fines into

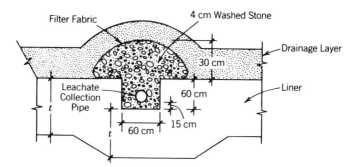

FIG. 8.1. Leachate collection trench detail.

the trench from the waste. Designs for both soil and geotextile filters are discussed below.

Soil Filter. Although several criteria are available for the design of granular filters, actual variations among them are minimal (Koerner, 1986). The following approach for the design of a soil filter may be used (Cedergren, 1977).

FIRST CRITERION

$$\frac{D_{15} \text{ of the filter}}{D_{85} \text{ of the overlying soil}} < 4 \text{ to } 5 \tag{8.1}$$

SECOND CRITERION

$$\frac{D_{15} \text{ of the filter}}{D_{15} \text{ of the overlying soil}} > 4 \text{ to } 5 \tag{8.2}$$

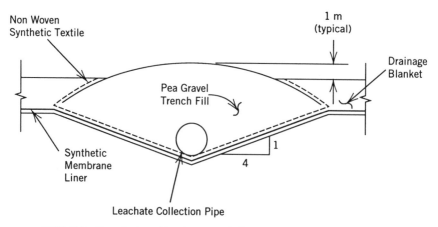

FIG. 8.2. Leachate collection trench for synthetic membrane liner.

in which D_n = the particle size of which $n\%$ of the soil particles are smaller. The first criterion is aimed at preventing migration of overlying soils into the filter layer and the second criterion is aimed at ensuring sufficient hydraulic conductivity of the filter layer to maintain proper drainage. A critique of several design criteria can be found in Sherard et al. (1984a,b) along with their suggestions for alternative criteria.

Geotextile Filters. Filter design criteria for geotextiles have been frequently discussed in the literature (Cedergren, 1977; Chen et al., 1981; Giroud, 1982; Horz, 1984; Lawson, 1982; Carroll, 1983).

The approach for filter fabric design primarily consists of comparing the soil particle size characteristics with the Apparent Opening Size (AOS; also called the Equivalent Opening Size or EOS) of the filter fabric. The following simple procedure is recommended by Koerner (1986)

1. For soils in which ≤50% of the particles pass through a 0.074-mm sieve (P200) the AOS of the filter fabric should be ≥0.59 mm (US No. 30 sieve).
2. For soils in which >50% of the particles pass through a 0.074-mm sieve (P200) the AOS of the filter fabric should be ≥0.297 mm (US No. 50 sieve). The criteria proposed by Carroll (1983) are more restrictive and the criteria proposed by Giroud (1982) are probably the most conservative.

The AOS or EOS of the filter fabric is found by sieving glass beads of known size through the fabric (Corps of Engineers, 1977). The test has several problems but is widely used for lack of a better test (Koerner, 1986). The EOS values for several commercial filter fabrics are available elsewhere (Corps of Engineers, 1977).

Leachate Pipe. As indicated in Section 8.1.1, a leachate pipe may fail due to clogging, crushing, or faulty design. Design and maintenance of leachate pipes for each of these situations are discussed below.

Clogging. Cleaning the pipe on a regular basis is an effective way of minimizing clogging due to biological or chemical processes. A clean out port along with an exit point design detail is included in Fig. 8.3. Essentially two methods are available for cleaning leachate lines: mechanical and hydraulic. At present all the cleaning equipment mentioned below are not used for leachate line cleaning, but are used for sewer line cleaning. However, they may be adapted (redesigned) to suit leachate line cleaning.

Three different types of mechanical equipment are available for cleaning purposes: rodding machines, cable machines, and buckets. In rodding machines, a series of rigid roads are joined together to make a flexible line that is pushed or pulled through the line. In cable machines, an attachment is

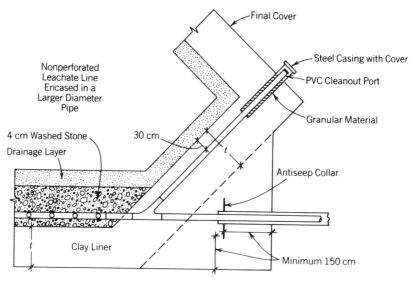

FIG. 8.3. Typical cleanout port.

spinned at the end of a cable that is pushed or pulled through the line. Both machine use a rotary motion of various attachments to clean the line. Several attachments for both types of machines are available (Bass, 1986). The disadvantage of the rodding/cable machine is that it cannot remove the dislodged debris and large quantities of water are required to flush the debris after rodding. Rodding/cable cleaning is advantageous for situations in which hydraulic methods are not capable of dislodging chemical or biological buildup or jetting may damage the pipe. Rodding/cable cleaning is less expensive than hydraulic methods in some cases. Machines fitted with a bucket that opens when pulled from one direction but closes when pulled from the other direction may be used to remove large quantities of debris from collection pipes. The bucket is pulled through a line in between two manholes, by a cable connected to a power winch (Hammer, 1977). Several accessories are available to remove materials not cleaned by a bucket (Foster and Sullivan, 1977). One disadvantage of the use of a bucket is that a manhole at each end of the pipe must be available to run the equipment. Buckets are available for use in 15-cm (6-in.) pipes and can be used to clean lines up to 230 m (750 ft) long (Foster and Sullivan, 1977).

Hydraulic equipments are almost exclusively used for cleaning leachate pipes. Two different types of hydraulic equipment are available for leachate line cleaning: jetting and flushing. In jetting, water is pumped through a nozzle that is self-propelled. Water pressure of up to 13.8 MPa (2000 psi) can be generated normally. It was observed that a high-pressure jet may damage the drainage layer (Ford, 1974); however, only a 5-cm (2-in.)-

thick drainage layer was used for the experiment. The nozzle loses its self-propelling motion at low pressure. An on-site experiment may be performed to determine the maximum allowable pressure that will not damage the drainage layer. The section used for the experiment must have sufficiently thick overburden so that a realistic equivalent pressure is exerted on the drainage layer. A low pressure (in excess of the self-propelled pressure) may be specified for the first few years to reduce the risk of damaging the filter layer. However, use of extremely high pressure is not recommended. Jetting is effective in cleaning most types of clogs, is easy to use, and needs access from only one end of the pipe. However, jetting may damage the filter layer and collection pipe, and may not be effective in removing heavy debris. A vacuum device may need to be used to clean debris dislodged by the jet. The nozzle size and type should be based on size, length, and expected clogging condition in the pipe. Jetting can clean up to a maximum length of 303 m (1000 ft); however, the capability of jetting equipment may vary (Foster and Sullivan, 1977). A field study using different pipe diameters undertaken by Babcock and Graham (1993) indicated that (1) 303 m (1000 ft) of 10-cm (4-in.)-diameter pipes will take 4–6 hr to clean where as same length of 15-cm (6-in.) or 20-cm (8-in.)-diameter pipe can be cleaned in less than an hour; (2) negotiating a bend in a 10-cm-diameter pipe is relatively difficult compared to 15- or 20-cm-diameter pipes; (3) 20-cm-diameter pipes should be used if the length of the collection line is more than 303 m; (4) it is preferable to limit leachate collection pipe length to 303 m; and (5) it is preferable to use 20.7 MPa (3000 psi) of jet pressure for easy and effective cleaning.

Collection lines can also be cleaned using a hose connected to a high-pressure water source (e.g., a fire hydrant). Sewer balls (inflatable rubber balls that reduce the flow area so that water flows around the ball at higher velocity, increasing the cleaning ability) and sewer scooters (or hinged-disc cleaners) are two attachments that may be used with the hose for better cleaning of the line (WPCF, 1980). The debris are washed downstream, and should be removed using a vacuum device. Although access to only one end of the pipe is needed for flushing, access from both sides of the pipe is preferable for proper cleaning. Access from both ends of a pipe is required if a sewer ball or sewer scooter is used. Flushing is simple and is useful only in the initial years or in certain waste types (e.g., foundry waste) landfills, where heavy biological or chemical buildup is not expected.

Crushing. Leachate collection pipes may crush during the construction of or during the active life of the landfill. To safeguard against crushing, leachate pipes should be handled carefully and brought on the liner only when the trench is ready. Running of heavy equipment over a pipe must be avoided. A pipe can be installed in either a positive or negative projection mode. Every effort should be made to install it in a negative projection mode (Figs. 8.1 and 8.2), although at times it may be necessary to install a pipe in a

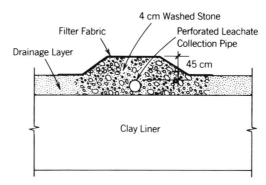

FIG. 8.4. Leachate collection pipe in a positive projection mode.

positive projecting mode (Fig. 8.4). The strength of a pipe must be checked to ascertain whether it will be able to withstand the load during both pre- and postconstruction periods. Usually two types of pipes are used, PVC and HDPE. The pipes are considered as flexible pipe. The design approach consists of calculating the deflection of the pipe, which should not exceed 5% [Uni-Bell Plastic Pipe Association (Uni-Bell, 1979]. Two approaches are available for checking pipe strength; (1) approach using Iowa formula; (2) approach developed by Paruvakat (1993).

Either of the following formulas, commonly known as modified Iowa formulas, can be used to estimate pipe strength:

$$\% \frac{\Delta}{D_p} = \frac{DBP(100)}{0.149(E/\Delta Y) + 0.061 E'} \tag{8.3}$$

$$\% \frac{\Delta}{D} = \frac{DBP(100)}{[2E/3(DR - 1)^3] + 0.061 E'} \tag{8.4}$$

in which D = the deflection lag factor, B = the bedding constant, P = the pressure on the pipe (psi), D_p = the pipe diameter (in.), Δ = deflection, E = the modulus of elasticity of the pipe material (psi), E' = the modulus of the soil reaction (psi), DR = the dimension ratio = outside diameter/ wall thickness (both in in.), $F/\Delta Y$ = the pipe stiffness (psi). Equation (8.3) is used when the pipe stiffness ($F/\Delta Y$) is known and Eq. (8.4) is used when DR is known. Note that these formulas are in F.P.S. units so all metric dimensions must be converted before using the formula. Pipe stiffness is measured according to ASTM D-2412 (ASTM) (Standard Test Method for External Loading Properties of Plastic Pipe by Parallel-Plate Loading). The modulus of elasticity of the pipe material depends on the compound used. The values of the variables in Eqs. (8.3) and (8.4) can be found in the Uni-

Bell handbook (Uni-Bell, 1979), from experiment, and from the published literature. The approximate range of values of some variables are included in Table 8.1, which provides a general guideline. Accurate values of these variables should be used for deep (20 m or more) landfills. The following formula may be used to calculate pipe stiffness (Uni-Bell, 1979):

$$\frac{F}{\Delta Y} \simeq 0.559 \, E\!\left(\frac{t}{r}\right)^3 \tag{8.5}$$

in which t = the wall thickness (in.) and r = the mean radius of the pipe (in.). For pipes with SDR numbers (e.g., SDR 35), the following formula may also be used for pipe stiffness (Uni-Bell, 1979):

$$\text{Pipe stiffness} = 4.47 \, \frac{E}{(DR - 1)^3} \tag{8.6}$$

Laboratory and field tests on buried HDPE pipe showed that the Iowa formula or its variations highly overpredicted pipe deflection (Watkins, 1990). Based on laboratory study Watkins (1990) concluded that for pipes enveloped in select Pipe-Zone Backfill (PZB), incipient failure condition will not occur up to a ring deflection of 18 to 21% (with angle of internal friction of the PZB between 30° and 35°) irrespective of the height of the landfill. The PZB is typically placed above the pipe to a height of 30–60 cm (Fig. 8.5). When waste is placed over the PZB the pipe will deflect. This deflection of the pipe will cause a redistribution of stresses within the PZB adjacent to the pipe.

TABLE 8.1. Approximate Range of Values of Different Variables of Eqs. (8.3) and (8.4)

Variable	Range	Remarks
B	0.08–0.1	Pipes embedded in gravel or sand
D	1.5–2.5	If the soil in the trench is not compacted, then the higher value of D should be used
E' (in psi)		If the soil in the trench is not
Crushed rock	1000–3000	compacted then the lower
Sandy soil and rounded gravel	~100–400	value should be used
Stiffness		
Schedule 40	~129	Values are for 6-in-diameter pipes
Schedule 80	~700	
SDR 35	46	
SDR 26	115	

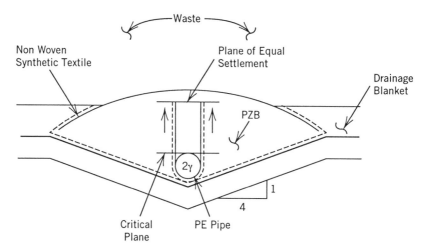

FIG. 8.5. Settlement of PE pipe under the waste load.

Paruvakat (1993) proposed a new approach for designing leachate collection pipes that considers the stress redistribution; his approach essentially combines the Iowa formula and Marston's formula. Paruvakat's (1993) method can be used for both positive and negative projection modes.

Marston's formula assumes that relative movements take place along imaginary vertical planes extending upward from the sides of the pipe (Spangler and Handy, 1982). For the landfill situation, Marston's formula is given by

$$W_c = \gamma(C_c)(B_c)^2 \tag{8.7}$$

$$C_c = \frac{e^{\pm 2kM(H_e/B_c} - 1}{\pm 2kM}$$

$$+ \left[\frac{H}{B_c} - \frac{H_e}{B_c}\right] e^{\pm 2kM(H_e/B_e)} \tag{8.8}$$

in which W_c = load per unit length of the pipe, γ = unit weight of soil above the pipe, B_c = outside width of the pipe, H = height of fill above pipe, H_c = height of plane of equal settlement above the critical plane, k = earth pressure coefficient, M = tan ϕ, ϕ = angle of internal friction, e = base of natural logarithm.

Figure 8.6. shows the positive projection mode where the relative stiffness of the pipe is smaller than the surrounding backfill and, therefore, shortening of the vertical height of pipe is more than the deformation of the adjacent backfill. Thus, an incomplete ditch condition will occur provided the plane of equal settlement is within the PZB. The height of the plane of equal settlement above the critical plane can be calculated from Eq. (8.9); all relevant terms are depicted in Fig. 8.6:

$$\pm r_{sd}(P_p) \left[\frac{H}{B_c}\right] = \left[\frac{1}{2kM} \pm \left\{\frac{H}{B_c} - \frac{H_e}{B_c}\right\} \pm \frac{r_{sd} P_p}{3}\right] \frac{e^{\pm 2kM(H_e/B_c)}}{\pm 2kM}$$

$$\pm 1/2 \left[\frac{H_e}{B_c}\right]^2$$

$$\pm \frac{r_{sd} P_p}{3} \left[\frac{H}{B_c} - \frac{H_e}{B_c}\right] e^{\pm 2kM(H_e/B_c)}$$

$$- \frac{1}{2kM} \left[\frac{H_e}{B_c}\right] \mp \left[\frac{H}{B_c}\right] \left[\frac{H_e}{B_c}\right] \tag{8.9}$$

in which r_{sd} = settlement ratio and P_p = ratio of pipe projection above the natural ground surface to the pipe diameter; r_{sd} is given by

$$r_{sd} = \frac{(S_m + S_g) - (S_f + d_c)}{S_m} \tag{8.10}$$

in which S_m = compression deformation of the backfill column adjacent to pipe of height PBC, S_g = settlement of natural ground surface adjacent to

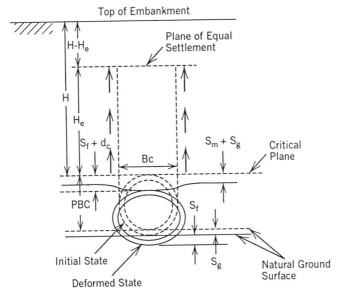

FIG. 8.6. Case of an incomplete ditch condition for a positive projecting conduit. After Paruvakat (1993). Reprinted with permission from the author and Industrial Fabrics Association International.

the pipe; S_f = settlement of the pipe into the liner, d_c = shortening of the vertical height of pipe.

(Note: In Eqs. (8.8) and (8.9), for ± signs, use the upper sign when r_{sd} is positive and use the lower sign when r_{sd} is negative.) For PE pipes in landfill liners, S_g and S_f are same whereby r_{sd} reduces to

$$r_{sd} = \frac{S_m - d_c}{S_m} \tag{8.11}$$

d_c is larger than S_m because the PZB materials are stiffer than the pipe. So r_{sd} is negative, indicating an incomplete ditch condition. Equations (8.8), (8.9), and (8.11) need to be solved simultaneously to find the deflection d_c for the design load of W_c. The following is a step-by-step procedure for calculating d_c (Paruvakat, 1993):

STEP 1. Assume a value of r_{sd}.

STEP 2. Calculate S_m based on the estimated vertical stress using the following formula:

$$S_m = \frac{(\text{vertical stress})\ (\text{mean pipe diameter})}{E_s} \tag{8.12}$$

in which E_s = deformation modulus of the PZB material [732.6 MPa (15.3×10^6 psf) approx].

STEP 3. Using the assumed value of r_{sd} and the calculated value of S_m, find d_c from Eq. (8.11).

STEP 4. Using the calculated value of d_c find W_c from the following, which is a variation of Iowa formula:

$$d_c = \frac{DBW_c}{(EI/r^3 + 0.061E'} \tag{8.13}$$

in which D = deflection lag factor, B = bedding factor, W_c = load per unit length of pipe, E = modulus of elasticity of the pipe material, I = moment of inertia of the pipe wall, E' = modulus of soil reaction, r = mean radius of pipe.

STEP 5. Using Eq. (8.7) and W_c calculate C_c.

STEP 6. Using Eq. (8.8) and C_c calculate H_e/B_c.

STEP 7. Use the calculated value of the $\dfrac{H_e}{B_c}$ in Eq. (8.9) to find r_{sd}. Steps 1 through 7 should be repeated until the calculated and assumed values are equal.

STEP 8. Verify that PZB extends up to at least H_e to satisfy the original assumption made in deriving the equations. The percent deflection ($d_c/2r$ in percent) can be checked against allowable percent deflection, or the actual stress on the pipe wall due to W_c can be checked against the allowable stress to calculate the factor of safety, which should be 2 or more.

In general, for shallow landfills (20 m for municipal or other wastes with similar unit weight and 10 m for sandy type waste) the Iowa formula may be used. For deeper landfills, Paruvakat's method described above should be used because it will provide a realistic pipe deflection.

Example 8.1 (in F.P.S. units)

Determine the suitability of 6-in. schedule 40 PVC pipe placed in a trench (Fig. 8.1) filled with rounded washed gravel (dumped in the trench). The landfill is 70 ft. deep and has a waste-to-cover ratio of 5:1. The final cover will be 4 ft thick and the drainage blanket 1 ft. thick. The unit weights of waste and soil are 50 and 110 lb/ft^3, respectively. The maximum leachate head is expected to be 3 ft.

$$\text{Thickness of daily cover} + \text{final cover} = (70 - 4) \times \frac{1}{(5 + 1)} + 4 = 15 \text{ ft}$$

$$\text{Pressure on the pipe } (70 - 15) \times 50 + 15 \times 110 + 3 \times 62.4$$
$$+ 1 \times 110 = 4697.2 \text{ psf} = 32.6 \text{ psi}$$

Using Eq. (8.3), the percentage deflection can be calculated:

$$\% \frac{\Delta}{D_p} = \frac{2 \times 0.09 \times 32.6 \times 100}{0.149 \times 129 + 0.061 \times 200} = 18.6\%$$

Since the percent deflection is more than 5%, a thicker pipe should be used. The percentage deflection for a schedule 80 pipe will be 5.03%. The maximum percentage deflection allowable is usually 5, however, manufacturers of PVC pipe may allow a value of up to 7.5%. Use of the lower value of 5% deflection is recommended if approximate values of variables are used in Eq. (8.3) or (8.4).

Pressure due to live load should be added to the pressure P to account for compaction machinery running in the landfill. The effect of live load is maximum for the first lift. Deflection of the pipe may be checked due to pressure of a compactor running on 1-ft. (30-cm) lift of waste. The effect of live load will be reduced with increases in the depth of the landfill. The Boussinesq equation (Sowers and Sowers, 1970) may be used to determine the effect of live load at various depths. The approximate pressure of equipment at various depths (Fig. 8.7) may be obtained by multiplying by the appropriate factor given in Table 8.2. In Fig. 8.7, W_p = the width of the compactor wheel in contact with the soil, q is the contact pressure of the equipment, and z is the waste thickness.

Example 8.2 (in F.P.S. units)

Assume $q = 20$ psi and $W_p = 12$ in. Find the total pressure on the pipe when the equipment is compacting the first lift of refuse 1 ft. thick (use appropriate values from the landfill in Example 1).

$$0.1 \times 20 + \frac{1 \times 110}{144} + \frac{1 \times 50}{144} = 3 \text{ psi} \tag{8.14}$$

This value of $P = 3$ should be added to the value of P in Example 8.1 to estimate percentage deflection. This should be checked, especially for landfills in which lower stiffness pipes (e.g., 6 in SDR-35) are proposed. It may be noted that the Boussinesq equation assumes an elastic soil condition, whereas most waste types are nonelastic; hence the values given in Table 8.2 are conservative.

Faulty Design. The bends in the leachate line should be smooth. Cleaning equipment cannot negotiate sharp bends. Criss-crossing of leachate lines should be avoided. Sometimes secondary leachate lines are connected to a

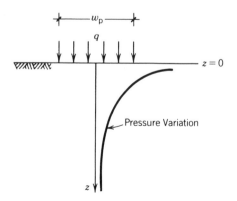

FIG. 8.7. Variation of pressure with depth due to a load at the surface. After Weiler (1993). Copyright 1993 by National Solid Waste Management Association. Reprinted with permission from *Waste Age* Magazine, August, 1993.

TABLE 8.2. Approximate Values of Pressure Coefficients at Various Depths

Depth	Coefficient
W_p	0.4
$2\,W_p$	0.1
$3\,W_p$	0.055
$4\,W_p$	0.03
$5\,W_p$	0.02
$6\,W_p$	0.015

header line to carry the entire leachate generated in a landfill. The diameter of the header pipe should be designed to handle total peak flow of leachate. T-joints should not be used to connect header pipes to a secondary line. A smoother 45° or lesser bend should be used to facilitate cleaning activities.

A minimum number of manholes should be used in a landfill. A flexible connection should be used for entry and exit points of leachate lines (Fig. 8.8). The manholes may settle, particularly if they are more than 7.5 m (25 ft.) in height. The settlement should be estimated for deep manholes. A lightweight material (other than concrete) should be considered for deep manholes.

The holes in a leachate line should be made as shown in Fig. 8.9. The pipe should be laid such that the holes are at the lower half of the pipe. Holes close to the springing line reduce the strength of the pipe and hence should be avoided. Slotted pipe may also be used for leachate collection.

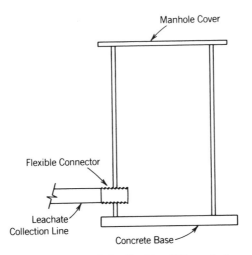

FIG. 8.8. Typical detail of landfill manhole.

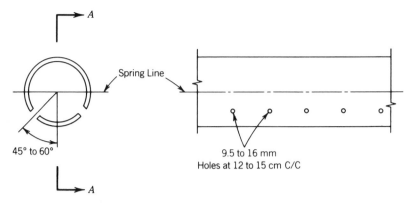

FIG. 8.9. Perforations in leachate collection pipes.

8.1.3 Leachate Line Clean Out Port

A typical clean out port is shown in Fig. 8.3. The clean out pipe must be well guarded at the exit point. A shallow concrete manhole may be constructed to provide additional safety against runover. Usually the pipe is laid along the side slope. However, if it is laid on a nearly vertical slope, a smooth bend should be used to connect it to the leachate line.

8.1.4 Leachate Collection Pump and Lift Station

The pump capacity must be calculated carefully for proper functions. In choosing a pump both suction and delivery head must be considered; it should be noted that the density of leachate is somewhat higher than water. A typical detail of a lift station and pump is shown in Fig. 8.10. Usually automatic submersible pumps are used in a lift station. Positioning of the starting and shut off switches should be such that the pump can run for a while; frequent start and stop may damage the pump. The shut off switch must be located at least 15 cm (6 in.) below the leachate line entry invert. Guide rails must be provided so that the pump can be lifted and lowered easily for maintenance. For larger landfills arrangements should be made for a standby pump, either available on site or easily accessible. A valve operable from the ground surface should be installed on the incoming leachate line(s). This is useful during periodic maintenance of the pump. If the lift station is constructed outside the landfill then it should be encased in clay or synthetic membrane to minimize the potential of leachate leakage into the ground. The lift station must be made leakproof from inside. The settlement of the lift station should be checked; in addition to the self-weight of the lift station, the weight of the pump and guide rail and of the leachate ponded inside must

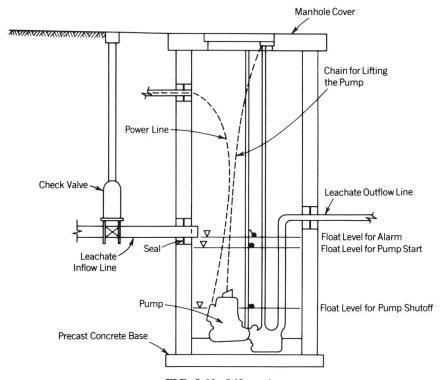

FIG. 8.10. Lift station.

be taken into consideration in calculating settlement. The leachate collection line entry connection should be made flexible.

Most pumps are efficient if run continuously; however, since the leachate generation rate varies, intermitant operation of the pump becomes essential. A 12-min cycle is considered satisfactory (Bureau of Reclamaton, 1978), an issue which should be verified from the prospective pump manufacturers. The sump size is estimated based on the volume of leachate inflow in one-half cycling time; this would mean equal on and off time. The storage capacity (or sump size) (S_s) is given by

$$S_s = V_c t \tag{8.15}$$

in which V_c = the rate of leachate collection per minute and t = one-half the cycling time. The pumping rate (P_1) is determined by

$$P_1 = \frac{V_c t + V_c t}{t}$$

$$= 2V_c \text{ volume/min} \tag{8.16}$$

The maximum volume of leachate expected in a day should be used while estimating the storage capacity and pumping rate. The maximum leachate level in the sump (for starting the pump) should be 15 cm (6 in.) below the inlet pipe invert. The minimum leachate level in the sump (at which the pump will stop) should be 60–90 cm (2–3 ft) from the bottom of the sump. The difference, D_f, between the maximum (pump start level) and minimum (pump shut off level) elevation should be kept low [60–90 cm (2–3 ft)] so that the sump dimension is reasonable. The cross-sectional area, A_r, of a sump is calculated by assuming a value of D_f and from the known value of S_s:

$$A_r \times D_f = S_s \tag{8.17}$$

From the known difference in head, h_d, between suction and delivery, the pump break horsepower (BHP) can be calculated as follows:

$$BHP = \frac{\gamma_1 h_d P_1}{550E} \tag{8.18}$$

in which γ_1 = the leachate density (note: if the density of leachate is higher than water, a 10–15% higher value may be used for a conservative estimate) and E = pump efficiency. Both the delivery and suction head should be taken into consideration when choosing a submersible pump.

Example 8.3 (in F.P.S. units)

Design a circular sump for a leachate collection system for the following case: V_c = 1.25 ft³/min, cycling time $(2t)$ = 12 min, D_f = 2 ft, and h_d = 16 ft. Assume a pump efficiency of 80% and the leachate density to be 10% higher than water.

$$\text{Sump storage capacity} = 1.25 \times 6 = 7.5 \text{ ft}^3$$

For a sump of diameter d

$$\frac{\pi d^2}{4} \times 2 = 7.5$$
$$d = 2.1 \text{ ft}$$

Use a 2-ft-diameter sump:

$$\text{Pumping rate} = 2 \times 1.25 = 2.5 \text{ ft}^3/\text{min}$$
$$BHP = \frac{62.4 \times 1.1 \times 16 \times 2.5}{550 \times 0.8} = 6.24$$

8.1.5 Leachate Holding Tank

Leachate tanks should have enough volume to hold leachate for a period of time (usually 1–3 days) during the peak leachate production season. The regulatory agency should be contacted to find out if a minimum holding capacity is mandated. The holding volume will depend on frequency of pumping out and maximum allowable discharge rate to a treatment plant.

Both double-wall and single-wall leachate holding tanks may be used. Single-wall leachate tanks should be installed within a clay or synthetic membrane encasement (Fig. 8.11). Arrangement should be made to monitor the inside of the encasement. The monitoring well will detect tank leakage at an early date. The well should be monitored once a month for indicator parameters. This type of encasement is not needed for double-wall tanks; however, provision for monitoring the space in between the two walls should be available.

Both metallic and nonmetallic tanks may be used. Metallic tanks must be protected against corrosion. The inside of the tanks must be coated with suitable material so that leachate does not damage the tank. The tank(s) should be pressure tested before installation to check for leaks. Manufacturer's guidelines regarding leak testing and installation should be followed. A long-term performance warranty should be obtained for the tank(s) from the manufacturer.

Tank(s) should be installed properly. Improper handling or backfilling may damage tanks during installation. Tanks should be anchored properly with the base concrete if the water table is expected to rise above the base level of tank. Anchoring is important for tanks installed at shallow depth below

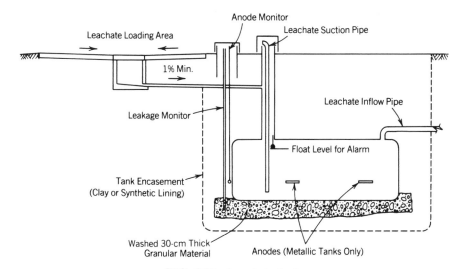

FIG. 8.11. Leachate tank.

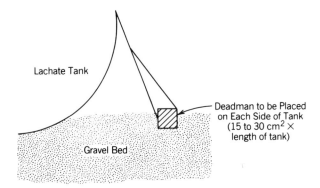

FIG. 8.12. Deadman to hold down tank.

the water table. Figure 8.12 shows a strapping arrangement; the strapping interval should be 1.8 m (6 ft) or per the manufacturer's design. The tank trench should be backfilled with pea gravel or crushed stone (3–15 mm size). A manhole is constructed above the tank that houses the pump (Fig. 8.13). A guiderail should be provided so that the pump can be lifted easily for maintenance and repair. The pipe exiting the manhole should be encased in another pipe of larger diameter to provide secondary containment. Some tankers are equipped with a pump so that the pump in the manhole is not needed. The capability of the tanker should be checked to find out whether a pump is necessary. A concrete loading pad should be constructed away from the tank. The pad should be sloped toward a sump connected to the tank so that any spill during loading will go back to the tank.

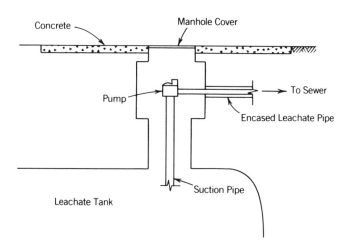

FIG. 8.13. Leachate tank manhole with outflow pump.

8.1.6   Leachate Removal Systems

Leachate from a landfill may be removed either by gravity flow or by using a side slope riser. Either system has its advantages and disadvantages. Each system is discussed separately in the following sections.

8.1.6.1   Gravity Flow. This system of leachate removal is used where the landfill base is at grade or at shallow depth. This is a popular system used in clay-lined landfills. The header pipe, which connects all the leachate from the landfill, exits through the side. The number of header pipes should be minimized to minimize liner penetration. The header pipe either discharges directly into a sewer line or into a sump with a lift station (see Section 8.1.4).

An antiseep collar should be constructed on all leachate line exit points to ensure that leachate does not seep through the hole. Figure 8.14 shows a typical antiseep collar design. The soil around the collar should be hand compacted. Care should be taken when compacting berm around a leachate exit point.

The advantage of gravity flow are (1) the method is economical, especially if direct discharge to a sewer line can be made; (2) the operational cost of a gravity flow is very low; (3) not much maintenance of the system is required; and (4) the system is in use for more than a decade, so that design and construction details are well established.

The disadvantages of the gravity flow system are (1) the chance of leakage at the exit point is relatively high, especially in landfills where synthetic membrane liners are used, and (2) the integrity of the antiseep collar cannot be tested and leakage through the liner cannot be monitored.

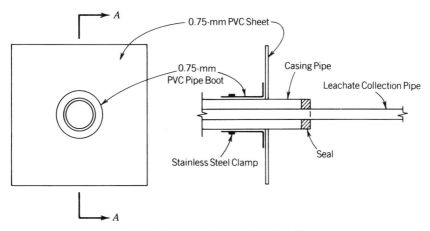

FIG. 8.14. Antiseep collar detail.

8.1.6.2   Side Slope Riser. This system of leachate removal is used primarily where the landfill base is deep (in excess of 20 m). In most cases this system is used mostly for landfills with synthetic membrane liners. Figure 8.15 shows a typical side slope riser detail. A 30–45 cm (12–18 in.) diameter HDPE pipe is used that houses the leachate withdrawal pipe. A sump pump is installed at the lower end that pumps the leachate. A side slope riser is installed at the end of each leachate line. Thus the need for a header pipe within the landfill is eliminated; however, the leachate from each side slope riser is discharged into a header pipe that runs outside of the landfill. Thus it eliminates T-joints within the landfill; T-joints are a major source of problems in leachate line cleaning. The leachate from the collection line is collected in a sump within the landfill that is withdrawn by the pump. The sump is filled with gravel. Enough redundancy should be built into a side slope riser. In addition to a steel chain attached to the pump, a hook should be attached to the pump to retrieve it if the chain is broken. In addition, the diameter of side slope riser pipe should be large enough (45 cm is suggested) so that a second pump can be installed in case the first one fails. In addition, a steel plate should be placed in the sump area. The purpose of this steel plate is to provide a guide in case a cassion needs to be installed in the event of side slope riser failure. The failure of a side slope riser means that leachate from the entire cell will be ponded, which in turn will increase leakage. Failure of the leachate collection line in a cell is also a possibility.

The advantages of the side slope riser are (1) it can be used for deep landfills; (2) the chances of leakage at the exit point is minimal; (3) it eliminates the need for a header pipe within a landfill; (4) additional redundancy can be built into the design; and (5) leachate removal from only one cell will be

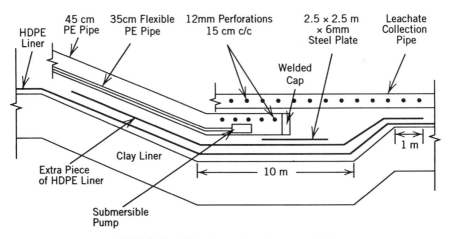

FIG. 8.15. Side slope riser (not to scale).

affected in case of pump failure in a side slope riser. The disadvantages of the side slope riser are (1) this system is difficult and costly to construct; (2) leachate removal from a cell will stop if a pump fails; and (3) no direct access to the collection point is possible, making maintenance of the pump and electrical system difficult.

8.2   STORMWATER ROUTING

Routing of stormwater on and around the landfill is essential to reduce leachate generation. All run-on water should be diversified away from the landfill by constructing drainage ditches. Usually landfills are not located on natural drainage ways with few exceptions. Precipitation falling on a landfill also needs to be routed to natural drainage swales or toward a sedimetation basin. The following sections mainly address routing of precipitation falling on a landfill; similar approaches can be used to route run-on water.

8.2.1   Design of Stormwater Ditch

The design of a stormwater ditch (also called a drainage swale) uses principles of open channel flow. There may be several ditches running over and around a landfill; in many instances one or more secondary ditches are connected to a primary drainage, which carries the entire runoff from the landfill area. In designing these ditches care should be taken to estimate proper volume of runoff water flowing through each section. Ditches running over the landfill should have low base slope to minimize erosion (note: the recommended maximum is 10%). Even though short-term maintenance is expected, long-term maintenance of drainage ditches cannot be ensured. The Manning formula is used to design a channel section:

$$V = \frac{1.486}{n_r} r_h^{2/3} S^{1/2} \tag{8.19}$$

in which V = the mean velocity of water (fps), r_h = the mean hydraulic radius (ft, obtained by dividing the cross-sectional area by the wetted perimeter), S = slope of the energy line (= slope of channel for small slopes), and n_r = the roughness coefficient.

The greatest difficulty in applying the Manning formula is to choose the proper value of n_r. Typical values of n_r based on experience and observations are available (Chow, 1959). Some restrictions should also be imposed on the allowable velocity to minimize scour. Table 8.3 provides some suggested velocity ranges and n_r values based on information obtained from several sources (Chow, 1959; SCS, 1971; Schwab and Manson, 1957; Bureau of Reclamation, 1978). The initial values are to be used for estimating the channel dimensions, which should be checked for adequacy in the long term.

TABLE 8.3. Recommended Values of n_r and V for Design of Drainage Swales in Landfills

Variable	Values
n_r	Initial: 0.02–0.03
	Long term: 0.1–0.14
Maximum permissible V in fps	Initial: 3–5
	Long term
	Clay: 4–5
	Sandy loam: 1.5–2.5

Chow (1959) provides a detailed procedure for designing a grassed channel. However, for most landfills a simpler approach shown in the example below may be used for designing drainage swales on a landfill. Usually either a trapezoidal or triangular section is used for landfill drainage swales. A typical drainage swale arrangement is shown in Fig. 8.16, and a typical cross section of a primary drainage ditch is shown in Fig. 8.17. Additional erosion protection measures such as lining with an erosion mat or riprap or construction of check dam(s) should be undertaken if higher velocity is used. In some cases drop inlets (Fig. 8.18a and b) may be needed to route surface water from the landfill surface to a nearby surface drainage swale.

Example 8.3 (in F.P.S. units)

Design a drainage swale to carry a flow of 5.8 ft³/sec (0.15 m³/sec) from a landfill.

For all drainage design, a trial-and-error method is used to find the dimensions of the section. For most cases a slope of the base is assumed and kept constant throughout the trial-and-error process. A triangular cross section with 3 : 1 side slope and 1% base slope is assumed for the example.

For the initial design:

1. Using the Manning formula, find the dimensions of the section capable of allowing the design flow (use initial n_r values given in Table 8.3).

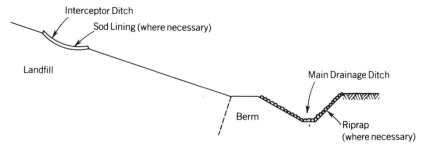

FIG. 8.16. Typical drainage swale arrangement in landfills for surface water routing.

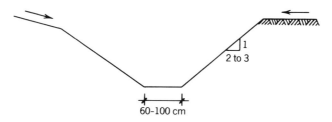

FIG. 8.17. Typical cross section of drainage ditch.

2. Check whether the velocity is within the maximum permissible value. The trial computations are given in Table 8.4. The long-term design consists of verifying whether the channel is large enough to route 1.5 times the design flow in the long term. The retardance to flow will increase due to growth of vegetation and hence the high n_r values suggested in Table 8.3 should be used. The entire channel depth of 1.8 ft is used for the check, which obviously reduces free board. However, such a reduction in free board is acceptable only for secondary swales on the landfill. For primary swales the final freeboard should be at least 80% of the initial freeboard. (Thus, for the example the revised depth of channel could be 1 not 1.8 ft.)

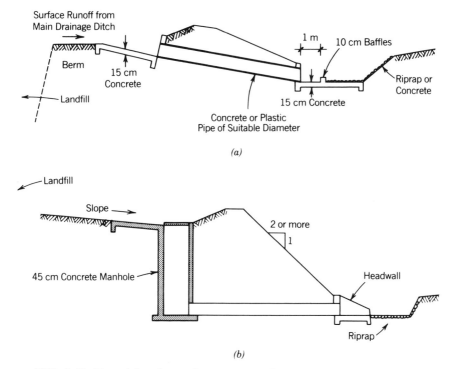

FIG. 8.18. Drop inlets for surface water routing: (a) inclined; (b) vertical.

TABLE 8.4. Trial Computations for the Example on
Drainage Swale Design

Trial	Depth ft	r_h ft	V ft/sec	ft³/sec
1	0.6	0.29	2.6	2.8
2	0.8	0.37	3.06	5.87
3	1.8	0.47	3.59	10.77
Depth of channel = 0.8 + 1 (free board) = 1.8 ft				

r_h for a 1.8-ft-deep triangular section is 0.85 ft. $V = 0.95$ ft/sec and $Q = 9.25$ ft³/sec. Since the capacity of the ditch is more than 1.5 times the design flow the swale dimension is acceptable.

8.2.2 Design of Culvert

Circular or rectangular culverts are used to drain water below a road. The culvert inlet and outlet should provide a smooth transition to minimize erosion at entrance and exit points; concrete should be used for entrance and exit. In many instances, maintenance of the culvert in the long term is not envisioned. Therefore, a concrete culvert is preferable over a metal culvert, which needs to be replaced more often. As an alternative, arrangements should be made to replace the culvert with an open section by cutting the road at the end of the long-term care period (usually 30–40 years after closure of landfill); this alternative should be resorted to only if lack of maintenance after the long-term care period is envisioned. A culvert should be over-designed because in most cases long-term maintenance is not expected. A culvert can flow full or partly full. The flow characteristics depend on inlet geometry, slope, size, roughness, approach, tailwater condition, and so on. Although the use of nomographs is suggested for high design flows, 45- to 50-cm (18- to 20-in.) circular section culverts with a minimum base slope of 1% can be safely used for flows up to 0.28 m³/sec (10 ft³/sec).

8.2.3 Design of Stormwater Basins

Sedimentation basins may need to be constructed before allowing the surface water to enter a natural flowage. The purpose of the sedimentation basin is to reduce the total dissolved solids (TDS) from the surface water. The sediment collected at the bottom of the basin should be cleaned periodically and disposed of in the landfill. Settling velocity of a particle is calculated using Stoke's law:

$$V_s = \frac{g(\rho_s - \rho_w)d^2}{18\mu} \tag{8.20}$$

in which V_s = the settling velocity of a particle, g = the gravitational constant, ρ_s = the density of the particle, ρ_w = the density of the water, d = the diameter of the particle, and μ = the absolute viscosity of water. Ideally the surface area of a basin (A) capable of settling all particles with the settling velocity of V_s is given by

$$A = \frac{Q}{V_s} \tag{8.21}$$

in which Q = the surface loading or overflow rate. In reality the sedimentation basin does not perform according to the above theory and the two following factors influence basic performance: currents and particle interactions. Four types of currents are identified: surface currents induced by wind, convection currents arising from temperature differences within and outside the basin, density currents that develop due to different densities of the incoming water and basin water, and eddy currents produced by the incoming water. Three different particle interactions are identified: settling is increased due to flocculation of two or more particles, settling of a particle is hindered due to the upward movement of the water particle displaced by the particle, and settling of a particle is hindered due to high concentrations of sediment near the bottom of the basin. The design of inlet and outlet structures also influence sedimentation. Particles of 100–10 μm (0.0039–0.00039 in.) and above are usually removed from the runoff water. The areas of the basin for 1 m^3/sec (35.937 ft^3/sec) flow for different particle sizes are included in Table 8.5. Performance of a sedimentation basin depends very little on the depth of the basin. Normally the depth of basin used is 1.5 m (5 ft). The design approach discussed above (and enumerated in the example below) is considered adequate for most landfill designs. Charts and nomographs are also available (Gemmell, 1971; Fair and Geyer, 1954) for more detailed design. Figure 8.19 shows a typical plan and cross section of a sedimentation basin. Usually the length-to-width ratio of 2:1 is used for basin sizing. The size of the basin depends on particle size distribution and Total Suspended Solids (TSS) of the runoff water. Usually removing the particles that are 40 μm and above provides an acceptable TSS of the effluent. However, regulations may dictate removal of lower particle size.

Example 8.4

Design a sedimentation basis for removing particles 40 μm and above for a landfill in which the expected peak flow is 1.5 m^3/sec (54 ft^3/sec).

From Table 8.5 the required base area of the basin is

$$1.5 \times 476 = 714 \text{ m}^2$$

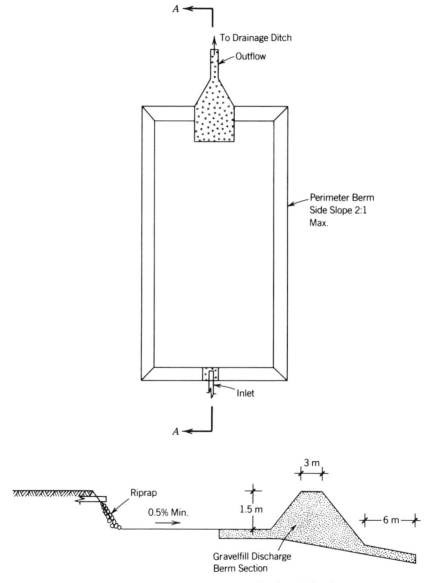

FIG. 8.19. Typical sedimentation basin detail.

TABLE 8.5. Sedimentation Basin Surface Area for 1 m³/sec Flow

Particle Size (mm)	Area (m²)
0.1 (100 μm)	125
0.06 (60 μm)	263
0.04 (40 μm)	476
0.01 (10 μm)	6.7×10^3
0.001 (1 μm)	6.7×10^5

Assuming the width to be A and length-to-width ratio as 2 : 1:

$$2A^2 = 714$$

or

$$A = 18.9 \text{ m}$$

Use a basin size of 38 × 19 × 1.5 m.

8.3   SYNTHETIC MEMBRANE

Several design issues are discussed. A simple approach using fundamental principles of mechanics are used in deriving the equations. The values used in the examples are approximate and may be different for a product used in a site. New products with better physical properties are marketed from time to time, hence physical properties for the synthetic membrane should be obtained from the manufacturer or from experiments whenever necessary. Dimensions of the trench and so on shown in Fig. 8.20 have been used for deriving equations in this section.

8.3.1   Anchor Trench Design

A trench is due on the berm and the membrane is inserted in the trench for anchoring. The dimensions of the trench need to be calculated so pullout does not occur. The maximum allowable pull (F_p) at point A is given by Eq. (8.22), which assumes total mobilization of yield stress.

$$F_p = \sigma_y t 1 \tag{8.22}$$

in which σ_y = the yield stress of the synthetic membrane and t = the thickness of the synthetic membrane. The total force resisting the pullout (F_R)

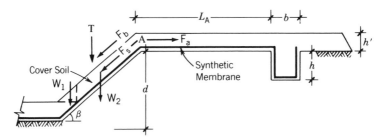

FIG. 8.20. Synthetic membrane design variable.

for a case in which there is no cover soil but the trench is filled with soil (note: sandbags are sometimes used until the trench is filled with soil) is

$$F_R = \gamma_s hb \tan \delta + wL_A \tan \delta \qquad (8.23)$$

in which γ_s = the unit weight of the soil, b = the trench width, h = the trench depth, δ = the friction angle between the membrane and soil, and w = the weight per square area of the synthetic membrane. For empty landfills the yield stress is not mobilized totally, hence a safety factor (1.2–1.5) may be used on F_p to design the trench. The total force resisting the pullout when cover soil is applied (F_{rc}) is

$$F_{rc} = \gamma_s(h + h')b \tan \delta + (w + \gamma_s)L_A \tan \delta \qquad (8.24)$$

w is negligible compared to γ_s. A safety factor (1.2–1.5) may be used to calculate F_{rc}.

Example 8.5

Design an anchor trench for a 40 mil HDPE liner ($\sigma_y t = 1440$ kg/m); the cover soil is 60 cm thick and has a unit weight of 1.7 g/cm^3. Assume a run on length of 1 m and a trench width of 1.2 m. Use $FS = 1.3$ and $\delta = 18°$:

$$F_{rc} = 1.7 \times 10^3(h + 0.6) \times 1.2 \tan 18 + 1.7 \times 10^3 \times 0.6 \times \tan 18$$

$$h(\text{obtained by equating } F_p \text{ and } F_{rc}) = 0.57 \text{ m}$$

Use a 60-cm-deep trench.

8.3.2   Allowable Weight of Vehicle

To avoid pullout, rupturing, and undesirable elongation leading to failure, heavy vehicles cannot be allowed on synthetic membranes. The maximum allowable weight of the vehicle (T) can be calculated using Eq. 8.25. A cover soil should be applied before allowing any vehicle into the landfill.

$$T \sin \beta + W \sin \beta = W \cos \beta \tan \delta + F_{rc} + \frac{\sigma_y t}{FS} \qquad (8.25)$$

in which T = the allowable weight of the vehicle, W = the total weight of the soil on the slope ($= \gamma_s h'/d/\sin \beta$), β = the slope angle of the liner, and FS = the factor of safety.

Example 8.6

Calculate the maximum weight of a vehicle for the case in Example 8.5. The landfill is 6 m deep and has a side slope of 3.5 : 1 ($\beta = 15.9°$).

From Eq. (8.25)

$$T \sin 15.9 + 1.7 \times 10^3 \times 0.6 \times 6$$

$$= 1.7 \times 10^3 \times 0.6 \times (6/\sin 15.9) \times \cos 15.9 \times \tan 18$$

$$+ 1.7 \times 10^3 \times (0.6 + 0.6) \times 1.2 \tan 18$$

$$+ 1.7 \times 10^3 \times 0.6 \times 1 \times \tan 18 + 1440/1.3$$

$$= 11{,}253 \text{ kg}$$

8.3.3   Check for Sliding of Cover Soil

The cover soil may slide down the slope if adequate friction is not provided by the membrane. Sometimes a geotextile is placed below or above the membrane for protective purposes. In such cases, the friction angle between the synthetic membrane and geotextile (which are usually lower) or the lowest friction angle between the materials should be used for design purposes. Williams and Houlihan (1986) studied the friction angle for three separate interfaces: soil to synthetic membrane (range: 27–17°), soil to geotextile (range 30–23°), and geotextile to synthetic membrane (range 23–6°). The study showed that the friction angles depend on the materials involved. Even though the soil to synthetic membrane and soil to geotextile friction angles were somewhat close, the geotextile to synthetic membrane friction angles are drastically low; HDPE friction angles were lowest in all cases. In the absence of experimental data, the following values may be used for sandy cover soil and HDPE liners:

Soil to synthetic membrane = 17°. Soil to geotextile = 23–25° (depends on type of geotextile). Synthetic membrane to geotextile = 6–8°.

Proper friction angle should be used for designing a combined layer of geotextile and synthetic membrane. The factor of safety against sliding (*FS*) is given by (see Fig. 8.21)

$$FS \times \frac{W \cos \beta \times \tan \delta}{W \sin gb} = \frac{\tan \delta}{\tan \beta} \qquad (8.26)$$

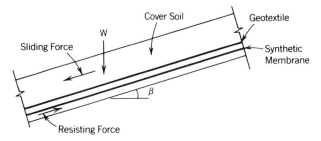

FIG. 8.21. Force schematic for synthetic membrane design.

Example 8.7

Calculate the factor of safety against sliding of a 60-cm soil cover placed over a geotextile; the geotextile is in turn placed over an HDPE synthetic membrane. The unit weight of soil $= 1.6$ g/cm^3 and the slope of the liner $= 18.4°$. The following friction angles may be assumed: geotextile and soil, 23°; synthetic membrane and soil, 18°; synthetic membrane and geotextile, 10°.

In this case the geotextile is most vulnerable to sliding over the synthetic membrane, hence

$$FS = \tan 10/\tan 18.4 = 0.53$$

To avoid failure the liner slope may be reduced to 5 : 1 (not a very practical approach though); the use of geotextile may be avoided and the slope is reduced to 3.5 : 1 (the $FS = \tan 18/\tan 15.9 = 1.1$); an anchor trench is designed to withstand a higher pull from the geotextile ($\delta = 10°$ should be used in the design) and the strength of the geotextile and synthetic membrane should be checked [Eq. (8.25) can be used (assuming $T = 0$ and $FS = 1$) to find the stress and Eq. (8.24) can be used for the anchor trench design].

8.3.3.1 General Equation for Cover Soil Sliding. An equation corelating effects of slope angle, strength properties and thickness of cover soil, seepage forces, and equipment load with the properties of the synthetic membrane has been proposed by Druchel and Underwood (1993). The maximum allowable pull (F_p) is given by (see Fig. 8.20)

$$F_p = F_b + F_s + \frac{(W_e + W_2)\sin(\beta - \delta)}{\cos \delta} - \frac{W_1\sin \phi}{\cos(\beta + \phi)} \qquad (8.27)$$

in which $F_b =$ equipment braking force, $F_s =$ seepage force; $W_e =$ weight of vehicle on the side slope; $W_1 =$ weight of soil anchorage at base,

W_2 = weight of cover soil on the side slope, β = slope angle of berm or liner. F_b can be assumed to be 30% of W_e; thus

$$F_b = 0.3 \ W_e \tag{8.28}$$

and

$$F_s = \frac{\rho_w h_s^2 \tan \phi}{2 \sin \beta \cos \beta} + \frac{\rho_w h_s}{\sin \beta}[d - \frac{h_s}{2 \cos \beta}]\cos \beta \tan \delta \tag{8.29}$$

in which h_s = seepage thickness (this will be maximum after a rain event or during spring thaw when the entire side slope soil will be wet), ρ_w = unit weight of water. W_1 and W_2 can be calculated using the following equations:

$$W_1 = \frac{\gamma_s(h')^2}{2 \ \sin\beta \ \cos \ \beta} \tag{8.30}$$

$$W_2 = \frac{\gamma_s h'}{\sin \ \beta} [d - \frac{h'}{2 \ \cos \ \beta}] \tag{8.31}$$

8.3.4   Check for Uneven Settlement

Uneven settlement may occur in landfill cover. However, the size of the settlement is an unknown that cannot be predicted. The maximum permissible size of curvilinear settlement may be estimated using the theory of cables. However, synthetic membranes are not elastic material and the stress developed due to a deflection is expected to be somewhat less than that predicted by Eq. (8.32). The shape of a cable deflected due to a uniform load is parabolic. The maximum stress would occur at M and N (Fig. 8.22) and its magnitude is given by (Oden and Ripperger, 1981)

$$\sigma_y = H(1 + 16S_r^2)^{1/2} \tag{8.32}$$

in which

$$H = \frac{(w + \gamma_s)h'L^2}{8d_f}$$

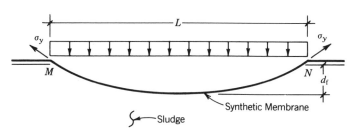

FIG. 8.22. Force schematic for the deflection of synthetic membrane.

S_r = the sag ratio = d_f/L, L = the diameter of a depression, and d_f = maximum deflection.

Equation (8.32) can be used to estimate the permissible size of a depression or if the depression size is known than σ_y can be estimated. Every effort should be made to avoid the formation of a depression when using a synthetic membrane.

8.4   BERM DESIGN

Berms around a landfill should be checked for stability. Various approaches are available for checking stability of a berm, which essentially deals with slope stability. One or more chapters will be needed to do justice to the topic, which makes it somewhat beyond the scope of this book. A basic concept and a few references are cited that should be considered as a complement to this section. Designers dealing with a difficult situation should consult the current literature on slope stability analysis.

Failure along a circular arc is assumed for most analysis (Fig. 8.23). The berm within the circular arc is subdivided into slices and analyzed by equilibrium (Bishop, 1955; Bishop and Morgenstern, 1960; Skempton, 1964). A graphical method known as the friction circle method is also available (Taylor, 1948). The method of slices using a modified Bishop's method (Bishop, 1955) provides good accuracy. Analysis can be made easier by using a hand calculator or computer (Whitman and Bailey, 1967; Little and Price, 1958). A computer disk for use in a personal computer is also available (Bosscher, 1987). Summary discussions on the topic can be found elsewhere (Morgenstern, 1992; Lambe and Silva-Tulla, 1992). The height of the berm, climatic condition, effective angle of internal friction (ϕ), effective cohesion (C'), and unit weight of the berm material influence the stability of the side slope. The effect of earthquakes on the structural stability of a berm constructed in earthquake-prone regions should also be investigated (Marcuson et al., 1992). Although rigorous analyses using correct material properties is suggested for checking the stability of berms, a 2 to 2.5 horizontal to 1 vertical side slope may be considered structurally safe for 3–4 m (10–13 ft.) high berms made of sandy soil (not applicable for earthquake-prone regions).

Apart from ensuring the structural stability of inboard and outboard side of a berm, the outboard side of a berm should be protected from erosion due to wind and water. Freeze–thaw also causes deterioration of the slope face. Soil erosion depends on rainfall intensity, length (along the slope) and gradient of the slope, status of vegetation on the berm, and soil type. The length and slope of a berm should be designed to minimize erosion. The erosion of bare soil is much higher for a 3H : 1V compared to a 4H : 1V slope (SCS, 1972). The length of the slope may be reduced by constructing a horizontal bench on the (outboard side) berm and by constructing drainage ditches. These measures should be undertaken if the slope length exceeds 20 m (66 ft.). Growth of vegetation retards soil erosion (Gray and Leiser,

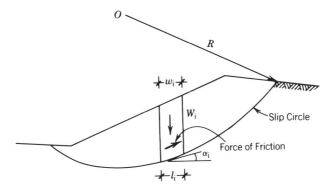

FIG. 8.23. Geometry and forces in "method of slice" analysis.

1982). The effect of freeze–thaw is prominent in berms with no vegetation. To account for soil erosion the outboard side of berm slopes should be made less steeper than what is calculated by structural stability analysis. A 3V : 1H slope for a 3–4 m (10–13 ft.) high berm (made of sandy soil) may be used in the absence of a detailed analysis. Note that the suggested side slope is less steep than mentioned in the above paragraph; the reduction in slope is to account for the erosion losses.

8.5   STABILITY OF WASTE SLOPE

The stability of the waste slope also needs to be assessed. The following are the four modes of failure of landfill slopes (Mitchell and Mitchell, 1992) two of which have already been discussed in previous sections: (1) failure of the inboard side of the berm (see Section 8.4), (2) failure of the synthetic membrane liner on the side slope (see Sections 8.3.1, 8.3.2, 8.3.3, and 8.3.3.1), (3) failure of the landfill foundation leading to waste slope failure, and (4) failure of the waste face.

Landfill foundation may fail due to the presence of a soft layer below the liner. Failure could be due to excessive settlement or due to lack of shear strength. Both trench type (Figs. 1.1 and 1.2) and at grade (Fig. 1.4) landfills should be checked for foundation failure if a soft soil layer is encountered during subsoil investigation. A check for both settlement and bearing capacity failures should be made. A case history of landfill foundation failure has been reported in the literature (Divinoff and Munion, 1986).

Waste face can fail due to lack of stability of the slope. Sliding of the waste mass as a whole may also take place. A case history of landfill failure due to sliding of the waste mass has been reported in the literature (Byrne et al., 1992). Both short-term and long-term analysis should be undertaken to check stability of the waste face (Howland and Ladva, 1992). Stability of the waste face should be checked both from an operational as well as a design stand point.

In a Canyon type landfill (Fig. 1.3), in addition to checking stability of the waste face during operation and at closure, the stability of the canyon itself should also be checked.

The problem in analyzing a waste slope is to obtain proper shear strength parameters for the waste and to develop an applicable analysis. For relatively homogeneous waste, such as foundry sand, the standard tests may be used to obtain the value of ϕ' and C'. However, for most waste types the standard tests are not applicable for the following reasons: (1) many sludge type wastes (e.g., papermill sludge) release gas during the test, which will influence the test results; (2) putrescible wastes may deteriorate in the landfill leading to a change in the shear properties; (3) some wastes are extremely heterogeneous (e.g., municipal garbage) and hence small samples cannot provide the shear properties of the waste as a whole; and (4) it is difficult to compact a waste in the laboratory.

Since most of the wastes are heterogeneous, strictly speaking the theories mentioned in Section 8.4 are not applicable. Thus, analyzing the stability of a waste slope is somewhat difficult. A combination of theoretical analysis and field study should be undertaken to investigate the stability of waste slopes (Bagchi, 1987a). The stability analysis theories mentioned in Section 8.4 may be used if a higher factor of safety (1.5–2) or lower values of ϕ' and C' are used in arriving at a stable slope angle. While checking the final waste face care should be taken to see that the entire slip circle passes through the waste only and no part intercepts the berm (Fig. 8.24). The shear strength properties (ϕ' and C') obtained from standard laboratory tests should be reduced by 15–25%. Observations regarding stable waste slopes of existing landfills for the same waste type (if available) should be taken into consideration. Several papers discussing various issues of this topic can be found elsewhere (ASTM, STP1070, 1990). The following waste slopes have been observed to be stable:

Municipal waste: 3H : 1V to 4H : 1V

Sludge type waste with minimum 40% solid content: 8H : 1V

Fly ash-treated sludge: 7H : 1V to 6H : 1V

Sandy foundry waste and fly ash: 3V : 1H to 4V : 1H

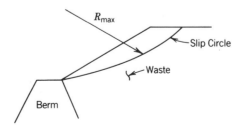

FIG. 8.24. Maximum allowable slip-circle for analyzing a waste slope.

The regulatory agency should be contacted to check whether the rules call for a mandatory maximum permissible waste slope. Proper geotechnical investigation should be undertaken at all sites before siting a landfill (refer to Chapter 2 for details).

8.6   COMMENTS ON SEISMIC DESIGN OF LANDFILLS

Seismic design of landfill elements is an emerging field in which standards of practice are evolving rapidly. The analytic techniques used are involved and are somewhat beyond the scope of this book. Although it appears that seismic design of landfill elements is undertaken in some cases, there is a lack of literature dealing with the topic. The purpose of this section is to familiarize readers with the current trends in design approaches rather than providing methods of analyses. A designer needs to address the following issues that are associated with seismic analysis:

1. Selection of appropriate design ground motion.
2. Evaluation of the effect of selected ground motion on the proposed site.
3. Estimation of the dynamic properties and shear strengths of waste fills, liner materials, and subgrade materials.
4. Checking the seismic stability of the landfill elements due to the selected ground motion.

A regulatory agency should be consulted to choose the proper ground motion. In general, a landfill should not be located on or near (within 60 m) a known fault. In an earthquake-prone zone seismic investigation should be undertaken to ensure the absence of a fault under the proposed site. Although primary faults are mapped, secondary faults are not well mapped. U.S.E.P.A requires that landfills located in seismic impact zones should be designed to resist the Maximum Horizontal Acceleration (MHA) for the area. A seismic impact zone is defined as an area that has 10% or more probability of exceeding MHA of $0.1g$ in 250 years. The dynamic response analysis should be performed to check the stability of subgrade strata and/or natural slope faces. The estimation of dynamic properties of soil materials can be easily done using available laboratory tests. Pseudostatic stability analyses (Newmark, 1965; Makdisi and Seed, 1978) is generally used to check dynamic stability. Use of a computer program (Schnabel et al., 1972) for this purpose has also been reported (Seed and Bonaparte, 1992).

Although Singh and Murphy (1990) reported dynamic characteristics of waste fill, in general there are few data on this important issue. Seed and Bonaparte (1992) suggested use of back analysis from field recordings of waste fills to obtain reliable data. Anderson et al. (1992) reported a case history using this approach.

Seismic design of landfills involves checking the stability of both the waste face and the liner. Stability of the liner (especially synthetic liners) dictates whether a failure will occur. Use of same shear strength properties of the waste mass for analyzing both static and dynamic loadings is reasonable, although the approach is conservative. The performance of 10 natural attenuation type landfills after earthquake in California (acceleration range 0.1–0.45g) was studied by Orr and Finch (1990). Relatively minor surface damage was observed, indicating good attenuation and damping properties of waste fills in general. However, subtle failure such as localized rupture of clay or synthetic membrane liner or final cover can lessen the integrity of the landfill.

The mode of failure due to the ground motion during an earthquake will induce a sliding block type failure along the liner–waste interface (Singh, 1992). This drag will cause a strain on the clay liner resulting in rupture. For synthetic membrane liners the dynamic properties of interface materials can be established using conventional laboratory tests. However, for clay one needs to find the limiting permanent deformation beyond which the clay will crack or rupture. The development of cracks in clay will depend on the magnitude of the sliding shear force and stress–strain characteristics of the compacted clay. The differential settlement of subsoil due to an earthquake will make a clay liner act like a beam, resulting in tension cracks at the

TABLE 8.6. Selected Techniques to Improve Landfill Stability in Seismic Impact Zones

Problem Area	Suggested Solutions
Liquefaction of foundation soils	Densify foundation soils
	Dynamic compaction
	Vibrocompaction
	Compaction grouting
Inadequate slip resistance in liner/leachate collection system	Increase strength of soil liner components (use different soil, revise compaction specifications)
	Increase interface strength (geomembranes)
	Reduce grading
	Change overall bottom geometry
Potential slip in waste	Alter operations to increase strength of waste (increase compaction or soil bulking)
	Alter cell configuration
	Institute interior soil berms
	Institute or increase size and/or strength of perimeter dikes
Slip of landfill cover	Increase strength of soils/materials
	Increase interface strength (geomembranes)
	Reduce landfill slopes

After Weiler et al. (1993). Copyright 1993 by National Solid Waste Management Association. Reprinted with permission.

bottom of the liner. Thus from a seismic design standpoint, use of a more plastic clay in constructing the clay liner appears to be preferable. It should be noted, however, that highly plastic clay is difficult to compact and chances of developing desiccation cracks are also higher. Table 8.6 provides a summary of possible earthquake design problems and suggested solutions for avoiding failure of the landfill.

8.7 ACCESS ROAD DESIGN

Roads, both within and outside of a landfill, are important in maintaining the smooth operation of a landfill. The road within the landfill should be designed so that dumping vehicles can move in and out easily. A typical cross section of a road within the landfill is shown in Fig. 8.25. Both geometric and structural design of road(s) outside of the landfill area should be undertaken. Sufficient sight distance, merging, and an exit lane to and from the primary public road should be provided. The vehicles entering and exiting from the landfill do not maneuver at high speed so a rigorous design calculation is not needed. A typical design for a landfill entrance is shown in Fig. 8.26. Caution sign(s) should be posted on the public road regarding the existence of a landfill and/or heavy vehicle entering the road.

Some arrangements should be made to minimize spreading of waste on the public road and off the landfill area, which sticks to the wheels of dump trucks. Washing of tires may be needed in hazardous waste landfills. Running the trucks over a rough gravel road or constructing a grate or a series of bumps in a portion of the exit road are some of the approaches used with partial success to reduce waste spreading.

The structural design drawing of the access road should indicate the thickness of the base and subbase coarse material and the type of surfacing. Flexible pavements are used for the access road. Palmer and Barber (1940) and Barber (1946) developed a method for designing pavement thickness using Boussinesq's equation (Kansas State Highway Commission, 1947). The pavement thickness (t_p) is given by

$$t_p = \left[\sqrt{\left(\frac{3A_t mn}{2\pi E_s \Delta}\right)^2 - a^2} \right] \left[\sqrt[3]{\left(\frac{E_s}{E_p}\right)} \right] \tag{8.33}$$

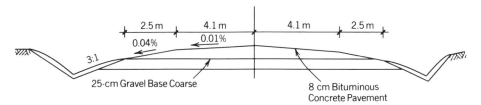

FIG. 8.25. Typical structural design of a landfill access road.

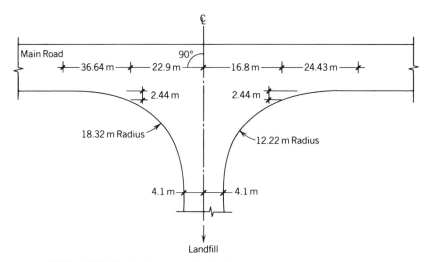

FIG. 8.26. Typical geometric design of a landfill entrance road.

in which A_t = the total weight of the vehicle, m = the traffic coefficient, n = the saturation coefficient, E_s = the modulus of elasticity of the sub-grade, Δ = the allowable deflection (= 0.1 in.), a = the radius of contact of the tire, and E_p = the modulus of elasticity of the pavement or surface course. In the absence of experimental values the following approximate values (in F.P.S. units) may be used for the design: m = 0.5, n = 0.5 for an average annual rainfall of 15.0–19.9 in. and 1.0 for a rainfall of 40–50 in. n increases by 0.1 for each 5-in. increase in average annual rainfall (i.e., n = 0.6 for an average annual rainfall of 20–24.9 in.); E_s = 5000 psi for gravel and 1500 psi for clay, E_p = 15,000 psi, and a = 6 in.; a granular base coarse material reduces frost heave.

If a granular base coarse material is to be used for a portion of the bituminous mat, then the thickness of the base coarse material (t_b) can be calculated as

$$t_b = (t_p - t_s) \sqrt[3]{\frac{E_p}{E_b}} \qquad (8.34)$$

in which t_s = thickness of wearing surface (usually 5–15 cm (2–6 in. thick) and E_b = modulus of elasticity of the base coarse material. Several other design approaches are available for the design of flexible pavements (Yoder, 1967).

Example 8.8 (in F.P.S. units)

Design a landfill access road for a region having an annual average rainfall of 30–34.9 in. and a single wheel load of 20,000 lb. The modulus of elasticity of a base coarse gravel is 5000 psi and that of the subgrade clayey soil is 500 psi.

From Eq. (8.33)

$$t_p = \left[\sqrt{\left(\frac{3 \times 20000 \times 0.5 \times 0.8}{2\pi \times 500 \times 0.1} \right)^2 - 6^2} \right] \left[\sqrt[3]{\left(\frac{500}{1500} \right)} \right]$$

$$= 11.5 \text{ in. of bituminous mat directly on the clayey subgrade}$$

To include a 2-in. surface and 6-in. of gravel base, the thickness of subgrade coarse material is calculated as follows from Eq. (8.34):

$$\text{thickness of base coarse gravel} \times (11.5 - 2) \sqrt[3]{\left(\frac{1500}{5000} \right)}$$

$$= 13.7 \text{ in.}$$

Hence, the required thickness of subgrade is

$$= 13.7 - 6) \sqrt[3]{\left(\frac{5000}{1500} \right)}$$

$$= 11 \text{ in.}$$

Thus the road section will consist of 2 in. of wearing surface, 6 in. of base coarse gravel, and 11 in. of subgrade.

8.8 LANDFILL COVER DESIGN

The purpose of a landfill cover is to minimize infiltration of water into the landfill. A multilayer configuration is used for cover in which each layer has a task to perform (Fig. 8.27). The first layer, called the grading layer, should consist of coarse-grained material and is usually 15–60 cm (6–24 in. thick). The thickness depends on the stability of the waste and gas collection system design. For an unstable waste surface, a layer of bark or a geotextile may be used below the grading layer. The purpose of the grading layer is to provide a stable surface on which the low permeability layer can be constructed and to facilitate landfill gas venting. The second layer, called the barrier layer, provides a barrier for water infiltration. The material used can be clay, synthetic clay liner, or synthetic membrane. This is an important layer that must be constructed carefully. A low permeability layer must be constructed below a synthetic membrane layer. The purpose of the third layer, called the protective layer, is to protect the barrier layer from freeze–thaw and

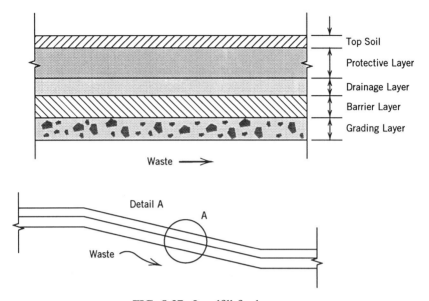

FIG. 8.27. Landfill final cover.

desiccation cracks and to provide a medium for root growth. The thickness of this layer should be sufficient to perform both the above tasks; the thickness of this layer should be between 30 and 105 cm (1 and 3.5 ft.) depending on geographic location. Although incorporation of this layer in the final cover may increase infiltration, it is considered as an important element for the tasks it is supposed to perform.

As indicated above, a low permeability layer (i.e., either a clay or synthetic clay layer) should be used below a synthetic membrane in the barrier layer. Usually a 60-cm-thick clay layer or a layer of synthetic clay is used for this purpose. However, only a 60-cm-thick clay layer may be used as the barrier layer. If the liner is a composite liner then the barrier layer should also be a composite layer.

The drainage layer may be constructed using coarse sand (5 to 15 cm thick) or synthetic net. The drainage layer is an important component of the final cover if a synthetic membrane is used at the top of the barrier layer. The purpose of this drainage layer is to provide better drainage of the protective layer so that the interface of the protective layer and the synthetic membrane does not become saturated. In general, the friction angle of the fine grained soil–synthetic membrane interface decreases due to saturation of the fine grained soil. A decrease in the interfacial friction angle may result in unstable soil conditions leading to failure or increased erosion of the protective layer. If the permeability of the protective layer is high (1×10^{-3} cm/sec or more) then it will drain quickly, whereby reduction in the interfacial friction angle due to saturation is not expected. So use of the drainage layer to avoid a reduction in the interfacial friction angle due to saturation in landfills with a high permeability protective layer is not justifiable. However, use of a layer of sand or synthetic net will protect the synthetic membrane from damage during placement of the protective layer. If the soil used for the protective layer is sandy then a separate drainage layer may not be essential. If the soil used for the protective layer has relatively low permeability and a synthetic membrane is used below the protective layer then a drainage layer should be used.

Example 8.9 (in F.P.S. units)

Check the stability of the various components of the final cover configuration shown in Fig. 8.27. The thickness of the various layers is as follows: (1) top soil: 6 in.; (2) protective layer of silty sand: 2 ft; (3) barrier layer: 60 mil textured HDPE over a 2-ft compacted clay layer; (4) grading layer: 6 in. of sand; (5) drainage layer is a geocomposite net (a synthetic net with a synthetic textile on either side). The interfacial friction angles are (1) protective layer/geocomposite, 25°; (2) geocomposite/textured HDPE, 32°; (3) textured HDPE/clay layer, 30°; (4) clay layer/grading layer, 34°. Undrained cohesion and friction angles of the clay samples compacted to 90% modified proctor were 500 psf and 15°, respectively. The maximum slope of the final cover is 3.5H : 1V. The length of the slope is 200 ft. The average unit weight of the soil is 130 pcf. The undrained friction angle of the silty sand is 28° and the cohesion is zero.

For drained conditions the factor of safety (F.S.) of the final cover components can be calculated using the infinite slope method. The infinite slope method neglects passive resistance near the toe. This method is valid for long slopes:

$$\text{F.S.} = \frac{\tan R}{\tan A} \tag{8.35}$$

in which R = the required friction angle, and A = the slope angle = 16° for a 3.5H : 1V slope. Usually an F.S. between 1.25 and 1.5 is used to calculate stability. The lower value is used when confidence on the data (e.g., interfacial friction angle is high and the consequences of failure are not catastrophic. For important structures a high F.S. is used. Using an F.S. of 1.4, R is calculated from Eq. (8.35) as 21.2°. Since all the interfacial friction angles are higher than 21.2° the final cover components are stable under drained conditions.

Undrained conditions may occur either in the protective layer or in the clay layer. The shear strength of soil for this case is given by

$$S_u = \sigma_n \tan \phi_u + C_u \tag{8.36}$$

in which S_u = the undrained shear strength of the soil layer, ϕ_u = the undrained friction angle of the soil, σ_n = normal pressure. The F.S. for this situation is given by

$$F.S. = \frac{S_L \times S_u}{S_L \times \gamma_s t \sin A}$$

$$= \frac{S_u}{\gamma_s t \sin A} \tag{8.37}$$

F.S. = 1.4, t = 2.5 ft (2 ft protective layer + 6 in. top soil) A = 16°, γ_s = 130 pcf.

$$1.4 \times 130 \times 2.5 \sin 16 = 2.5 \times 130 \cos 16° \tan \phi_u + C_u$$

or

$$125 = 312 \tan \phi_u + C_u$$

Table 8.7 is developed for various combinations of ϕ_u and C_u. Since the undrained friction angle of the silty sand protective layer is 28° a failure is not expected.

To check the failure of the clay layer a t value of 4.5 needs to be used. In this case from Eq. (8.37)

$$S_u = 1.4 \times 4.5 \times 130 \sin 16 = 226$$

now

$$S_u = \sigma_n \tan \phi_u + C_u$$

$$= 4.5 \times 130 \cos 16 \tan 15 + 500$$

$$= 650.7$$

TABLE 8.7. Combination of ϕ_u and C_u Values for Example 8.9

ϕ_u (in degrees)	C_u (in psf)
0	125
4	103
8	81
12	58
16	35
21.8	0

Since the actual S_u is more than the required S_u, the clay layer will be stable. (Note: Usually the interfacial friction angle between a synthetic material and a soil is 90% of the internal friction angle of the soil.)

Finally, 10–15 cm (4–6 in.) of organic soil should be spread on top of the protective layer to facilitate seed germination and help the growth of vegetation. Necessary lime and nutrient should be applied at least for the first 5 years to help vegetative growth. Vegetation is helpful in reducing soil erosion, increasing structural stability of the cover, and reducing infiltration by increasing evapotranspiration. A horticulturist should be consulted to help choose proper species, which can grow vigorously without cultivation.

8.8.1 Clay or Amended Soil as Barrier Layer

The properties of these materials are discussed in Chapter 7 and construction-related issues are discussed in Chapter 9. The thickness of this layer depends on its ability to provide a low permeability layer to reduce infiltration. From principles of soil mechanics it can be determined that a thin barrier layer is enough to provide overall low equivalent permeability of the cover. However, a thicker layer is needed for construction-related issues. A 60-cm (2-ft.)-thick layer for clay and a 30-cm (1-ft.) layer for bentonite amended soil is considered to be the minimum. A synthetic clay liner may also be used as a barrier layer. A thicker layer will perform better in case of differential settlement of the waste.

Installation of a filter layer, to minimize migration of fines from the clay layer to the underlying coarse-grained soil, should be considered, especially if the layer is thin (30 cm). The soil filter criteria mentioned in Section 8.1.1 may be used to design this layer. As an alternative, a geotextile layer may be laid over the grading layer where needed.

8.8.2 Synthetic Membrane as Barrier Layer

The properties of these materials are discussed in Chapter 7 and construction-related issues are discussed in Chapter 9. A thickness of 1 mm (40 mil) is considered minimum but 1.5–2 mm (60–80 mil) is preferable. Provision should be made in the membrane regarding installation of a gas extraction system in the future (refer to Section 9.2). If the soil used for the protective layer has a significant amount of 150-mm (6-in.) particles then use of a geotextile is recommended. In such cases, the side slope should be reduced or a special synthetic membrane with high surface friction should be used. A low permeability layer must be constructed below a synthetic membrane liner to minimize infiltration in case of a leak in the membrane.

8.9   GAS VENTING SYSTEM DESIGN

Only venting systems are discussed in this section; design of systems for collection and use of landfill gas is not included although the extraction system design included in Section 8.8.2 can provide some input. It is important to decide at the beginning whether an active or passive system will be utilized for venting the gas. Contrary to popular belief, a passive venting system cannot be converted to an efficient active venting system simply by connecting the vents with a header pipe and by installing a blower at the end of the header pipe. Only a series of deep gas extraction wells, each connected to a header pipe and placed suitably in the landfill, can provide efficient withdrawal of gas from a landfill. The following issues need to be considered for choosing one system or the other:

1. Landfill design: chances of gas migration are higher from natural attenuation type landfills than from containment type landfills.
2. Type of soil surrounding the landfill: gas migration can occur more easily through sandy soil than through clayey soil.
3. Distance of usable closed space (homes, warehouses, etc.) near the landfill. Landfill gas can migrate 150 m (500 ft.) or more. Any usable closed space within 300 m (1000 ft.) of a landfill should be monitored for methane gas concentration.
4. Possibility of future use of the landfill.
5. Regulatory mandate: the regulatory agency may mandate the type of gas venting system to be used in a landfill.
6. Waste type: gas generation depends on waste type. Gas venting is essential for putrescible waste.

8.9.1   Passive Venting System

Such systems are installed where gas generation is low and off-site migration of gas is not expected. Essentially passive venting is suitable for small municipal landfills (40,000 m^3) and for most nonputrescible containment type landfills. The system may consist of a series of isolated gas vents (Fig. 8.28). No design procedure is available to calculate the number of vents required, but one vent per 7500 m^3 (~10,000 yd^3) of waste is probably sufficient. Sometimes these isolated vents are connected by a perforated pipe embedded in the grading layer (Fig. 8.29).

8.9.2   Active Venting System

An active venting system consists of a series of deep extraction wells connected by a header pipe to a blower that either delivers the gas for energy reuse purposes, or to an on-site burner or simply releases it to the atmosphere.

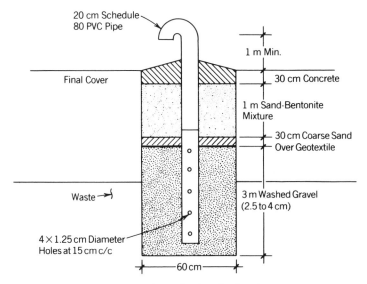

FIG. 8.28. Typical detail of an isolated gas vent.

Whether the gas can be released to the atmosphere without burning depends on the following:

1. Chemical constituents of the gas. If hazardous air contaminants such as vinyl chloride or benzene are present then burning the gas is the preferred option. If such contaminants are absent, releasing the gas to

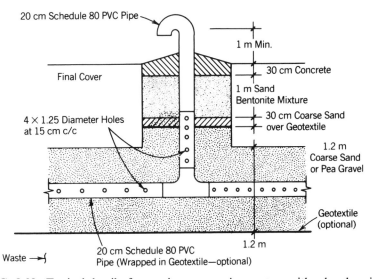

FIG. 8.29. Typical detail of a passive gas venting system with a header pipe.

the atmosphere may be acceptable in some (but not all) situations. In addition the regulatory agency should be contacted to determine whether burning landfill gas is mandatory.
2. Landfill location. If the landfill is located near/within a community then burning is necessary because methane has an odor of its own that may create a nuisance condition.

A typical layout of an active venting system is shown in Fig. 8.30, which includes the elements of the system. Designs of these elements are included in the following paragraphs. A detail design is included in Example 8.9.

Extraction Well. Spacing of extraction wells is a key issue in extracting landfill gas efficiently. They should be spaced such that their zone of influence overlaps. As shown in Fig. 8.31, a 27% overlap can be obtained by installing the extraction wells on the corners of a equilateral triangle of side 1.73R and a 100% overlap can be obtained by installing the extraction wells on the corner of a regular hexagon of side R. A square array would provide a 60% overlap. Thus, spacing of extraction wells is given by

$$\text{Spacing} = (2 - O_1/100)R \qquad (8.38)$$

in which R = the radius of influence of gas extraction wells and O_1 = the required overlap.
The zone of influence of a gas extraction system should be determined from actual field study. An extraction well should be installed within the landfill with gas probes at regular distances from the well (Fig. 8.32). Short-term and/or long-term testing is done to design an efficient withdrawal system. Short-term extraction tests usually runs for 48 hr to several days. A short-term test is sufficient where the intension is to design an extraction system to minimize landfill gas migration.

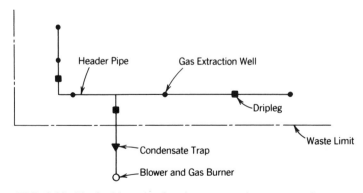

FIG. 8.30. Typical layout of active gas venting system elements.

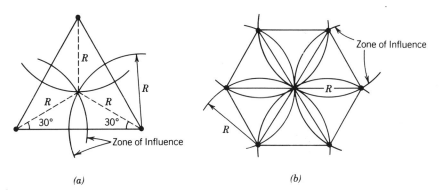

FIG. 8.31. Positioning of gas extraction well for complete overlap: (a) triangular array; (b) hexagonal array. Solid circles indicate locations of gas extraction wells.

A long-term test is used to simulate full recovery project conditions. The suggested probe spacing is shown in Fig. 8.32. The extraction wells should penetrate 80–90% of the refuse thickness and lower 70–80% of the well should be perforated. The well should be pumped for at least 48 hr and then pressure at all the probes should be monitored for 3 consecutive days, at

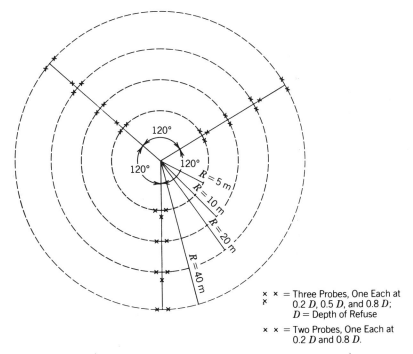

FIG. 8.32. Gas extraction well and probe cluster configuration for zone of influence determination.

least twice a day. The probes nearest to the well show highest negative pressure, which drops rapidly with distance. The radius of influence is that radius at which the negative pressure is nearly zero. In the absence of test data a radius of influence between 45 and 67 m may be used. The radius of influence depends on the cover type, depth of landfill, age, and composition of waste. The lower value is used for relatively shallow landfills [15 m (50 ft) or less] where the final cover does not have a synthetic membrane barrier layer. Figure 8.33 shows details of a typical gas extraction well. A well with arrangements for separating leachate is used where a relatively high leachate head buildup is expected [1.2 m (4 ft.) or more].

Header Pipe. Nonperforated 15- to 20-cm diameter plastic pipes are used to connect the extraction wells to a blower. The diameter of the pipe may be increased to reduce head loss due to friction. These pipes are embedded in

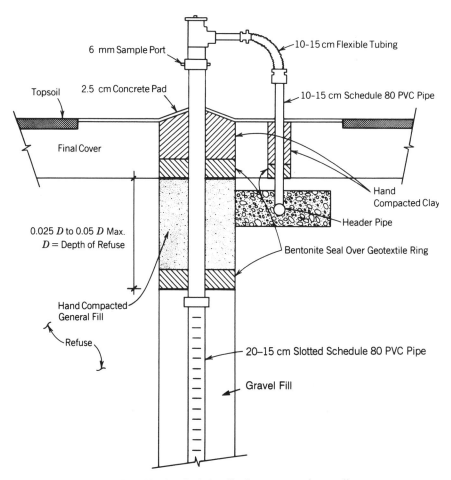

FIG. 8.33. Typical detail of gas extraction well.

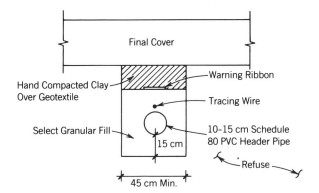

FIG. 8.34. Typical detail of gas extraction header pipe and trench.

trenches filled with sand (Fig. 8.34). If leachate is extracted from the landfill along with gas, then the leachate discharge pipe can also be embedded in the same trench. PVC or HDPE pipes are used for header pipes. The header pipe must not be perforated. The head loss through these holes is high and a significantly high capacity blower will be needed without much increase in extracted gas volume.

Blower. The blower should be installed in a shed at an elevation slightly higher than the end of the header pipe to facilitate condensate dripping. The size of the blower is designed based on total negative head and volume of gas to be extracted. Note that a three-phase electrical connection is needed for most motors with horsepower of 5 or more. If a three-phase electrical connection is not available on site then the system should be designed with multiple blowers each requiring less than a 5-horsepower motor. An approach for calculating the blower size is enumerated in the example at the end of this section.

Condensate Removal. Landfill gas is saturated, which forms condensate due to a reduction of temperature while moving through the header pipe. Arrangements need to be made to remove the condensate before the gas enters the blower. A typical dripleg used for condensate removal is shown in Fig. 8.35. Usually driplegs are spaced at 150–230 m (500–750 ft.). Condensate may be allowed to drip back into the landfill or collected and treated with leachate. A study by Cook et al (1991) indicated that collection of condensate from a (natural attenuation type) landfill gas extraction system reduce Volatile Organic Compounds (VOC) concentration in surrounding groundwater. Thus condensate removal will decrease VOC concentration in leachate. It should be noted, however, that the concentration of several parameters in condensate usually exceeds the hazardous waste concentration

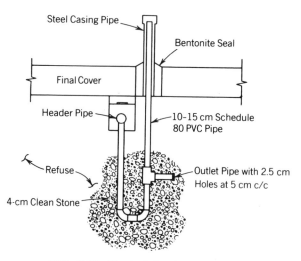

FIG. 8.35. Typical detail of a dripleg.

limit; thus the condensate may need to be handled as hazardous waste if dictated by the regulatory agency. The drip leg exit should be designed in such a way that the condensate can be directed to the leachate tank or collected in a separate tank if needed. This option would allow the collection of condensate if the reduction of VOC in the leachate becomes necessary. The drip leg should be designed such that the seal is maintained at all times.

Burner. Landfill gas may have enough methane to burn once ignited but for complete burning of hazardous air contaminants an additional methane supply may be needed. The operating temperature and residence time in a burner should be enough to destroy the contaminants completely. An operating temperature of 815–900°C (1500–1650°F) and a residence time of 0.3–0.5 sec is needed to destroy most hazardous air contaminants. A flame arrester should also be installed on the landfill gas line to arrest flames going back into the blower.

Example 8.9 (in F.P.S. units)

Design an active gas venting system for a landfill of 30,000 T waste volume using HDPE pipes for the layout shown in Fig. 8.36. Assume a gas production rate of 0.3 ft³/lb/year and equal production rate from each well. A minimum negative head of 2 in. of water column is to be maintained at the farthest well. (Note: For formulas in this example, refer to Perry, 1976.)

Total volume of gas generated from the landfill $= \dfrac{30000 \times 2000 \times 0.3}{365}$

$$= 45{,}315 \text{ ft}^3/\text{day}$$

$$\text{Reynolds number} = \frac{D_i V \rho_g}{\mu} \tag{8.39}$$

where D_i = inside diameter of the pipe = 5.7 in. = 0.475 ft
    V = velocity = 0.58 ft/sec
    ρ_g = density of gas = 0.0808 × 1.05 = 0.08484 lb/ft³
    μ_g = viscosity of gas = 0.093 lb/ft-hr

(Note: Values of ρ_g and μ_g are estimates based on a gas of molecular weight 30 and specific gravity of 1.05. Use of laboratory values is suggested for major/critical projects.)

The Fanning friction factor is a function of the roughness of the pipe and Reynolds number. For plastic pipes that may be categorized as drawn pipe, a roughness factor (rf) of 0.000005 may be used, thus (rf)/d = 0.000005/0.475 = 0.0001. An appropriate friction factor for each segment of the header pipe was found. The loss of pressure (Δp) in each pipe was calculated using

$$\Delta p = \frac{32\rho_g f L_p q^2}{\pi^2 \, g_c D_i^5} \tag{8.40}$$

in which f = the Fanning friction factor, L_p = the length of the pipe (ft), q = the volumetric flow rate (ft³/sec), g_c = the dimensional constant (= 32.17), and D_i = the internal diameter of the pipe (= 0.475 ft).

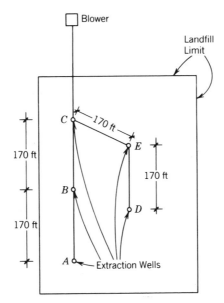

FIG. 8.36. Layout of active gas collection system for Example 8.9.

Loss of pressure in each fitting and valve can be calculated by knowing the velocity and the friction loss factor (K). In the absence of better estimates for K the following values can be used: 45° elbows, 0.35; 90°bends, 0.75; tee, 1.0; gate valve (½ open), 4.5; Eq. (8.41) is used to estimate the pressure loss in fittings and valves:

$$\Delta p = \rho_g K \frac{V^2}{2g} \tag{8.41}$$

in which V = the velocity at the fittings (ft/sec) and g = the gravitational constant.

The velocity at each fitting must be calculated. Frictional losses at different pipe sections and fittings are included in Table 8.8. Loss of head at each extraction well is given by

$$\Delta p_s = \frac{\mu_g F_g D_r}{2 K_g g}\left[R^2 \ln\left(\frac{R}{r_w}\right) + \frac{\gamma_w^2}{2} - \frac{R^2}{2} \right] \tag{8.42}$$

in which μ_g = the gas viscocity (= 0.0293 lb/ft-hr = 8.14 × 10^{-6} lb/ft-sec), F_g = the gas production rate (= 0.3 ft³/lb/year = 9 × 10^{-9} cft/lb/sec), D_r = the density of the refuse (= 1000 lb/yd³ = 37.03 lb/ft³), K_g = the intrinsic permeability of the refuse (permeability × μ/ρ_w = 1 × 10^{-7} cm² = 1.08 × 10^{-10} ft²), R = 100 ft, and r_w = the radius of the extraction well (assume 2 ft).

Putting proper values in Eq. (8.27) $\Delta\rho_w$ = 0.093 psi

Total loss of head in five extraction wells = 5 × 0.093 = 0.465 psi

Total loss of head = 0.465 + 0.02131

Total loss of head = 0.48631 psi

TABLE 8.8. Friction Loss in Different Sections of the Header Pipe and Fittings

Pipe Section	Friction Loss (psi)
DE	0.00006
EC	0.00019
AB	0.00006
BC	0.00019
CF	0.00094
Bend at E	0.00397
Bend at C	0.0159

It is common practice to maintain a minimum negative head of 2 in. of water ($=$ 0.0722 psi) in the header pipe. Thus the

$$\text{Total suction head required} = 0.48631 + 0.0722$$

$$= 0.55831 \text{ psi}$$

The horsepower of the suction motor

$$= \frac{\text{pressure in psf} \times \text{flow rate in ft}^3/\text{sec}}{550}$$

$$= 0.08 \text{ HP} \tag{8.43}$$

The brake horsepower with an 80% efficiency $= 0.08/0.8 = 0.1$ HP

An alternative design procedure for designing gas extraction systems is discussed below:

1. Estimate gas flow from each well; assume 1 cfm of gas/vertical foot of the extraction well (note: use the entire well depth, not the slotted length).

2. Estimate the flow in each segment of the header pipe. The gas flow in the header pipe increases in the direction of flow due to addition of flow from each well in the line. For example, the flow in section EC (Fig. 8.36) is the sum of flows from extraction wells D and E.

3. The header pipe diameter is calculated using Spitzglass formulas (Spitzglass, 1912) given in Eq. (8.44). Choose a diameter of the pipe such that the maximum velocity in the header pipe is 50 ft/sec and the maximum head loss/100 ft of pipe is 1 in. of water column.

$$Q = 3550 \, K \, [\Delta_p / SL_p]^{1/2} \tag{8.44}$$

in which $Q =$ flow rate (ft^3/hr) and $S =$ specific gravity of the gas.

$$K = \frac{D_i^5}{[1 + (3.6/D_i) + (0.30D_i)]^{1/2}} \tag{8.45}$$

4. For a looped header pipe, the pipe diameter may need to be varied to balance head loss at the point where the loop meets (e.g., point C in Fig. 8.36). Start with an external pipe diameter of 6 in.

5. To the head loss in the header pipe add the head loss in each bend and pipe fittings and each well as described in Example 8.9. To this head loss add the head loss due to the elevation difference between the header pipe and the blower and the head loss at entry to the blower.

6. Use the total head loss and flow rate to choose a blower.

7. From the total gas flow rate and an average temperature range of 100° to 70°F (note: gas temperature may vary depending on landfill age, waste

composition, and ambient temperature) estimate condensate volume. To estimate condensate volume find the vapor content of the saturated gas at the maximum and minimum anticipated temperature using a standard gas chart (e.g., Natural Gas Engineering Handbook). Condensate volume is the difference in water vapor of saturated gas at maximum and minimum temperature.

Manufacturers' charts relating flow rate, head loss, diameter and rpm of the wheel, and outlet diameter may be used to choose a suction pump. A higher pump rating should be used to account for uncertainties in the refuse, higher than estimated friction loss due to a change in gas properties, and so on.

Usually a larger blower is installed to compensate for the reduction in efficiency through time, higher gas density due to saturation, and reduction in performance during cold weather.

Performance Monitoring. After an active system has been installed, its performance should be monitored to determine whether the system is working as designed. For this purpose the pressure and gas composition in each extraction well and the offsite gas probes are monitored twice a day for 2–3 days. The monitoring is done 7 days after the shakedown. During the shakedown period, the opening of valves in the extraction wells needs to be adjusted to arrive at the design pressure at the farthest well. Any serious leakage/blockage in the header pipe or malfunctions of extraction well valves and the blower assembly can be detected through this performance monitoring.

8.9.3 Temporary Gas Collection Systems

Temporary gas collection systems are used to mitigate odor problems in landfills. In some waste type (e.g., sludge) odor is a significant operational problem that might affect health (e.g., nausea) of workers or of the population living close to a landfill. If a change in operational practice, such as covering the waste with soil at the end of the day, is not a practical alternative or if odor persists even after instituting such an operational practice then temporary gas collection should be undertaken.

In this system the exposed waste is covered with one or more synthetic membranes seamed together. Attempts should be made to minimize the exposed area by covering it with 15 cm of soil. The floating cover of membrane is connected to a system of flexible pipe manifold that will extract gas from under the membrane (Fig. 8.37). A low radius of influence of 15–20 m should be assumed in spacing the collection points. The laterals are connected to a header pipe, which in turn is connected to a blower of sufficient capacity. The blower may be sized by using the method described in Section 8.9.2. The system is not efficient, so loss of head will be rather high. For design purposes 10 cfm of gas from each extraction point may be assumed. A drip

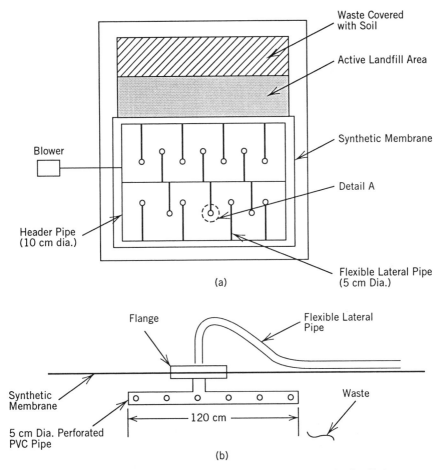

FIG. 8.37. Temporary gas collection system (a) plan (b) detail A.

leg should be used on the header pipe before it enters the blower. The leachate/condensate from the drip leg may be directed back to the landfill or collected separately and treated. The membrane should be secured in place with a sufficient number of sand bags. The membrane and the flexible pipe assembly should be designed such that it can easily be moved within the landfill to accommodate active disposal area. Usually a 40-mil-thick synthetic membrane is used. The membrane should be disposed in a landfill after use.

8.10 CONVERTING EXISTING NATURAL ATTENUATION LANDFILLS TO CONTAINMENT LANDFILLS

Many times existing natural attenuation landfills have to be converted to containment landfills. Before proceeding with such a design, it is necessary to confirm that groundwater contamination is nonexistent at the site. The purpose of such conversion is to retrofit old landfills for a change in regulatory rules or to reduce future groundwater contamination potential. Retrofitting must not be confused with remedial actions, which are undertaken when groundwater contamination has been detected or is highly likely to occur. Retrofitting may not be possible in all cases. It should be attempted only at sites having suitable geology and hydrogeology. Two approaches are normally used: construction of a base leachate collection system, and construction of a perimeter leachate collection system. A third option is to use leachate extraction wells within a landfill that will reduce groundwater impact.

8.10.1   Base Collection System

If a sizable area of the landfill has not been used, then a regular base collection system can be constructed on that part and covering the existing landfill with a barrier layer as shown in Fig. 8.38. A 60- to 90-cm (2- to 3-ft) layer of sand compacted to 85% relative density on the inside of the existing landfill will provide a stable support for the liner. If needed, a 30-cm (1-ft) layer of gravel may be compacted below the sand layer to provide additional strength. The waste slope on which the liner is to be constructed should be checked for stability. Groundwater should be below the subbase of the new portion of the landfill.

8.10.2   Perimeter Collection System

If the entire base area is already utilized and if a suitable low permeability layer exists at reasonable depth below the existing landfill, then a perimeter collection system may be constructed as shown in Fig. 8.39. A cutoff wall, which should extend at least 90 cm (3 ft) into the low permeability layer, has to be constructed. Then a perimeter leachate collection line should be installed inside of the cutoff wall. A submersible pump needs to be used to pump out leachate from the trench. The design can significantly reduce leachate seepage into the groundwater.

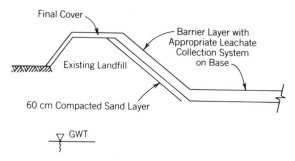

FIG. 8.38. Retrofitting of partially full existing landfill with a basal leachatte collection system.

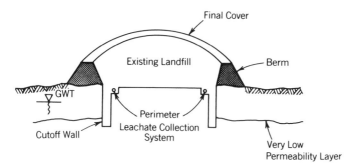

FIG. 8.39. Retrofitting of existing landfill with a perimeter leachate collection system.

8.10.3 Leachate Extraction Wells

To minimize groundwater impact leachate extraction wells may be installed within natural attenuation type landfills or in containment type landfills in which leachate collection systems for one or more cells have failed. Use of these wells is successful if leachate is ponded within the landfill or if the waste is saturated (e.g., sludge type waste). Usually 10-cm-diameter wells with slotted screens are installed with the waste. The well is installed within a gravel pack similar to a gas extraction well (see Fig. 8.33). The yield from a well depends on the amount of leachate present and on the pump capacity. A radius of influence between 10 and 20 m may be used if installation of more than one well is planned. The leachate pumped from each well is collected by a header pipe (in case of multiple wells) and discharged into a collection tank or directly into a sewer. The approach used for groundwater well design may be used for designing these wells. The optimum extraction schedule should be determined by recovery rate of each well. 10 to 15 cycles per day may be used for optimum performance. A detail draw down test may be undertaken to optimize extraction (Hentges et al., 1993).

LIST OF SYMBOLS

D_n	= particle size of which $n\%$ of the soil particles are smaller
D	= deflection lag factor
B	= bedding constant
P	= pressure on the pipe
D_p	= pipe diameter
Δ	= deflection
E	= modulus of elasticity of the pipe material
E'	= modulus of the soil reaction
DR	= dimension ratio = outside diameter/wall thickness
$F/\Delta y$	= pipe stiffness

t	= wall thickness of pipe
r	= mean radius of the pipe
W_p	= width of the compactor wheel in contact with the soil
q	= contact pressure
z	= waste thickness
V	= mean velocity of water
r_h	= mean hydraulic radius
S	= slope of the energy line (= slope of channel for small slope)
n_r	= roughness coefficient
V_s	= settling velocity of a particle
g	= gravitational constant
ρ_s	= density of particle
ρ_w	= density of water
d	= diameter of particle
μ	= absolute viscosity of water
Q	= surface loading or overflow rate
A	= surface area of a sedimentation basin
F_p	= maximum allowable pull
F_R	= total force resisting pullout
γ_s	= unit weight of soil
b	= trench width
h	= trench depth
δ	= friction angle between the membrane and soil
F_{rc}	= total force resisting the pullout when cover soil is applied
w	= weight per square area of the synthetic membrane
σ_y	= yield stress of synthetic membrane
t	= thickness of synthetic membrane
T	= maximum allowable weight of vehicle
W	= total weight of soil on slope
β	= slope angle of the liner
h'	= thickness of landfill cover
d	= depth of landfill
L	= diameter of a depression
S_r	= sag ratio = d_f/L
d_f	= maximum deflection
A_t	= total weight of vehicle
m	= traffic coefficient
n	= saturation coefficient
E_s	= modulus of elasticity of the subgrade
Δ	= deflection
a	= radius of contact of tire
E_p	= modulus of elasticity of the pavement or surface course
t_p	= total pavement thickness
t_s	= thickness of wearing surface
t_b	= thickness of base coarse material

S_s = storage capacity of sump
V_c = volume of leachate collection
t = one-half the cycling time of pump
P_1 = pumping rate
D_f = difference between maximum and minimum elevation
γ_1 = leachate density
A_γ = area of sump
h_d = delivery head of sump
D_i = inside diameter of the header pipe
V = velocity of gas in the header pipe
ρ_g = density of gas
μ_g = viscosity of gas
Δp = loss of pressure in header pipe
f = Fanning friction factor
L_p = length of pipe
q = volumetric flow rate
g_c = dimensional constant
K = friction loss factor in fittings and valves
F_g = gas production rate
D_r = density of refuse
K_g = intrinsic permeability of refuse
R = radius of influence
r_w = radius of extraction well
O_1 = overlap of radius of influence
Δp_w = loss of head at each extraction well
W_c = load per unit length of the pipe
γ = unit weight of soil
B_c = outside width of pipe
H = height of fill above pipe
H_e = height of plane of equal settlement above the critical plane
k = earth pressure coefficient
M = tan ϕ; ϕ = angle of internal friction
S_f = settlement of the pipe into the liner
r_{sd} = settlement ratio
P_p = ratio of pipe projection above the natural ground surface to the pipe diameter
S_g = settlement of natural ground surface
S_m = compression deformation of backfill column adjacent to the pipe of height P_{BC}
E_s = deformation modulus of the PZB material
I = moment of inertia of the pipe wall
r = pipe radius
F_p = maximum allowable pull
F_b = equipment braking force
F_s = seepage force

W_e = weight of vehicle on the side slope
W_1 = weight of soil anchorage at base
W_2 = weight of cover soil on the side slope
h_s = seepage thickness
ρ_w = unit weight of water
Q = flow rate
S = specific gravity of the gas

9 Landfill Construction

This chapter includes construction and quality control tests for liners and final cover construction. Details of many landfill appurtenances and their construction have already been included in Chapter 8. The functional success of liner and final cover elements depends on the correct choice of material and proper quality control tests during construction. Discussions regarding properties of various liner materials are included in Chapter 7, which will be helpful in making the correct choice of material for liner or final cover construction relative to waste type.

Two types of construction specifications are usually used. In a "work type" specification the contractor is told what to do and how to do it. A performance specification on the other hand requires a specific end result. The bid for the work type specification will be lower compared to performance specification because the contractor does not have to spend time in researching how to do the job or hire a technician/engineer for quality control purposes. As far as liner construction goes it may be somewhat risky to use a detailed "work type" specification because all the construction and quality control issues are not yet standardized. For example, the number of passes for compacting a clay liner lift specified in the contract may or may not be sufficient to obtain the desired density. Thus, this type of detail "work type" specification should be avoided in preparing a bid document. It is essential that independent quality control personnel are utilized for maintaining construction quality.

9.1 SUBBASE CONSTRUCTION

The subbase for a liner system refers to the ground surface on which the liner is constructed. Compaction and grading of the subbase are necessary so that the actual liner can be constructed easily. If the subbase is not compacted then it becomes difficult to compact the first one or two lifts to 90 or 95% modified Proctor's density. The subbase should be compacted to the same degree as the actual liner. If the subbase material is sandy then it should be compacted to 85–90% relative density. It should be noted that the Proctor's density versus moisture relationship does not hold good for coarse grade materials (ASTM D698-78). So the percentile relative density should be specified and the standard relative density test (ASTM D4253-83) should be used. The density should be checked at 30-m (100-ft.) grid points. A smaller or larger grid may be chosen depending on the soil properties, and reliability of the contractor. A vibratory roller may be used for compacting sandy soil. However, if the sand is already at a dense state application of vibration will loosen the sand. Therefore, the *in situ* density of the sand should be checked prior to choosing compaction equipment. A nuclear device or other standard method (e.g., sand cone) may be used to determine the in-place density. Consolidation and rebound characteristics of the subbase should be studied during design of the site. If the subbase is predominantly sandy then checking consolidation/rebound is not as important as when the subbase is clayey. Based on a national survey Peirce et al. (1986) reported that the maximum possible overburden pressure in hazardous waste landfills can be 370 kPa (7575 psf). Settlement of the subbase for such high pressure should be checked. Rebound of subbase due to excavation of overburden should be checked if the subbase is clayey or if a thick clayey stratum is identified at a short distance below the subbase grade. Rebound of the subbase may cause uneven heave of the liner resulting in failure of the leachate collection system. When estimating rebound, the weight of the liner may be considered as a weight-restricting rebound if quick construction of the liner is planned; however the total weight of the waste should not be taken into account because sufficient time elapses between liner construction and disposing of waste up to the final grade, which may be enough for total rebound to take place. This issue should be investigated during landfill design where necessary.

9.2 LINER CONSTRUCTION

As mentioned in Chapter 7, four types of materials are used for liner construction: clay, bentonite-amended soil, synthetic clay liner and synthetic membrane. The techniques used for constructing a liner using clay or bentonite-amended soil are very similar. Therefore construction and quality control issues for these two materials are grouped together.

9.2.1 Construction of Clay and Amended Soil Liner

Because the construction and quality control used for both base liner and the barrier layer in the final cover system are the same, the construction of the barrier layer in the final cover will not be discussed separately.

Construction. Laboratory studies have shown that clod size influences the permeability of the clay (Daniel et al., 1984; Benson and Daniel, 1990). So it is important to reduce clod size. Clod sizes recommended by engineers vary between 4.6 and 25 mm and are dependent on the thickness of the lift (Goldman et al., 1986). The clod size becomes a more important issue if the preconstruction moisture content of the soil is less than the wet of optimum moisture at which compaction is proposed. Wetting of clay takes time. If the clod size used in the field for permeability testing is larger than what was used in the laboratory, then the entire clay mass may not attain the same moisture content within the short time period usually allowed in the field. The type of equipment used for compaction also plays a role in developing low permeability. It is recommended to start compaction with a small clod size of approximately 2.5 cm (1 in). Tiller or disks may be used for breaking up clay, however, high-speed pulvi mixtures, which are used for breaking asphalt pavement, are expected to reduce clod size of clayey soil to less than 3.8 cm (1.5 in.) after two passes (Goldman et al., 1986).

Compaction of clay must be done at "wet of optimum" moisture. Adequate time must be allowed to ensure uniform distribution of moisture. Nonuniform moisture distribution in the clay liner can be due to larger clod size, uneven water distribution by sprinklers, and insufficient "curing time" allowed for moisture penetration (Ghassemi et al., 1983). The moisture content of compacted portions of the liner should be maintained during inactive periods to prevent drying (which may lead to desiccation cracks) or overwetting (which will need long period to dry). Moisture can be maintained by "proof rolling" the liner with a rubber tire or smooth steel drum vehicle or covering the liner with plastic. Once started, a clay liner project should be finished entirely. If construction is halted due to onset of winter (in cold regions) then scarification and recompaction should be done for the top 30 cm (1 ft) of the liner.

Three main categories of rollers are available for compaction: sheep's foot roller (self-propelled or towed), vibratory (smooth drum or sheep's foot), and rubber tire. A nonvibratory type sheep's roller is recommended for constructing a clay liner because it can provide kneading compaction and has the ability to break down clods further. A vibratory type roller or rubber tire rollers may not provide the right type of particle orientation desired for constructing a low permeability liner. Vibration may increase pore pressure within clay clods temporarily, which increases its shear strength; as a result more pressure will be needed to construct a uniform, homogeneous well-compacted layer (Bagchi, 1987c). Static compaction produces a flocculated structure with large pores (Mitchell et al., 1965). At wet of optimum the particle groups are weaker; high shear strain will help in breaking down the flocculated structures (Seed and Chan, 1959). Comparisons of the effect of kneading versus static compaction for various water content on soil permeability show that the lowest permeability can be achieved by using kneading compaction at wet of optimum moisture (Mitchell et al., 1965). The use of a heavy weight sheep's foot roller is preferred by many for compacting clay (Hilf, 1975; Department of the Navy, 1971). The size of the landfill should also be considered in choosing equipment. Because large compaction equipment needs a large turning radius, suitable smaller equipment should be selected for small landfills. However, small equipment may not provide the pressure required to achieve a low permeability liner. Thus, construction of a small containment type clay-lined site poses a construction-related problem, which should be considered during design of such sites.

The slope of the sidewall should also be considered when selecting equipment. Two types of construction techniques are used to construct a liner on a side slope: flat and stepped construction (Fig. 9.1a and b). Usually stepped construction is used for slopes steeper then 3 : 1. Another approach is to construct a much flatter slope to facilitate construction (Fig. 9.2). The final slope of 3 : 1 is maintained by constructing a much thicker sand drainage layer. In addition to the ease of compaction using a sheep's foot roller, the thicker sand drainage layer provides better protection for the clay liner on the side wall from the effects of freeze–thaw and desiccation. However, a larger area is required to construct a landfill using a flatter side slope.

Quality Control. There are several issues involved in providing proper quality control during liner construction using clay or bentonite amended soil: quality control before and during clayliner construction, quality control personnel, and documentation report. Each item is important in providing proper quality control for the successful construction and functioning of a landfill. Construction of containment type landfills involves a substantial amount of money. In addition, failure of a landfill would contaminate the groundwater; groundwater clean up may cost millions of dollars. Therefore sufficient importance should be given to quality control during construction of such structures. Spending at least 10–20% of the cost of the project for quality control purposes is routine for all civil engineering projects.

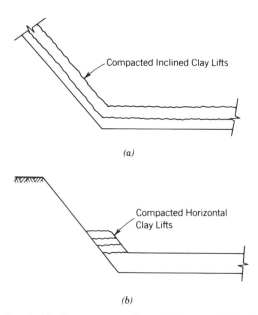

(a)

(b)

FIG. 9.1. (a) Inclined side liner construction. (b) Stepped side liner construction.

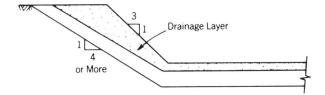

FIG. 9.2. Flatter sidewall construction.

Quality Control before and during Clay Liner Construction. The steps involved in choosing the borrow source for a clay liner, discussed in Section 2.5.2, must be followed. The quality of soil, percentage compaction, and moisture content at compaction are thus known before liner construction is undertaken. Proper testing is required to implement the proposed compaction during construction. The items that need attention are lift thickness, number of equipment passes, frequency of testing compaction density and moisture content, and permeability of the compacted liner.

Usually a lift thickness of 20–25 cm (8–10 in.) before compaction [which becomes 15–20 cm (6–8 in.) after compaction] is recommended. This lift thickness provides enough kneading for the entire lift and provides good bonding with the lift below. Tieing consecutive lifts is an important issue for those side liners that are constructed using horizontal lifts (Fig. 9.1b). An improperly tied lift may exhibit higher permeability in the horizontal direction along the lift boundary. A sheep's foot tends to mesh the boundary between successive lifts (Johnson and Sollberg, 1960) and thus is a better choice of equipment for compaction.

The total number of passes depends on the compactive effort, which is expressed as kg-m/m^3 (ft-lb/ft^3). Foot contact pressure is an important issue for achieving the desired permeability at the proposed density and moisture content. Two approaches are available. In the first approach the foot contact pressure and number of passes are specified; in the second approach the numbers of passes are calculated based on compactive effort, drawbar pull, and so on. DM-7 (Department of the Navy, 1971) specifies four to six passes of sheep's foot rollers for fine-grained soil, foot contact pressure of 250–500 psi (1722–3445 kPa) for PI greater than or equal to 30 and 200–400 psi (1378–2756 kPa) for PI less than 30. (Note: Foot contact pressure should be regulated such that the soil is not sheared on the third or fourth pass.) The specified foot contact area varies between 5 and 14 in.^{2} (32.25 and 90.3 cm^2). The second approach is to calculate the number of passes based on the following formula:

$$CE = (P \times N \times L)/ (R_W) \times (L_T) \times (L) \qquad (9.1)$$

in which CE = the compactive effort, P = the draw bar pull, N = the number of passes, L = the length of the pass, R_W = the roller width, and

L_T = the lift thickness. The usual field approach is to measure the density more frequently at the beginning of a compaction project. The number of passes required for a particular equipment to achieve the specified density at wet of optimum is standarized based on the initial test runs (Note: Use of the DM-7 guideline for foot contact pressure is suggested.) Usually the foot shears the soil initially, but after a few passes the foot "rides on the clay" (i.e., no longer shears it). If the foot contact pressure is excessively high and the layer continues to be sheared even after seven or eight passes, then foot contact pressure should be reduced by reducing the weight of the drum. However, if the foot does not shear the clay lift to at least two-thirds its thickness from the beginning then the weight of the drum should be increased. The weight of the drum can be adjusted by adjusting the volume of sand within the drum.

The number of density, moisture content, and permeability tests to be done for quality control purposes must be specified. Clay liners are a heterogeneous medium in a microscale but may be considered homogeneous if the scale of observation is enlarged. A statistical approach using representative elementary volume (REV) has been attempted to determine the number of quality control tests necessary during clay construction (Rogowski and Richie, 1984). The REV for a hydrologic property is defined as the smallest volume above which the decrease in variance of property of the medium is not significant (Bear, 1972). REV is difficult to define in the real world, however, according to Benson (1990) a 1500 cm^2 area is large enough to incorporate the variability. So when small samples of soil liners are measured to predict the overall permeability of the liner, they are assumed to be representative of the entire compacted volume. The probable variability is highest for permeability, average for water content, and lowest for compacted density (Rogowski and Richie, 1984). The frequency of the quality control tests specified by design engineers varies widely (Goldman et al., 1986). Usually a large number of density and moisture content tests are specified during construction to ensure uniformity. Fewer permeability tests (laboratory or field) are usually specified. A detailed study of compacted density, moisture content, and permeability of a clay liner indicated some variability of these properties within the liner (Rogowski et al., 1985), which were within reasonable limits. Although a shelby tube sample represents a small liner area, proper compaction technique and quality control can ensure uniformity whereby sample size becomes a redundant issue. Significant difference in permeability was not observed between shelby tube samples and large 30-cm-diameter samples collected from Wisconsin landfills (Bagchi, 1993). However, depending on the confidence in the quality control tests and the importance of a project, *in situ* permeability tests may need to be performed. Refer to Section 7.1.6.2 for a detail discussion on *in situ* permeability tests.

In the absence of a specified testing frequency the following testing frequency may be used: Dry density and moisture content tests on every 30 m (100 ft) grid points for each lift subject to a minimum of 12 tests per hectare (5/acre) per lift. Density and moisture content testing frequency should be increased for confined areas in which equipment movement is restricted or hand compaction is necessary. Grain size analysis, liquid limit, plasticity index, and permeability should be performed on samples collected at 25% of the 30 m (100 ft) grid points subject to a minimum of 5 tests per hectare (2/acre) per lift permeability tests must be performed on undisturbed samples. The grain size analysis and Atterberg limits tests are necessary to document that the soil originally proposed is being used for liner construction. In addition, the grain size distribution, Atterberg limits and Proctor's curve should be checked during construction for every 3800 m^3 (5000 yd^3) of soil placed to check the variability of the source. The color of the soil should also be checked in every truck load; a significant change in soil color usually indicates a change in soil properties.

Usually clay is required to be compacted to 90% of maximum modified proctor's density at wet of optimum (2–3% more than optimum) moisture. Daniel and Benson (1990) argue that this method may not ensure the specified permeability if the compactive effort is varied in the field. They suggested a new approach for controlling permeability during liner construction. The recommended procedure involves (1) compaction of 5–6 specimens each with modified, standard, and reduced proctor (note: a reduced proctor compactive effort procedure is similar to a standard proctor except the number of blows is reduced to 15) compactive efforts in the laboratory; (2) determination of saturated hydraulic conductivity of each specimen; and (3) specifying the acceptable zone of density and moisture content on a dry unit weight vs moisture content plot. While plotting the density vs moisture content data a different symbol should be used for the specimens whose hydraulic conductivities were equal to or less than the maximum allowable value. A typical plot with acceptable zone specified by this method and traditional acceptance zone is shown in Fig. 9.3. Because the zone of acceptance is larger, construction operations are expected to accelerate.

Quality Control Personnel. The presence of technical personnel with some experience and knowledge about clay compaction projects is essential during landfill liner construction. The person(s) must be capable of advising the field crew, performing the quality control tests in the field (e.g., compaction tests, moisture content tests, and collection of undisturbed shelby tube samples for permeability tests), and surveying. The field technician(s) must be under the direct supervision of a qualified engineer capable of supervising the technician(s). Many regulatory agencies require that the quality control tests are signed off by certified or registered engineers. In those situations it is essential the supervising engineer is certified or registered by an authority acceptable to the regulatory agency. Although the construction contractor may arrange for on-site technical supervision, the owner should hire independent personnel for quality control purposes. A conflict of interest will exist if the quality control personnel are employees of the construction contractor or have an interest in the construction firm. The owner should realize that the landfill owner is ultimately responsible for any faulty construction.

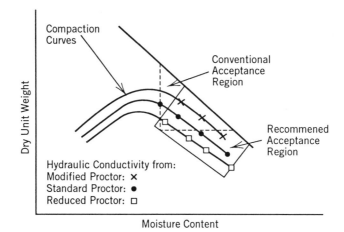

FIG. 9.3. Typical compaction curves showing conventional and recommended acceptance regions. Based on Daniel and Benson (1990).

For controlling the quality of a clay construction project the quality control technician(s) should be present at all times when compaction is done. The homogeneity of the compacted clay cannot be ensured by performing quality control tests alone, no matter how many or what type of tests are done. In addition, the supervising engineer should arrange for routine and surprise visits during the construction of the landfill. Although the interests of the contractor and the quality control team are in conflict, a good professional relationship is essential for the smooth running of the project. A preconstruction meeting attended by representatives from the contractor, quality control team, owner, and regulatory agency is suggested. In many cases the contractor insists on using compaction equipment other than the sheep's foot roller because the required compaction can be easily obtained by using a rubber tire vehicle. The contractor and the quality control team must be convinced that compaction and moisture content specifications are not the goal but are only indirect tests to achieve a low permeability clay liner. Perhaps a technical presentation arranged by the owner (or the regulatory agency) regarding landfill clay liner construction and quality control tests is an effective means of bringing all parties to a common understanding and mutual cooperation.

The quality control tests are usually arranged by the owner and performed with equipment (and the laboratory) owned by the quality control team or their firm. For major projects, especially large municipal waste and hazardous waste sites, the owner should arrange to perform at least 10–20% of the tests by an independent consulting firm not owned by the quality control team or the contractor. Sometimes this type of "split sampling" is performed by the regulatory agency. The owner should more than welcome such a "split sampling" program because this will provide additional quality control for the project at no extra cost. At no time should the quality control tests be performed in a laboratory owned by the contractor.

Documentation Report. All the items of construction must be well docu-mented. Clear and concise documentation provides construction details, notes any departure from the original proposal, and discusses reasons for such departures; it is also helpful in case of any litigation. So due importance should be given to the documentation report. The location of any test must be clearly indicated. Since compaction, moisture content, permeability, and other tests are done on different lifts it is a good idea to draw each lift separately and then show the appropriate test locations. The drawings and narrative should be clear and concise. A daily log of construction activities and quality control tests should be maintained at the site by the technician(s). Such field logs are important documents, especially in case of a dispute. Writing over and erasing must be avoided in the log book. In case of an error, the mistake should be simply crossed out. The report should be reviewed in detail by all the parties, i.e., the owner, contractor, and the quality control team, before submitting it to the regulatory agency if directed. Usually the consulting firm that designed the site has responsibility for quality control and documentation. If detailed documentation requirements are spelled out by a regulatory agency, then those must be followed to obtain a permit.

9.2.2 Construction of Synthetic Membrane Liner

Although the construction activities for synthetic membrane installation are not as time consuming as clay liner construction, the quality control tests are intensive. Although activities during installing them in the field are minimum, extreme care is necessary during their installation.

Construction. The subbase must be properly prepared for installation of synthetic membrane. The subbase should be compacted as per design speci-fications (usually 85–90% relative density for sand or 90% modified Proctor density for clay or amended soil). The subbase must not contain any particles greater than 1.25 cm (0.5 in.). Larger particles may cause protuberance in the liner, especially due to freeze–thaw effects during winter (Katsman, 1984). A herbicide may be used on the subbase below the synthetic membrane to inhibit vegetative growth. Use of herbicide can be avoided if the time interval between subbase construction and synthetic membrane installation is not large (e.g., one month). The panel layout plan should be made in advance so that travel of heavy equipment on the liner can be avoided. In no case should a vehicle be allowed on a completed liner. Only the seaming equipment, seam testing equipment, and necessary minimum number of personnel should be allowed on the liner. Seaming of panels within 0.9 m (3 ft) of the leachate collection line location should be avoided if possible; this issue can be finalized during the layout plan. The subbase must be checked for footprints or similar depressions before laying the liner. The seaming equipment tends to get caught in such small depressions, causing burnout and subsequent repair. A small piece of the synthetic membrane placed below the membranes that are being seamed (this piece is moved forward along with the seaming equipment) may reduce burnout due to small depressions. The crew should be instructed to carry only the necessary tools and not to wear any heavy boots (tennis shoes are preferred). Laying of the synthetic membrane should be avoided during high winds [24 kmph (15 mph) or more]. Seaming should be done within the temperature range specified by the manufacturer.

Synthetic membrane absorbs heat very easily because of the carbon black and becomes overheated very quickly during sunny days. In higher latitudes construction should be done within a temperature range of 4–38°C (40–100°F). In lower latitudes the higher limit should be lower than 38°F to avoid overheating. Wrinkles may develop due to seaming at various times of the day, especially if the daytime and nighttime temperature differential is high. Typically, seaming is done (especially extrusion type) after the dew evaporates from the liner during the early hours after sunrise. Welding cannot be done in the rain or if the membranes are wet after a recent rain. So the membranes to be seamed must be dried using a sponge before seaming. In case it becomes essential to continue seaming during rain (e.g., only a few minutes of seaming left) a portable protective structure may be used to cover the area being seamed. Such portable protective structures should be kept ready in the equipment shed at all times for emergency use.

Settlement hubs may need to be installed in the final cover over some waste types that are expected to settle (refer to Section 10.7 for numbers, etc.). The construction details of a settlement hub are included in Fig. 9.4. Proper care must be exercised in connecting the boot with the membrane. Gas vent risers are also needed in most landfills. Figure 9.5 shows details of a gas vent riser used in synthetic membrane final covers. In some instances (mostly for putrescible waste) active gas venting may become necessary at a future date. A passive gas venting system cannot be retrofitted to act as an active gas venting system. If it is felt that an active gas venting system needs to be installed at a future date, then provision for installing deep extraction wells should be made. A preliminary design regarding extraction

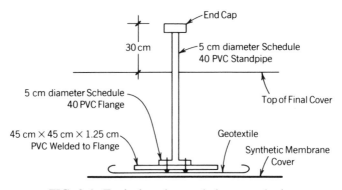

FIG. 9.4. Typical settlement hub on synthetic.

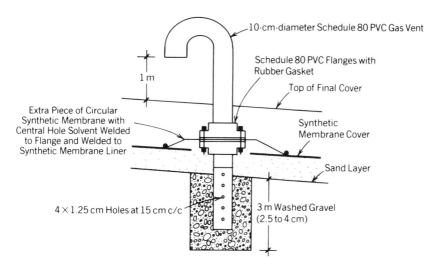

FIG. 9.5. Typical gas vent details for a synthetic membrane cover.

well location should be made so that holes on the membrane are made at proper locations during installation. A 15–20 cm-diameter riser pipe (closed at the top) should be installed at the proposed extraction well locations; connection details of such risers are the same as the details shown in Fig. 9.5. Punching holes and connecting the membrane with the extraction well may be an extremely difficult, if not impossible, task in the future.

Transportation of the membrane must be done carefully, both from the factory to the site and within the site. A carefully prepared panel layout plan can reduce handling. The panel length and mark should be handed over to the factory prior to their manufacturing. Necessary overlap [15–23 cm (6–9 in.)] should be used in preparing the panel layout plan. The seaming machines capability and seaming technique used dictate the overlap length. The marked panel can then be placed at a location where it can be laid without much movement.

Synthetic membranes deteriorate due to ultraviolet rays. They should be covered if kept on site, with a nontransparent sheet. If the estimated time of installation is high then the rolls of synthetic should be stored in a shed to avoid exposure to ultraviolet rays. If the ambient temperature is expected to be high then the shed should be well ventilated so that the temperature inside is not excessive. Manufacturer's guidelines on storage must be followed at all times.

Several types of seaming methods are available. The following are some of the commonly used seaming techniques: thermal-hot air, hot wedge fusion, extrusion welding (fillet or lap), and solvent adhesive. The manufacturer usually specifies the type of seaming to be used and in most cases provides the seaming machine. Manufacturer's specifications and guidelines for seaming must be followed. Seaming is more of an art even with the automatic machines. Only persons who are conversant with the machine and have some actual experience should be allowed to seam. For HDPE, hot wedge fusion and extrusion welding type seaming are commonly practiced. A detailed study on strength and durability of seams was conducted by Morrison and Parkhill (1985). The results indicate the following:

1. No direct correlation exists between the seam shear and peel strength.
2. The peel strength of a seam should be tested to evaluate the quality of the seam.
3. The dead load peel test is not a valid procedure for evaluating the quality of a seam.
4. Short-term (6 months or less) chemical immersion tests may not be enough to determine chemical compatibility of some seams.
5. Of the three field seaming methods (thermal, extrusion, and mechanical) used for seaming HDPE membranes, the extrusion lap weld produces the highest shear and peel strength.

Synthetic membranes must be covered with soil and or waste as soon as possible. The results of the quality control field tests are available immediately; however, results of quality control tests done in the laboratory are not available immediately. The turnaround time for these tests must be minimized by proper planning and project scheduling. A testing time should be reserved in the laboratory (owned by the manufacturer or independent third party), where the quality control tests are to be done so that the waiting period is minimized. Enough volume of soil should be stockpiled near the site so that it can be spread on the finished membrane as soon as the test results are available and the final inspection is over. Synthetic membranes can be damaged by hoofed animals. Bare membrane should be guarded against such damage by fencing the area or by other appropriate methods.

Usually liners are constructed in phases to minimize the length of time the liner is exposed to the elements. Such phasing reduces liner deterioration due to freeze–thaw and desiccation. Thus liners built in subsequent phases need to be connected with the previous phase. The splice detail for connecting liners is shown in Fig. 9.6 (note: Fig. 9.6 shows splice details for both synthetic and clay liners).

At least a foot of sand or similar soil should be spread on the membrane. The soil should be screened to ensure that the maximum particle size is 1.25 cm (0.5in.) or less. The traffic routing plan must be carefully made so that the vehicle(s) does not travel on the membrane directly. Soil should be pushed gently by a light dozer to make a path. Dumping of soil on the membrane should be avoided as much as possible. One or two main routes with 60–90 cm (2–3 ft) of soil should be created for use by heavier equipment

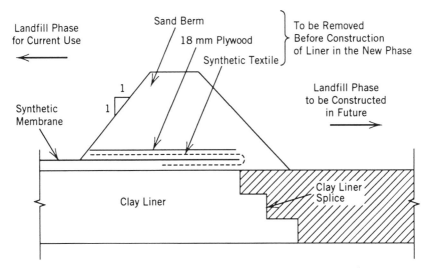

FIG. 9.6. Splice details for synthetic and clay liners.

for the purposes of soil moving. The damage of the membrane due to traffic can be severe, but will probably remain undetected. Even the utmost precaution and quality control during installation will be meaningless if proper care is not taken when covering the membrane. As a guideline, it may be assumed that the time required to cover the membrane with soil is 50–100% of the time spent on installation. "Slow" and "carefulness" are the key words during the soil spreading process. It is recommended that a maximum of two parties (each party consisting of one small dump truck and a light dozer) be allowed for soil spreading. Minor wrinkles form in synthetic membranes (note: this effect is relatively more pronounced in polyethylene membranes) due to expansion during the day and contraction during the night resulting from temperature differential. Sand should be hand shoveled to the back of the minor wrinkles to prevent them from being pushed ahead of the sand and becoming a "large air bubble" (Fig. 9.7). The first lift of waste should be spreaded and compacted with light vehicles. It is preferable not to compact the first foot of waste. No bulky items should be dumped in the first lift.

The bid specification should include warranty coverage for transportation installation and quality control tests. The cost of a project may increase due to the warranty. A survey of available manufacturers' warranties may be made by the owner to ensure that the owner is making the best choice. Unlike clay or amended soil, the manufacturer's warranty is an important issue in synthetic membrane bids. The manufacturers usually provide a warranty regarding strength and durability, field and factory seams, safe transportation of rolls from factory to site, and product quality. Comparison of warranties from different biders should be made along with the cost. The

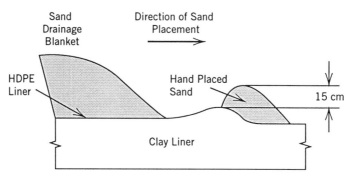

FIG. 9.7. Hand placement of sand to reduce formation of air bubble.

experience of the company (both in manufacturing and installation), quality control during manufacturing and installation, physical installation should be asked in the bid so proper comparisons among different biders can be made.

Quality Control. The quality control issues are the same as for liner construction. However, since comments on quality control personnel will be similar to those made for clay liner projects, they will not be repeated in this section. Only quality control before and during membrane installation and documentation report are discussed here.

Quality Control before and during Membrane Installation. Tests of several physical properties of the membrane must be performed before installation. Usually most of these tests are performed at the time of manufacturing in the manufacturer's laboratory. The owner may arrange for an independent observer to oversee the tests, conduct the tests in an independent laboratory, or use a "split sampling" technique. This issue of responsibility for preinstallation quality control tests must be clearly mentioned or resolved during the biding process. The following are tests (with the relevant test designations indicated within parentheses) used for quality control purposes: sheet thickness (ASTM D751), melt index (ASTM D1238), percentage carbon black (ASTM D1603), puncture resistance (ASTM D4833), tear resistance (ASTM D1004), dimensional stability (ASTM D-1204), density (ASTM D729), low-temperature brittleness (ASTM D746), peel adhesion (ASTM D4437), and bonded seam strength (ASTM D4437). The quality control tests that are performed during installation include the following:

1. Inspection of subbase grade. Both density and smoothness of the surface should be checked. As previously indicated, the subbase should not have any small depressions that tend to create more burnouts during seaming. The installation contractor and the quality control team should inspect the entire site.

2. Checking delivery tickets. The delivery tickets of all rolls delivered to the site must be inspected to verify that the site received proper shipments.

3. Verification of the proposed layout plan. The rolls may be marked at the factory or at the site. Usually the width of the roll is fixed but the length varies based on the layout plan.

4. Checking roll overlap. The overlap of each roll must be checked during and after sheet placement. Usually a 15 cm (6 in.) overlap is preferred, however, the seaming machine and the type of seaming dictate overlap width.

5. Checking anchoring trench and sump. The anchoring trench and sump dimensions must be checked prior to installation. No sharp objects or rocks should be present in any trench or sump. Material passing through a 1.25-cm (0.5-in.) sieve should be used to fill a trench or sump.

6. Testing of all factory and field seams using proper techniques over full length. Usually the vacuum box test (Fig. 9.8) is the most successful test for thick [0.75 mm (30 mil) and above] HDPE membranes. Air lance testing is not suitable for thick membranes. A dual hot wedge seam is tested by pressurizing the space in between the parallel seams. A lowering of pressure indicates leaking within the test section. Two newer techniques, the ultrasonic impedance plane (UIP) technique (also known as the wave resonant frequency technique) and the ultrasonic pluse echo technique, were also found to be useful for HDPE membranes (Morrison and Parkhill, 1985).

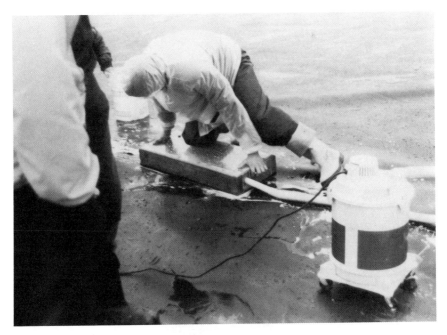

FIG. 9.8. Vacuum box test.

These techniques are still at the development stage and may be available for field use in the near future. It should be noted that these nondestructive tests check bonding between membranes, not seam strength. Only destructive testing for verifying seam strength is available currently.

7. Destructive seam strength test. Test seams at least 60 cm (2 ft.) long for each seaming machine should be made twice a day: once at the beginning of each work period. Each seamer should perform at least one test seam each day, if more than one person is handling one seaming machine. A portion of each test seam should be tested in the field and another in a laboratory. The manual field peel test is known to have produced erroneous results (Katsman, 1984) and hence should be avoided. The peel adhesion tests performed in the field can be standardized by specifying specimen size and a set displacement of the testing equipment; the displacement may be calibrated to reflect a tensile force. As an alternative a tensile testing machine may be used in the field. The test seam samples sent to the laboratory should be tested for peel adhesion (ASTM D4437) and bonded seam strength (ASTM D4437).

8. Patch up repair. Damage to synthetic membranes is unavoidable. Synthetic membranes can be damaged as a result of bad seams or burnouts, manufacturing defects, shipping problems, accidental puncturing (e.g., droping sharp object), wind damage during installation, and wrinkles. Every damaged area must be repaired (Fig. 9.9). Each patch up repair must be tested for proper seaming. There is no set standard or number of allowable

FIG. 9.9. Application of patch to repair a damaged area.

repairs. Based on experience and literature data (Katsman, 1984) it appears that an average of one or two repairs per 930 m^2 (10,000 ft^2) of installed synthetic membrane area is reasonable. This estimate assumes that the quality control team is very vigilant. Lapses on the part of the quality control team may show a lower repair rate but actually may represent a worse job. A detailed field report of all activities is very essential. The repair location, cause of failure (i.e., bad seam or manufacturing defect), time of detection, time of repair, and time of retesting must be recorded meticulously. Without such a detailed record, it is not possible to identify any problem in the future. A surprise visit by the supervising engineer should be arranged to ensure better quality control.

Documentation Report. The documentation report is in general similar to the documentation report discussed in Section 9.2.1 for clay construction. The report should include field and laboratory test results, all repair locations, warranties from the installer and manufacturer where appropriate, and a detailed field log of construction activities.

9.3 BERM CONSTRUCTION

Sandy soil is usually used to construct landfill berms. The compaction equipment most suitable for coarse-grained material includes rubber tire rollers, smooth wheel rollers, and a crawler tractor. It is suggested that lift thickness before compaction should be 25–30 cm (10–12 in.) and tire pressure should be 413–551 kPa (60–80 psi). The number of passes should be standardized in the field based on design relative density (usually 90%). Three to six passes are necessary to achieve the design density. Water may be added during compaction, but it is necessary to known whether the dry or wet method (ASTM D4253) was used in developing the relative density specifications and tests. Berm is constructed in horizontal lifts. The outer face and top of the berm should be covered with topsoil and vegetated as soon as possible. Surface erosion may be caused by a heavy rainstorm or freeze–thaw, resulting in failure of the berm. Vegetation helps in stabilizing slopes (Gray and Leiser, 1982).

Berms for natural attenuation type landfills may include a clay core and/or a riprap on the interior face. The design drawing should be studied carefully to develop a working drawing for construction of the clay core. For vertical extension of the berm, the upper berm should be keyed to the lower berm by a 90 × 90 cm (3 × 3 ft) to 150 × 150 cm (5 × 5 ft) key as shown in Fig. 9.10. The top surface of the existing berm must also be scarified. The key and the first lift should be compacted carefully. A power tamper or rammer may be used for constructing the key. A vibratory compactor may loosen the already compacted lower berm near the top. (Note: a dense sand loosens due to vibration.) Therefore, it is preferable that nonvibratory compacting equipment be used for constructing the upper berm. The junction of the

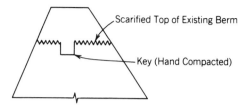

FIG. 9.10. Details for berm extension.

lower and upper berm may provide a channel for leachate to seep out, especially for natural attenuation type landfills. A 0.5- to 0.75-mm (20- to 30-mil) synthetic membrane may be placed on the inside face of the berm to minimize chances of leachate seepage.

Quality control tests should include density tests (any of the methods described in Section 7.1.3 may be used) at a regular frequency. The following frequency of testing is recommended: density tests should be performed on each compacted lift at every 30 m (100 ft.) for lift widths less than 4.5 m (15 ft). For lift widths greater than 4.5 m (15 ft) the density test should be checked at every 23 m (75 ft). The test location should follow a zigzag pattern so that density is tested both near the ends and at the middle of the berm. The grain size analysis of the berm material should be performed for each 765–1900 m³ (1000–2500 yd³) of material used; the higher limit is for a uniform borrow source and the lower limit is for a nonuniform borrow source. The material used for berm construction should not be too gravelly. The subbase grade on which the berm is constructed should be compacted. A 15-cm (6-in.) layer of 2-cm (0.75-in.) size gravel may be used to provide a firm base. The base of the berm should be at least 15 cm (6 in.) below the existing ground surface on the outer face to minimize chances of leachate seepage at the toe of the berm. If the geotechnical investigation indicates a soft clay lens below the site and the design calls for a preloading of the area before actual construction, then necessary arrangements should be made to preload the area. If preloading is undertaken, then additional quality control tests, such as the settlement history of the berm, should also be recorded during the preloading and postconstruction periods.

9.4 SAND DRAINAGE BLANKET CONSTRUCTION

Construction of a sand drainage blanket is not as difficult as construction of a liner, but more is involved than just dumping sand on the finished liner. The sand should be pushed carefully so that vehicles do not travel on the liner directly. A light dozer should be used for this operation. If the drainage layer is constructed on a synthetic liner then additional care must be taken to protect the liner. A sand drainage blanket on the side liner may erode due to heavy rainfall. Pea gravel is a more stable material on the side liner.

The 0.074 or less fraction (P200) content of the drainage blanket material should not be more than 5%. A clean coarse sand is the preferred material for the drainage blanket, however, pea gravel may also be used for this purpose. A layer of gravel should never be used as a drainage blanket; the fines from the waste may migrate and clog the blanket. A filtering medium design approach may be used in designing a graded filter over a gravel drainage blanket. The material specified in the design should be followed strictly when constructing the drainage blanket.

The quality control tests include tests for grain size analysis and permeability. Usually one grain size analysis for each 765 m^3 (1000 yd^3) and one permeability test for each 1900 m^3 (2500 yd^3) of material used is sufficient. For smaller volumes a minimum of four samples should be tested for each of the above properties. The permeability of the material should be tested at 90% relative density.

9.5   LEACHATE COLLECTION TRENCH CONSTRUCTION

The location of the leachate collection line, as shown in the design drawing, should be strictly followed. The pipe spacing is a critical item for minimizing leakage through the liner. A leachate collection pipe may be placed in a trench. When the collection pipe is to be installed in a trench care must be taken to ensure that the trench has the design slope (minimum of 0.5%) toward the collection manhole. A ditch must be excavated in the subbase (see Figs. 8.1 and 8.2) along the proposed leachate trench location. As previously indicated, seaming of the synthetic membrane liner should be avoided within 90 cm (3 ft) of the leachate collection trench. The gravel used in the trench and the leachate collection pipes should be brought in carefully. The collection pipes should be connected near the trench site so they do not have to be dragged on the liner. Dragging may rip the synthetic membrane. It is recommended that the drainage blanket be spread over the synthetic liner first, on the entire liner except 15 cm (6 in.) away from the collection trench. This will allow movement of light vehicles on the liner for placement of gravel and collection pipes.

When the leachate collection pipe is not installed within a trench, there is no need to thicken the liner below the collection pipe location. Typical details for this type of collection pipe installation are shown in Fig. 8.4.

All leachate collection lines that penetrate a liner should have an antiseep collar. A minimum of 1.5 m (5 ft) of compacted clay should be placed around the collar in all directions. Leachate transfer lines located outside the lined area should be encased in at least 60 cm (2 ft) of clay or a double cased pipe should be used.

Quality control tests for a leachate collection pipes and trenches should include the following:

1. Density testing (for nonsynthetic liners only) at 30 m (100 ft) centers.
2. A 0.074 mm or less fraction (P200) content testing of the gravel for every 76 m³ (100 yd³) of material use. The gravel should have a uniformity coefficient of less than 4, a maximum particle diameter of 5 cm (2 in.) [a particle size of 3.7 cm (1 1/2 in.) is recommended for a synthetic membrane liner], and should be subangular. Limestone and dolomite should not be used because they may dissolve in acidic leachate and clog the collection pipes. The integrity of the gravel for specific leachate types (especially from hazardous waste) may be checked in case of doubts.
3. Stiffness and strain of leachate collection pipes, in accordance with ASTM D2412. The test results should conform to specified standards (Uni-Bell, 1979).
4. Checking invert of collection trench. The invert of the leachate collection trench should be checked at 9.5-m (25-ft) intervals using a level instrument with good accuracy so that the 0.5% slope of the collection line can be checked. In addition, a dye test may be done to ensure the slope of the pipe. In a dye test a dye is injected into the collection line at a low rate (so that it does not leak out through the perforations) until it comes out through the other end. At steady state the rate of injection should be equal to the rate of outpour. This test may be done if there is any doubt regarding the survey work and in important landfills.
5. Cleaning of leachate collection lines. The leachate collection pipes should be cleaned using a water jet, sewer line cleaning equipment, or high-velocity gravity flow immediately after finishing construction. Water jetting will clean all construction debris inside a pipe and will detect any severe damage (e.g., crushing) to the pipe. The dye test should be performed after cleaning.

9.6 DOUBLE OR MULTIPLE LINER CONSTRUCTION

The construction and quality control are guided by each liner material. Discussions on the construction of clay liners and synthetic membrane liners are included in earlier sections of this chapter. The most difficult construction is the case in which both liners are synthetic membranes. As indicated in Section 6.2, the lower synthetic membrane is highly prone to damage, which is difficult or almost impossible to detect. The main source of damage is probably vehicular traffic, which is unavoidable during construction of the primary liner. Extreme care should be taken with traffic movement on a synthetic liner.

9.7 GROUNDWATER DEWATERING SYSTEM CONSTRUCTION

Usually this system is installed in a sandy environment. The construction of a dewatering system involves excavating trenches in which the groundwater pipes are to be laid, laying perforated groundwater collection pipes, and connecting them to a manhold or sump. Details of a groundwater collection trench is shown in Fig. 9.11. Usually the groundwater collection pipes are drained by gravity wherever possible and so the water is allowed to be discharged in surface water bodies via a ditch. However, for hazardous waste sites the groundwater collected may not be allowed to be discharged to surface water because of fear of contamination. Arrangements need to be made to direct the pipe to either a separate or a dual manhole.

In some landfills the groundwater collection system may need to be installed within a sand bed. Typical details of this type of installation are shown in Fig. 9.12. A geotextile should be laid over the sand bed. This will minimize migration of fines from the clay liner. It will also help the construction of the synthetic membrane liner.

9.8 LYSIMETER CONSTRUCTION

Lysimeters are constructed below liners. Typical details of lysimeters are given in Section 10.3.1. The critical construction events are laying the synthetic membrane and the boot on the collection pipe. The subbase should be cleared of all stones larger than 1.3 cm (0.5 in.). Overexcavation and recompaction of the subbase with soil passing U.S. no. 4 sieve is suggested. Leak testing of seams should be done very carefully. If possible, a membrane without a seam should be installed; otherwise the seam should be completed and tested in the factory to provide better quality control. The synthetic membrane must be laid on a low permeability layer (e.g., 60 cm clay layer, synthetic clay liner).

The lysimeter is usually tested for leakage. The lysimeter is filled with water and the top is covered as tightly as possible and left undisturbed for

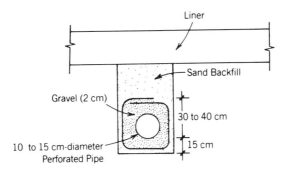

FIG. 9.11. Typical details of a groundwater collection trench and pipe.

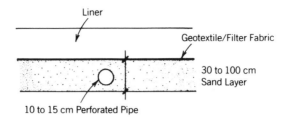

FIG. 9.12. Typical details of a groundwater collection pipe laid in a sand layer.

24–48 hr. The water elevation is measured at the beginning and end of the test. The test can successfully detect serious leakage problems but cannot detect small holes. The compatibility of the synthetic membrane with the leachate should also be tested.

9.9   LANDFILL COVER CONSTRUCTION

Landfill cover construction and quality control issues are similar to those for liner construction and therefore will not be discussed here. The layer below the low permeability layer, referred to as the grading layer or gas venting layer, should be constructed using poorly graded sand (SW or SP type sand per USCS classification). A grain size analysis for every 400 m^3 (520 yd^3) of material use is recommended for quality control purposes. The layer should be compacted to 85–90% relative density to provide a firm subbase for the low-permeability layer above. The density should be tested at 30 m (100 ft) grid points.

The layer above the low-permeability layer, referred to as the protective layer, should be SM, SC, or ML (per USCS classification). A finer soil mixed with sand may be used if silty soil is not available. A horticulturist or soil scientist should be consulted if such mixing is planned. A grain size analysis for every 765 m^3 (1000 yd^3) of material use is recommended for quality control purposes. This layer should not be compacted to help root penetration.

Laying of the topsoil layer should be done as soon as the protective layer construction is finished. Heavy construction equipment should not be allowed on the finished surface. The nutrient and liming requirements for the topsoil should be assessed from a competent agricultural laboratory. In the absence of a regulatory recommendation/requirement regarding seed mix, a horticulturist or soil scientist should be consulted. A combination of grass and bush type vegetation capable of surviving without irrigation water should be planted. At least five samples of topsoil per hectare (2.4 acres) should be tested for nutrient and liming requirements. Nutrient and seed mix application rates should be supervised on site for quality control purposes.

9.10 MATERIAL PROCUREMENT, CONSTRUCTION SCHEDULING AND SO ON

All construction material should be procured/ordered well in advance. Landfill construction must not be scheduled until specifications of all materials required for construction are approved by the proper authority and the contractor/supplier signs a specific delivery agreement. Construction scheduling is an important issue, especially in cold regions. Enough waste [usually 1–1.5 m (3–5 ft) lift] should be disposed of on a landfill liner before the onset of the winter/dry season to protect the liner from the natural elements. In general, because soil is compacted to a higher density in clay liner projects than their natural state, the shrinkage factor must be estimated. The following formula is used:

$$SF = (F_d - N_d)/N_d \qquad (9.2)$$

in which SF = the shrinkage factor, F_d = the proposed field density in kg/m^3 (lb/ft^3), and N_d = the natural density of the soil at the borrow source. The borrow source should have a volume of $(1 + SF) \times F_v$, in which F_v = the required volume of soil at the site. A certain percentage for spillage and so on should be allowed, varying from 2 to 10% of the fill volume (lower value for larger projects). The minimum required soil volume (R_v) at the borrow source with a 5% spillage is

$$R_v = (1 + SF)F_v + 0.05F_v = F_v(1.05 + SF) \qquad (9.3)$$

Example 9.1

Find the required volume of soil at a borrow source for a project in which a 5% spillage factor may be assumed. The following values are known: F_d = 1680 kg/m^3 (105 lb/ft^3), N_d = 1360 kg/m^3 (85 lb/ft^3), and F_v = 100,000 m^3.

9.11 EROSION CONTROL DURING LANDFILL CONSTRUCTION

An erosion control plan should be prepared before landfill construction is started. This will minimize erosion of soil from the construction site and sediment load in the surface water body (e.g., lake, stream). If planned properly the area disturbed during construction can be kept to a minimum. Figure 9.13 shows a typical erosion control plan. Straw bales and silt fences are used in drainage swales to minimize off-site migration of fines. Straw bales must be properly placed in trenches without any gaps between adjacent bales. The straw bales should be anchored in place using stakes. Typical details for straw bales and silt fences are shown in Figs. 9.14 and 9.15. The

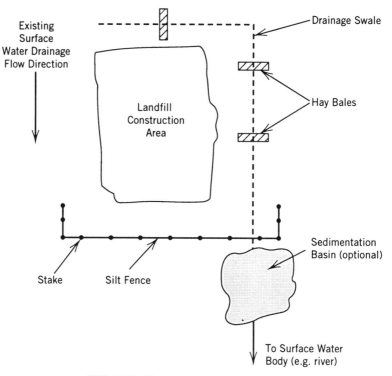

FIG. 9.13. Typical erosion control plan.

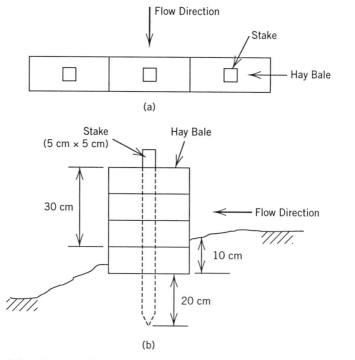

FIG. 9.14. Details of hay bales (a) plan (b) cross section (not to scale).

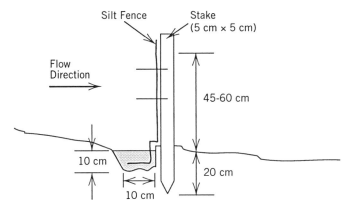

FIG. 9.15. Details of silt fence (not to scale).

straw bales and silt fences should be inspected at least once a week to ensure that they are still functional. Use of a sedimentation basin is recommended if the disturbed area is more than 2 acres. The disturbed area should be vegetated as early as possible after completion of construction. The straw bales and silt fences should be maintained until vegetative cover is established on the disturbed areas. The regulatory agency should be consulted to find out if they have any specific requirements regarding erosion control during construction.

9.12   CONSTRUCTION ON LANDFILLS

Although landfills do not provide firm foundations and at times are hostile environments, construction of structures on landfills are undertaken in many sites (Aderson and Hatayama, 1988; U.S.E.P.A, 1993). Figure 9.16 shows typical details for a building on a landfill. It is prudent not to construct dwelling units on putrescible waste because of possible fire hazard due to accumulation of explosive methane gas within the building. If construction of a closed space is unavoidable on a putrescible waste landfill then a continuous methane monitoring system should be installed in the building. If a house is built on a sandy area contaminated with VOC then a vacuum pump should be installed to vent contaminated vapor from the soil. Additional information regarding construction on landfills can be found elsewhere (Oweis and Khera, 1990).

Apart from environmental issues the structural stability and long-range integrity (due to exposure to corrosive chemicals) should also be studied. The materials used in constructing the foundation and so on should be such that they do not react with the chemicals present in the contaminated site. Dynamic compaction has been used to improve strength characteristics of sanitary landfills prior to road construction. In dynamic compaction a 8- to

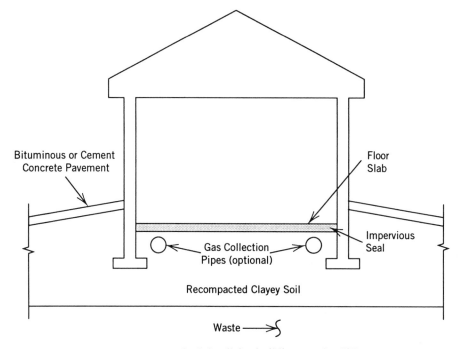

FIG. 9.16. Typical detail for buildings on landfill.

30-ton hexagonal or square steel weight is dropped repeatedly from a 40–80 ft. height using crawler cranes. Usually the pounding is done on 10–20 ft. grid points, which is repeated using a different grid location to ensure that the entire area is compacted. The repeated application of high energy impacts causes densification/compaction of the waste and soil mass to sufficient depth.

Stone columns may be used to provide structural support in landfills containing "soft" waste (e.g., municipal waste). Stone columns are compacted columns of aggregate interlocked with the surrounding soil/waste. A specially designed vibrator that is driven into the soil/waste like a pile is used to release stone radially within the subsurface creating 2- to 4-ft-diameter columns of stone. The number, depth, and spacing between columns will depend on soil conditions and loading. Stone columns can be used to support buildings, bridges, retaining walls, overhead tanks, and so on.

LIST OF SYMBOLS

CE = compactive effort
P = drawbar pull
N = number of passes

L	= length of pass
R_W	= roller width
L_T	= lift thickness
SF	= shrinkage factor
F_d	= proposed field density
N_d	= natural density of the soil
R_v	= required soil volume at the borrow source
F_v	= required volume of soil at the site

10 Performance Monitoring

There are essentially two purposes in monitoring a landfill: to find out whether a landfill is performing as designed and to ensure that the landfill meets all the regulatory standards. The performances of some of the monitoring instruments are well established and are well accepted, although many new instruments are being developed and are awaiting widespread acceptance. Usually hydrogeologists are assigned the task of developing a performance-monitoring program for groundwater around landfills; engineers conversant with air monitoring are entrusted with ambient air monitoring around landfills. However, landfill design engineers need to understand the basics of monitoring so that they can interact with the persons responsible for developing a monitoring program for a landfill and develop a proper perspective for overall design and management of a landfill. Although this chapter is meant mainly for engineers, hydrogeologists will also benefit from the information.

The following items are usually monitored in a landfill: leachate head within the landfill, head in the dewatering system (if installed), leakage through the landfill base, groundwater around the site, gas in the soil and the atmosphere around the landfill, head and quality of the leachate in the collection tank, and stability of the final cover. All of these items need not be monitored in every landfill. The following issues are considered when developing a monitoring program: (1) what equipment to use, (2) where to install the monitoring equipment(s), (3) how often monitoring should be done, and (4) what chemical constituents should be monitored. Mainly the first three issues are discussed in this chapter. Guidelines for item 4 are included in Chapter 4.

In addition to the above-mentioned issues the following three "management type" issues have to be resolved for an overall management of performance monitoring:

1. Identification of a laboratory for testing purposes. A laboratory capable of providing corect detection limits should be used. For example, let us assume that regulatory action is mandated for a certain chemical constituent at a concentration of 0.001 ppm. The laboratory chosen must be capable of detecting the chemical species at 0.001 ppm or less. In addition, the laboratory must follow standard procedures for storing and testing samples. So an inspection of the laboratory should be undertaken prior to contracting for a project.

266

2. Data aquisition and storing. Standard forms should be used that identify the chemical constituents to be tested for each monitoring point. This form should also indicate the highest allowable concentration of each chemical constituent, which are usually mandated by the regulatory agency. After obtaining the results from the laboratory, the data should be stored properly. A computer may be used to store and retrieve data, especially if the total amount of data to be handled is quite large.

3. Analysis of data. A set procedure should be used to analyze the data. A statistical method or other methods should be used to judge whether a medium (e.g., groundwater) has been impacted by the landfill.

10.1   LEACHATE HEAD MONITORING

The leachate head is monitored in a containment type landfill; leachate head monitoring is optional for a natural attenuation type landfill. Essentially two types of leachate head well design are available (Gear, 1988): horizontal and vertical. Typical designs of both well types are included in Figs. 10.1 and 10.2. Use of a wide concrete pad or stainless steel plate will allow reconstruction of another well if the first one is damaged. If a head well needs to be installed in the future after waste placement, then the exact location of the concrete pad/steel plate must be recorded in the drawing. Table 10.1 indicates the advantages and disadvantages of each well type. The leachate head within a landfill is expected to vary both with location and time. It is expected to be highest near the crest of liner and lowest near the collection pipes. At a minimum the leachate head level should be monitored near these two locations.

The leachate head level varies with time. Therefore monitoring of the leachate head level should be done frequently to see if the landfill is performing as designed. A weekly monitoring for the first 3–4 years of operation and monthly monitoring thereafter is suggested.

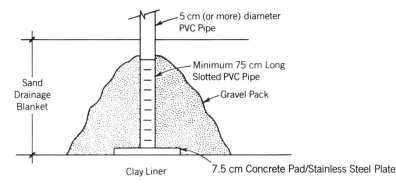

FIG. 10.1. Typical design for a vertical leachate head monitoring well.

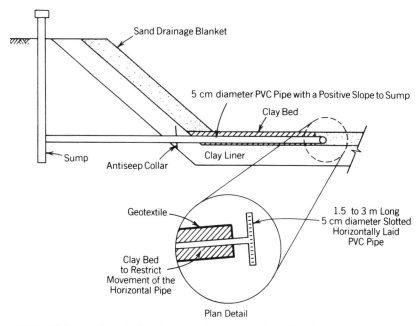

Plan Detail

FIG. 10.2. Typical design for a horizontal leachate head monitoring well.

10.2 MONITORING HEAD IN THE GROUNDWATER DEWATERING SYSTEM

Head within the groundwater dewatering system should be monitored to check whether the system is performing as designed. The head is also expected to vary both in terms of location and time. Only horizontal monitoring

TABLE 10.1. Advantages and Disadvantages of Vertical and Horizontal Type Leachate Head Wells

S1 Number	Vertical Head Well	Horizontal Head Well
1	Can be installed before, during, or at completion of waste placement	Can be installed before waste placement
2	Must be extended periodically if installed before or during waste placement	No extension is necessary
3	Chance of damage during waste placement is high	Chance of damage during waste placement is low
4	Hard to protect from waste shifting and settling, especially in sludge landfills	Waste shifting and settling has no effect

wells are used to monitor the head. The design of the head well is similar to that for the horizontal leachate head well. A monthly monitoring to study the seasonal variation is suggested for the first 3–4 years. Thereafter the frequency may be reduced to only the few months of the year in which seasonally high levels were observed prior to construction of the landfill.

10.3   LEAKAGE MONITORING

The unsaturated zone between the liner and the groundwater table (if the liner is completely above the seasonal high groundwater table) has to be monitored to detect leaks. The literature identifies two approaches:

1. Installation of instrumentation that can collect leachate exfiltrate. These are termed direct leakage monitors.
2. Installation of instruments that can detect water percolation, and so on. These are termed indirect leakage monitors.

Bumb et al. (1988) and McKee and Bumb (1988) had proposed models that can be used to design a leak detection monitor network for hazardous waste landfills in sensitive environments. It should be mentioned that although early leak detection provides an early warning, the results may not be used to enforce any legal remedial action if the laws/rules are applicable to groundwater. Water in the vadose zone is not normally considered groundwater unless the legal definition explicitly includes it. This is more of a legal issue and hence a lawyer should be consulted as to how the results may be used for enforcement purposes. However, the use of direct or indirect leakage monitors should not be restricted because of legal constraints. Remedial action costs less if a problem is detected at an early stage. A leakage monitor should be considered as an integral part of landfill design.

10.3.1   Direct Leakage Monitors

Two types of direct leakage monitors are available: suction lysimeters and basin lysimeters. The location and number of lysimeters depend on landfill design. Installation of more than one lysimeter below the subbase is preferable because performance of the landfill can be continued to be monitored even if one lysimeter fails. The lysimeters should be installed near the edge of a landfill so that the length of transfer piping is minimum; horizontal bends on transfer pipe should be avoided. Lysimeters can be installed almost anywhere below a natural attenuation landfill. However, for containment type landfills the following recommendations should be followed in selecting a lysimeter location. The leachate head is highest at the crest of the base liner, so this is the location where maximum leakage is expected. Minimum leakage is expected below the leachate collection trenches. If one lysimeter is to be installed per phase in a landfill, then it should be installed below a crest of the base liner. If two lysimeters are to be installed per phase then they should be located at the following points: one below a crest of the base liner and the other below the middle of a module. A third may be installed below the collection trench if necessary (Fig. 10.3). There should be at least one lysimeter below each phase of a containment type landfill.

Suction Lysimeter. Suction lysimeters should be installed below the liner area. To protect these lysimeters during landfill construction and operation, they should be installed several feet below the subbase grade. These lysimeters do not function very well if the moisture content of the vadose zone is extremely low.

Three different designs for suction lysimeters are available: vacuum operated, vacuum-pressure operated, and vacuum-pressure samplers with check valves (Everett, 1981). These have been used (in nonlandfill projects) for a number of years, both in laboratory and field studies (Parizek and Lane, 1970; Apgar and Langmuir, 1971; Johnson and Cartwright, 1980). Suction lysimeters may provide valuable information if installed and operated properly. Typically a series of suction lysimeters are installed at several depths in the same or immediately adjacent bore holes. A typical suction lysimeter is shown in Fig. 10.4.

Collection Lysimeter. These are also called basin type lysimeters. These have been used successfully in many landfills (Kmet and Lindorff, 1983). The lysimeter is to be constructed in the field. Figure 10.5 shows a collection lysimeter design. The exfiltrate collected can be monitored for rate of leakage and quality. The basin should be sufficiently large so that it collects an adequate volume of sample necessary for determining the quality of leachate. Usually 2 liters of leachate is necessary for analytical work. A testing laboratory should be contacted to verify the volume requirement.

The transfer pipe drains to a standpipe from which the leachate is withdrawn, either by pump or manually, on a regular basis. The standpipe should

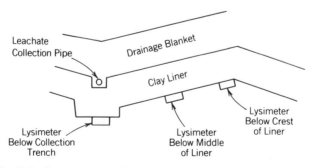

FIG. 10.3. Suggested location of lysimeters in order of preference.

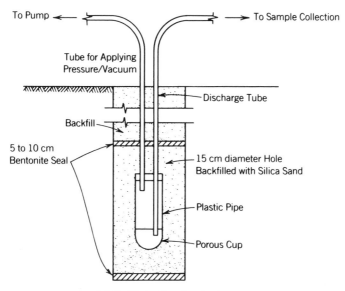

FIG. 10.4. Suction lysimeter.

be sized in such a way that it can hold leakage in between withdrawals. The standpipe should be large enough so that it can be cleaned occasionally.

The leachate head in the standpipe should be monitored more frequently (the suggested interval is 15 days) for the first 2 years initially. Leachate must be removed whenever the level reaches 15 cm (6 in.) below the invert

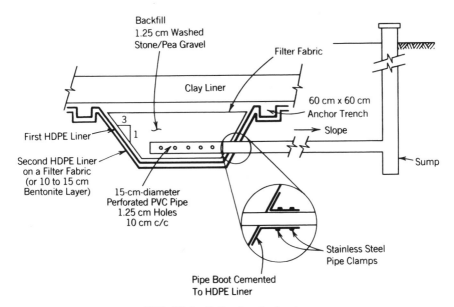

FIG. 10.5. Collection lysimeter.

of the transfer pipe. The frequency of head monitoring in the standpipe and withdrawal may be reduced when a stabilized leakage rate is observed for at least 2 subsequent years. The transfer pipe should be cleaned periodically (the suggested frequency is once every 2 years).

Example 10.1

Determine the area for a basin lysimeter installed below a 1.3-m (4.3-ft)-thick clay-lined site. Assume 100% spillage, and that sampling frequency = 30 days, liner permeability = 1×10^{-7} cm/sec (note: liner permeability is usually reduced due to clogging, etc., within 2–3 years of construction), and average leachate head over the liner = 30 cm.

$$\text{Leakage/sec} = (1 \times 10^{-8}) \times (130 + 30) \times A/(130)$$
$$= (1.23 \times 10^{-8}) \times A$$
$$\text{Leakage/30 day} = 0.03 \times A \ (= \text{basin area})$$

There should be at least 4 liters of liquid in a 30-day period (assuming a 100% spillage). Therefore

$$0.03A = 4 \times 1000$$

or

$$A = 125386 \text{ cm}^2$$

Assuming a length to breadth ratio of 2, the dimensions of the basin should be $(2B \times B = A = 125,386)$ 250 cm × 500 cm (8 ft × 16 ft). The size should be increased by 50–100% to account for lower permeability and additional spillage.

10.3.2   Indirect Leakage Monitors

Many researchers are attempting to develop new instruments or to adapt an existing instrument for landfill leakage detection purposes (Davis et al., 1984; Maser and Kaelin, 1986; Hwang, 1987; Burton, 1987; Daniel et al., 1981; Christel et al., 1985; Everett et al, 1982; Phene et al., 1971a,b; Thiel et al., 1963; Wilson, 1981, 1982, 1983). In most cases indirect leakage monitors were tested in an experimental setup or a conceptual use was discussed.

The indirect monitors can be subdivided into two groups: (1) instruments that detect changes in moisture content of the vadose zone, and (2) instruments that detect a change in the chemical concentration (salinity in most cases) in the vadose zone. These monitors detect changes of moisture content or chemical concentration in the vadose zone but not the quality at any particular time. Since some leakage is expected through a clay-lined landfill these monitors will provide information at the initial stage when changes are taking place. But after several years, when steady state is reached in the vadose zone, most of these monitors will probably fail to provide any valuable information. However, for landfills that are designed for zero or negligible leakage (double lined or landfills lined with synthetic membrane), use of these types of monitors may provide valuable information if a serious leak develops. Brief descriptions of several indirect monitors reported in the literature follow. In general, indirect monitors are not used for routine landfill monitoring but it appears that some of these monitors have potential for future use.

Neutron Moisture Meter. The meter essentially provides a profile of the moisture content of the soil (Wilson, 1980) below the landfill. These meters can be installed below a landfill or moved through a borehole (Fig. 10.6) next to the landfill (Daniel and Kurtovich, 1987).

Gamma Ray Attenuation Probes. These are used for detecting changes in moisture content (Schmugge et al., 1980). The degree of attenuation of gamma rays depends on the bulk density and water content of the medium. Two different methods are available: transmission and scattering. In the transmission method two wells are installed at a known distance. A single well is used in the scattering method. The method is usually limited to shallow depth because of difficulties in installing parallel wells.

Electrical Resistance Blocks. These also measure changes in water content. Electrode blocks embedded in porous material are installed in the soil. The electrical properties of the blocks change with the changing water content of the vadose zone (Schmugge et al., 1980). The water content is calculated from calibration curves after determining changes in the electrical properties of the blocks.

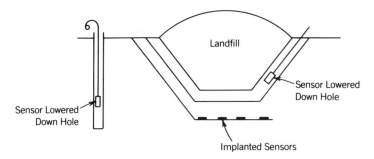

FIG. 10.6. Neutron moisture meter placement.

Thermocouple Psychrometers. These meters are used to detect changes in moisture content. The meter relies on cooling the thermocouple junction by the peltier effect (Merill and Rawlins, 1972). Two types of meters are available: wet bulb and dew point. The dew point method is more accurate (Wilson, 1981). The relationship between soil water potential and relative humidity is used in reading the water content of the medium.

Tensiometers. Tensiometers have been used for many years by soil scientist to measure the matric potential of soil. (Richards and Gardner, 1936; Marthaler et al., 1983). Tensiometers measure the negative pressure (capillary pressure) that exists in unsaturated soil. Many researchers (Christel et al., 1985; Long, 1982; Klute and Peters, 1962; Thiel et al., 1963) reported the use of an electrical pressure transducer instead of a conventional mercury manometers. Figure 10.7 shows a sketch of a typical tensiometer.

Heat Dissipation Sensors. These units also can monitor water content by measuring the rate of heat dissipation from the block to the surrounding soil. Calibration curves relating matric potential and temperature difference for the on-site soil need to be developed in the laboratory.

Electrical Probes. These probes can be used to determine the salinity of the vadose zone. Four probes are installed in an array (Wenner array) so that conductivity of the soil can be measured (Rhoades, 1979; Rhoades and Van Schilfgaarde, 1976). A calibration curve relating conductivity and soil salinity is used for predicting soil salinity.

Salinity Sensors. As the name indicates, these sensors can be used to monitor soil salinity. Electrodes attached to a porous ceramic cup are installed in the soil (Oster and Ingualson, 1967). Calibration curves relating specific conductance and salinity are used to interpret readings.

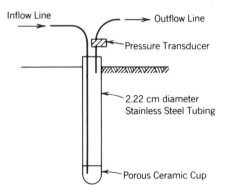

FIG. 10.7. Tensiometers.

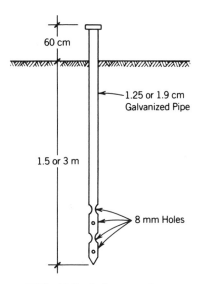

FIG. 10.8. Soil gas probe.

Soil Gas Probes. These probes monitor volatile organic compounds (VOC) in the soil. The gas probe consists of a galvanized pipe that can be driven into the ground easily. A typical soil gas probe is shown in Fig. 10.8. The gas may be analyzed *in situ* using a portable gas chromatograph or tested in a laboratory after absorbing it in charcoal (Silka, 1988; Daniel and Kurtovich, 1987).

Wave Sensing Devices. Use of both seismic and acoustic wave propagation properties for leak detection in experimental setup has been reported in the literature (Maser and Kaelin, 1986; Davis et al., 1984). In the seismic wave technique, the difference in travel time of Rayleigh waves between the source and geophones is used to detect leaks. In the acoustic emission monitoring (AEM) technique, sound waves generated by flowing water from a leak are utilized in leak detection.

Time Domain Reflectrometry (TDR). This technique is based on the difference in dielectric properties of water and soil. TDR measures wide-fequency bandwidth and short-pulse length that are sensitive to the high-frequency electrical properties of the material (Davis et al., 1984). TDR is reported to be insensitive to soil type, density, temperature, and pore liquid quality but is sensitive to changes in the water content of soil (Topp et al., 1980).

10.4 GROUNDWATER MONITORING

Two approaches are available to detect groundwater contamination: a direct method using groundwater monitoring wells and an indirect method using geophysical techniques. However, groundwater contamination detection using geophysical techniques is normally used for remedial actions but not for routine monitoring purposes. Hence only the direct method is discussed. The following are issues that should be considered for the design, installation, and sampling of a groundwater monitoring network: (1) coordinates of sampling points, (2) minimum number of sampling points required, (3) design and installation of wells, (4) groundwater quality status prior to landfilling, (5) frequency of sampling, and (6) collection and preservation of samples. The first two issues are discussed in some detail and the other issues are briefly addressed in the following section.

10.4.1 Coordinates and Number of Sampling Points

An approximate leachate plume configuration needs to be visualized for the design and installation of a monitoring network prior to disposal of waste in a landfill. This is a difficult task, especially if the soil below the groundwater table has extensive stratification, lenses of soil of different permeability, fractures, and so on. Such nonuniformity is a rule rather than an exception in glaciated regions.

A thorough site characterization is essential for the successful design of a monitoring well network. Some guidelines for site characterization are discussed in Chapter 2. When designing a monitoring network it is important to recognize that contaminant migration is a three-dimensional phenomenon. Thus, the vertical depth at which a well screen is placed is as important as its grid location. The three major factors that influence monitoring well location include chemical characteristics of the contaminants expected to seep into the aquifer (this item is waste specific), design of the landfill (i.e., whether the site is lined and if so the liner type; this item is landfill design specific), and the hydrogeological characteristics of the site (this item is site specific).

The movement of a contaminant in the subsurface depends partly on its solubility in water and on its diffusivity in and reactivity with the geologic material beneath the site. The mobility of different leachate constituents will be different. Light immiscible contaminants tend to float while transport of heavy miscible contaminants is governed by its viscosity and density. Nonreactive species (e.g., chloride, low-molecular-weight organics) are expected to be quite mobile, whereas heavy metals and high-molecular-weight organics are least mobile (Technical Enforcement Guidance Document, 1986). Thus, attenuation mechanisms of leachate constituents (discussed in detail in Chapter 5) should be given due consideration when designing a monitoring network.

The plume configuration is influenced by the velocity with which the exfiltrate (leachate after percolating through the landfill subbase) enters the aquifer (Paschke, 1982). This entry velocity is higher for natural attenuation landfills than for containment landfills. Therefore the leachate constituents that leak primarily due to advective transport are expected to penetrate deeper into the aquifer for natural attenuation landfills than for containment landfills. However, landfill design is not expected to have a significant influence on plume for leachate constituents that leak primarily through diffusion.

The hydrogeologic characteristics of a site can be heterogeneous or homogeneous. A monitoring network for a homogeneous situation is easy to understand and the concept can be helpful in visualizing monitoring network in heterogeneous situations. A three-dimensional plume geometry for a homogeneous aquifer is discussed in the following paragraphs, which is helpful in developing a concept for a three-dimensional array of monitoring points:

It is simpler to locate monitoring points in uniform aquifers. The following generalization regarding plume geometry is applicable to deep sandy aquifers. The discussion will provide some insight into a monitoring network design. It is recognized that the soil stratigraphy used is very simple, and is not usually encountered in the real world. It is introduced here with the hope that developing a concept for the design of a monitoring network in a simple case will help in visualizing a monitoring network in more complicated real world situations. The general trend of Concentration Distribution of Contaminants (CDC) and the shape of the plume (based on theory, laboratory experiments, and field observation) are summarized in this section (Bruch and Street, 1967; Paschke, 1982; Paschke and Hoppes, 1984; Kimmel and Braids, 1974; Nicholson et al., 1980; Bear, 1969; Bouwer, 1979; Cherry et al., 1981; Fattah, 1974; Pickens and Lennox, 1976). A circular landfill base is assumed for easy depiction and interpretation. A plume for a circular landfill base will have an elliptical cross section (Paschke and Hoppes, 1984); if the concentrations for a particular cross section are plotted, then all the points will ideally lie on a bell-shaped surface (Fig. 10.9). The highest concentration

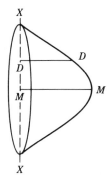

FIG. 10.9. Concentration distribution of contaminants.

(MM) at a particular section occurs slightly below $0.5d$, where d is the depth of the lower boundary of the plume at that section measured from the water table. The concentration MM will be higher for sections closer to the landfill. The spread of contamination increases with distance from the landfill. To visualize the concentration variation within a landfill plume, one may imagine that the ordinates of the bell-shaped surface change with distance from the landfill; the height of the bell decreases and the base area increases as the distance from the landfill increases. In Fig. 10.10 the dotted lines at X–X, Y–Y, and Z–Z illustrate the concentration distribution in a vertical plane within the leachate plume at these locations.

Let $(DD)x$ represent the concentration of a contaminant at section X–X and at some depth from the top of the plume, while $(MM)x$ represents the highest concentration of a contaminant at that section. At a subsequent section (e.g., y–y), $(MM)y$ will be less than $(MM)x$ and $(DD)y$ will be less than $(DD)x$. However, depending on the relative location of the planes x–x and y–y, $(MM)y$ could be less than, equal to, or greater than $(DD)x$. Hence, depending on the three-dimensional coordinates of a monitoring point, it is possible to find higher concentrations of a contaminant at a down-gradient monitoring point than in a relatively up-gradient monitoring point.

As shown in Fig. 10.10, the leachate plume for sandy aquifers can be enclosed approximately by straight lines. The angle in Fig. 10.10 is termed the divergence angle, which is the angle made by a straight line (with the

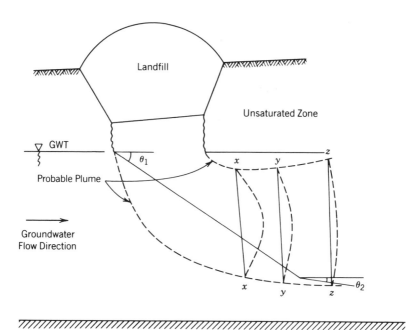

FIG. 10.10. Cross section of a landfill plume in a thick homogeneous sandy aquifer.

horizontal) obtained by joining the far end of the landfill and the boundary of the plume at an arbitarily fixed distance [a distance of 91 m (300 ft) was used]. The suggested plume geometry is based primarily on field plumes (Kimmel and Braids, 1974; Nicholson et al., 1980). The soil stratigraphy for the two field plumes was as follows: the base of the unlined NA type landfill was near the groundwater table, which occurred within a sandy aquifer consisting of coarse to fine sand with little or no fines. The groundwater flow was nearly horizontal in the vicinity of the landfills. For containment type sites the divergence angle is expected to be lower because the leachate entry velocity into the groundwater table will be lower.

Discussions in the preceding paragraphs can be used to design a network in homogeneous aquifers. Although homogeneous aquifers do exist, heterogeneity in aquifers is more common. Two common types of heterogeneity are interbedding (or alternate layers of sandy and clayey/silty materials) and clay lenses in a predominantly sandy stratum. General monitoring approaches for these two settings are discussed. Knowledge about the geology of the site is important in designing a monitoring network. Although contaminant transport modeling can be used to develop the probable plume configuration for heterogeneous subsoil conditions, in practice such modeling is not undertaken. In general, monitoring wells are placed on potential migration paths.

Interbedded Aquifer. Groundwater flow direction can be different in the upper and lower stratum. There may be different horizontal and vertical gradients in each stratum; there is a possibility of existence of vertical gradients in the upper strata in most cases. Proper assessment of the flow direction and gradients (horizontal and vertical) is the key to a successful monitoring network design. Figure 10.11 shows an example monitoring network for a natural attenuation type municipal waste landfill sited in an interbedded aquifer. Note that background water samples are collected from all three strata, and monitoring points are well dispersed in each stratum, because the plume is expected to penetrate into the lower sandy aquifer. For a containment landfill the plume configuration is expected to be different and hence the monitoring network should be adjusted accordingly. If light immiscible compounds are present in the leachate then additional monitoring points should be installed near the water table in the upper aquifer.

Aquifer with Clay Lenses. The size of the lenses is an important issue for such sites. Some times the lenses are quite large and may have perched water. Whether the perched water should be monitored depends on the groundwater rules applicable for the site. (Note: In some states in the United States perched water monitoring is required, whereas in others only the "usable aquifer" needs to be monitored.) If a big lens is present within an aquifer, it should be monitored. Figure 10.12 shows an example of a containment landfill for papermill waste sited in this type of aquifer. Note

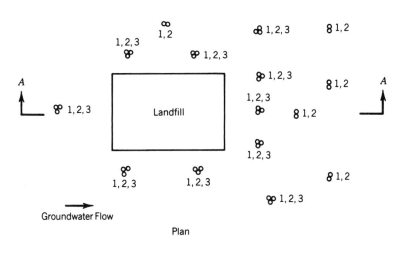

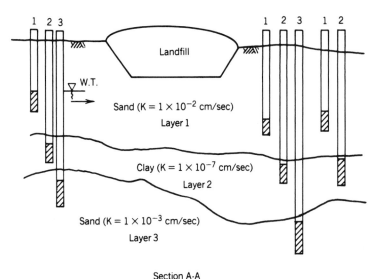

FIG. 10.11. Monitoring of an interbedded aquifer.

the background water is sampled at two depths and only the large clay lens is monitored separately.

Thus, a three-dimensional array of monitoring points is needed for proper monitoring of the groundwater down-gradient of landfills. The soil stratigraphy beneath the landfill base and the hydrogeology of the site will influence the plume geometry. Monitoring points for each landfill should therefore be custom designed. Data from down-gradient monitoring wells could be anomalous and are sometimes almost impossible to interpret without data from a background well that should be installed at a sufficient upgradient distance from the facility (Bagchi et al., 1980). It should be mentioned that leachate plumes do not always sink to the bottom of an aquifer, although leachate is denser than groundwater. The plume geometry depends on the leachate entry velocity, groundwater velocity, permeability of the aquifer, soil stratigraphy below the landfill, density of the leachate, *in situ* diffusion–disperson coefficient of leachate constituents, and the thickness of the aquifer (Paschke and Hoppes, 1984). Therefore, data from a number of deep monitoring wells will not necessarily represent the worst conditions of groundwater contamination from a landfill. A three-dimensional array of

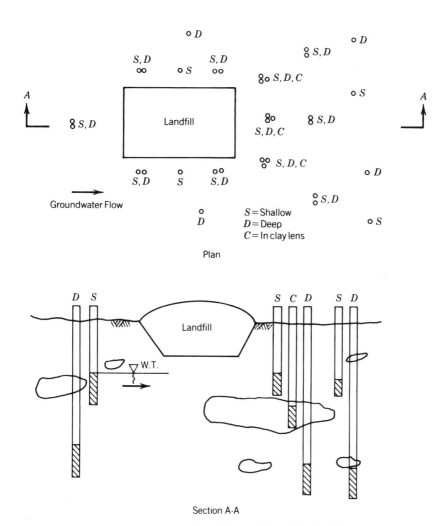

FIG. 10.12. Monitoring of an aquifer with clay lenses.

monitoring points intercepting the plume is necessary to determine the status of groundwater contamination due to landfill siting. The most difficult part is to predict the plume geometry.

10.4.2 Design of Wells and Sampling Frequency

Once the sampling point coordinates are fixed, properly designed monitoring wells need to be installed at these points. The design and installation of background wells and the status of groundwater quality are critical issues.

Usually an up-gradient well is installed to monitor background water quality. Monitoring at regular frequencies is needed to judge the change in quality of the groundwater down-gradient of the landfill. The quality of background water does change and it may be impacted due to a pollution source up-gradient of the landfill. If a known pollution source is absent then monitoring at a single point is enough. However, if a known polution source exists up-gradient then several monitoring points will be necessary to judge the background water quality. Usually a long-screen well (Fig. 10.13a) is used for monitoring background data. Multilevel monitoring is probably a better approach because this will indicate the CDC if one exists. However, several piezometers (Fig. 10.13b) installed at different depths can also be used. The above well designs are also used to monitor down-gradient water quality. It may be noted that in addition to collecting water samples at different depths, the depth of the water table must also be monitored for interpreting data.

The material for well casing depends on the chemical parameters monitored. PVC pipes can be used to monitor inorganic constituents. However, metalic pipe coated with a nonreactive chemical or stainless-steel pipe is recommended for monitoring organic constituents. For shallow wells 5-cm (2-in.)-diameter schedule 40 PVC casing may be used; however, for deeper wells 31-m (100-ft. or more) schedule 80 PVC pipe should be used. The well screen material should also be nonreactive with the expected leachate exfiltrate constituents. The slot size depends on the geologic material in which it is placed. For most material a 0.15-mm (0.006-in.) or 0.25-mm (0.01-in.) opening is suggested; a larger slot size of 0.5 mm (0.02 in.) may be used for a coarse sand or gravel environment. The slot opening should cover approximately 10% of the screen's surface area. Filter cloth should not be wrapped around the well screen (Bureau of Solid Waste Management, 1985). The material placed around the screen is known as filter pack. Properly sized filter pack filters out sediments and increases the effectiveness of the well. The material used should be nonreactive (e.g., clean silica sand) to chemical constituents monitored. The grain size of a filter pack depends on the grain size of the geologic material surrounding the well. Poorly graded (preferably single-sized particle) medium or coarse clean sand or pea gravel is generally used in a sandy environment. Clean fine sand is recommended

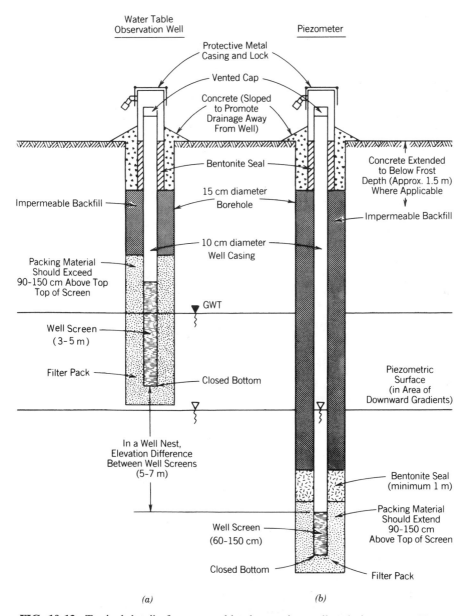

FIG. 10.13. Typical detail of a water table observation well and piezometer. Not to scale.

for a silty and clayey environment. The physical characteristics of the backfill material should exhibit a balance between strength, impermeability, continuity, and chemical compatibility. Bentonite, cement, and polymers are commonly used for backfilling and well sealing. Several types of equipment for drilling well boreholes are available. A drilling method that introduces the

least amount of foreign material to the borehole and causes minimum disturbance to the formation should be used.

After installation, a well must be developed (or cleaned) to remove fines
accumulated in the borehole during installation and to restore the natural
permeability of the surrounding soil. The best development method is one
that causes water to flow rapidly both in an out of the well screen in order
to dislodge and remove fine particles. Usually a pump is used to draw the
water out of a well. A bailer or surge block may be used for dislodging
sediments from the screen. A well should be developed until the water is
clear or until the conductivity, pH, and temperature of the pumped water
remain constant for a period of time.

A well must be protected at the ground surface. The log of well installation
must be documented properly. The following information should be included
in a well installation log: well number; well location (indicating site grid
point); date of installation; diameter and type of well casing; mean sea level
elevations of the top of the well casing, ground surface near the well and
top of the screen; length and material of the well screen; depth of the bottom
of the well; and the type of well (i.e., water level well or piezometer). The
quality of groundwater at the site prior to disposing of waste (known as
background quality) should be established by sampling on-site wells. The
following program for establishing background water quality may be used
in the absence of a program set by the regulatory agency: eight rounds of
monthly or quarterly sampling of all the wells including the background well
for all the parameters for which the groundwater is to be monitored after
disposing of waste in the landfill.

Usually groundwater is monitored quarterly, biannually, or annually, depending on the type of waste, size and design of the landfill, aquifer material,
and so on. In most cases a quarterly monitoring is undertaken; annual monitoring is undertaken for small landfills located in remote places far away
from any groundwater use source.

Collection, preservation, and testing of the groundwater sample are important to obtain representative data. The water level of each monitoring
well should be purged by removing four well volumes (internal radius of the
well × the height of the water column in the well) of water using a bailer or
a pump. Certain parameters (temperature, specific conductance, pH, color,
odor, and turbidity) of the water are measured in the field prior to filtering
the sample. Field filtering of groundwater samples should be avoided where
possible. Low flow pumps may be used for turbid samples. However, in
cold climates field filtering of turbid samples is unavoidable because use of
low flow pumps is not a practical alternative (Connelly, 1994). The collected
sample must be preserved using different chemicals for different parameters.
Bailer, pumps, and so on must be cleaned after sampling each well to avoid
cross-contamination (Lindorff et al., 1987). Groundwater samples should be
tested in a laboratory in which quality control measures are good and the
lowest possible detection limits are used.

10.5 GAS MONITORING

Gas around a landfill (both above and within ground) needs to be monitored. Although the possibility of the movement of gas through the soil for containment type landfills is low, it should be monitored on a routine basis. The air on and around landfills should also be monitored to check for hazardous air contaminants injurious to the health of landfill workers and people living around a landfill.

10.5.1 Underground Gas Monitoring

Gas probes similar in design to groundwater wells are installed around landfills. A study of the subsoil stratigraphy must be undertaken prior to selecting monitoring points. Usually migration occurs through sandy deposits, however, highly fractured formations can also serve as a conduit for gas migration. Gas can also migrate through gravel beds of utility lines running close to a landfill. So the possible conduits of migration should be identified before installing a gas probe. Either short-screen or long-screen probes (Fig. 10.14) may be used to monitor gas. The bottom of the probes should be above the highest seasonal elevation of groundwater. Sometimes a nest of three short-screen probes is used to detect gas.

Usually landfill gas is monitored for methane concentration; however, other hazardous air contaminants may be added to the list. Methane is explosive between 5 and 15% volume/volume concentration in air. Landfill gas migration seems to occur in pulses. Figure 10.15 shows a typical concentration variation of methane for a NA type landfill. Because of the high variability in gas concentration quarterly or even monthly monitoring may not detect the real status of migration because the time and date of sampling may not synchronize with the high concentration (Bagchi and Carey, 1986). It is suggested that gas monitoring be done twice a day (midmorning to early afternoon and late afternoon) for 7–10 consecutive days in the month(s) when migration is most likely to occur. Chances of gas migration are high when the ground is either frozen or saturated.

10.5.2 Landfill Air Monitoring

Sample collection and allowable concentration of air contaminants pose special problems for landfill air monitoring. The threshold limit values (TLV) issued by the American Conference of Governmental Industrial Hygienists (1987) are for indoor exposure. In a survey of landfill gas in Wisconsin, Chazin et al. (1987) used source sampling of active or passive gas vents. The sample collection method included a Gillian personal sampling pump and activated charcoal tubes with a sampling time of 60 min. The Wisconsin study included several landfills with different waste types: municipal cum industrial, municipal, combined municipal (papermill sludge, refuse, and

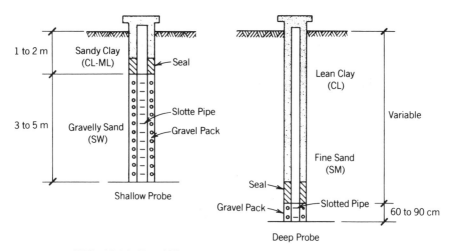

FIG. 10.14. Landfill gas probes. After Bagchi et al. (1986).

industrial), and papermill sludge. The study indicated that vinyl chloride (a known carcinogen) is expected to be present in all large municipal landfills (383,000 m³ or 500,000 yd³); benzene may also be present in large municipal landfills. In addition to hazardous air contaminants, dust concentration may also pose a health risk. The following possible causes for the presence of vinyl chloride in municipal landfills are suggested: (1) *in situ* formation by bacteria from other chemicals present in the landfill; (2) a by-product of the chemical decomposition of polyvinyl chloride plastic; and (3) formation of vinyl chloride from trichloroethylene by bacterial action. Of these three, the third is believed to be the most likely mechanism. Thus, discarding of plastic with solvents like trichloroethylene should be avoided to reduce the chance of hazardous air contaminants in landfill gas.

Several sampling techniques are available. The principal objective of sampling is to collect a polluted air sample to analyze the concentration of

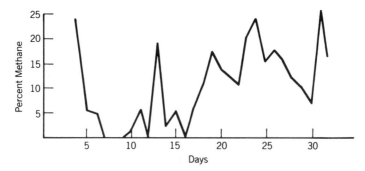

FIG. 10.15. Typical concentration variation of methane observed in landfill gas probes. After Bagchi et al. (1986).

pollutants. Gas sampling devices can be divided into three categories: passive, grab, and active. Passive sampling involves the collection of pollutants by the diffusion of gas to a collection medium. Although the passive sampling technique is simple and costs less, a long collection time is required (7–30 days) (Godish, 1987). Presently passive sampling is almost obsolete. An evacuated flask, gas syringe, or air collection bag made of synthetic materials is used to collect grab samples. An active gas sampling technique consists of a train of a pump, a tube, and a sampler, which may be a cassette, impinger, or sorbent tube. Particulate matter may be sampled by a hi-vol sampler (Axelrod and Lodge, 1977). In a hi-vol sampler a collecting glass fiber filter is installed upstream of a heavy duty vacuum pump with a high air flow rate of 1.12–1.68 m^3/min (40–60 ft^3/min). The filter is placed in a housing to protect it from impact of debris and so on. Paper tape samples draw ambient air through a cellulose filter tape. The particle concentration is measured by optical densiometers usually installed with the tape sampler. Because of difficulties in correlating optical reading with the gravimetric data, their use is limited (Godish, 1987). In impactors, air is drawn and then led through a deflecting surface whereby it can be fractionated to six or more sizes. The hi-vol samplers collect samples of all sizes. Since larger particles do not pose a health risk, data from a hi-vol sampler may not be reliable in assessing particulate density. Impactor samplers are better devices for assessing dust concentrations that pose health risks.

Table 10.2 includes time weighted average TLVs for some air contaminants. It may be noted that these are based on a 40 hr/week exposure in a workplace. Usually the allowable concentration in ambient air is much

TABLE 10.2. Threshold Limit Values of Selected Air Contaminants[a]

Contaminant	TLV
Dust	1 mg/m^3
Carbon monoxide	50 ppm
Asbestos	0.2 to 2 fibers/cm^3 (depending on asbestos type)
Benzene	10 ppm
Coal dust	2 mg/m^3
Cotton dust	0.2 mg/m^3
Grain dust	4 mg/m^3
Hydrogen sulfide	10 ppm
Nuisance particulates	10 mg/m^3
Phenol	5 ppm
Vinyl chloride	5 ppm
Wood dust	
Hard wood	1 mg/m^3
Soft wood	5 mg/m^3

[a] Values of TLV obtained from the American Conference of Governmental Industrial Hygienists (1987).

less (e.g., the allowable concentration of vinyl chloride in Wisconsin is 0.0006 ppb for a 24-hr exposure) because a 24-hr exposure is assumed. A health risk assessment should be undertaken to establish allowable concentrations of a contaminant in ambient air.

10.6   LEACHATE TANK MONITORING

A leachate tank should be monitored for level of accumulated liquid and leachate quality. The leachate generation rate is estimated during the landfill design. Therefore the rate of filling the tank can be anticipated. Based on this rate the head level in the tank and extracted volume should be monitored daily/weekly/monthly to ensure that overflowing of the tank does not occur. The quality of leachate should be monitored at least annually during the active life of the landfill and 2–5 years after closure. Leachate quality monitoring is needed to interpret groundwater data and revise a groundwater monitoring program. In addition, leachate quality monitoring is helpful in running the treatment plant. Sometimes BOD and volume of leachate are monitored daily prior to discharging the leachate into the intake point of the treatment plant.

10.7   FINAL COVER STABILITY MONITORING

The stability of the final cover should be monitored if a higher than usual waste slope is used for a landfill. Approaches for monitoring a synthetic cover and clay cover are discussed below.

10.7.1   Synthetic Cover Monitoring

Settlement hubs are installed over synthetic covers to monitor settlement. Excessive settlement may lead to shearing of the synthetic membrane. Usually settlement is monitored for sludge type waste. Settlement hubs should be placed at 30 m (100 ft) grid points (or closer for very unstable sludge) and may be monitored quarterly or biannually. They may be placed on the side slope to monitor its stability. Settlement hubs are usually installed on the top surface of sludge landfills, which is more likely to settle.

10.7.2   Clay Cover Monitoring

Settlement of clay-covered landfills may also be monitored. Usually settlement monuments are used to monitor settlement. Monuments may be installed on the side slope to monitor the stability of the slope. Monuments may be established at 30 m (100 ft) grid points (or closer) on the top surface and monitored quarterly or biannually. Usually slope failure occurs along a circular arc (Fig. 10.16). So if the side slope is to be monitored then a

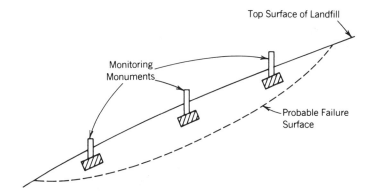

FIG. 10.16. Arrangement of monuments for slope stability monitoring.

minimum of three monuments along a slope line should be established. Both horizontal and vertical movement of these monuments should be monitored biannually or annually to judge stability.

10.8 GROUNDWATER DATA ANALYSIS

If an allowable concentration for a parameter is set by a regulatory agency then it is easy to know whether any of the monitoring wells has been impacted.

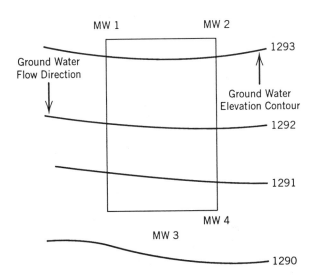

FIG. 10.17. Plan of an NA type landfill with monitoring points.

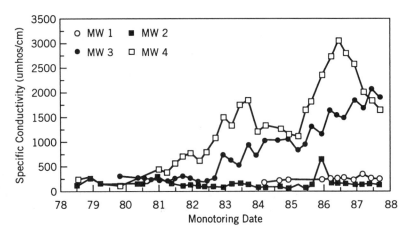

FIG. 10.18. Time vs. concentration plot.

However, in the absence of such set concentration, interpretation and assessment of groundwater data are difficult tasks. Although several statistical approaches are available (e.g., Student's *t* test), time vs concentration plots and box plots are two powerful tools for analyzing the data and detecting contamination. Figure 10.17 shows a plan view of an example NA type landfill and the groundwater monitoring points. The time vs concentration plots of the monitoring points are shown in Fig. 10.18. As is obvious from Fig. 10.17, MW 1 and MW 2 are background wells that were installed to monitor the water quality of the groundwater before getting contaminated by the landfill leachate. Groundwater sampled from MW 3 and MW 4 show an increase in specific conductance with time (Fig. 10.18) indicating impact. The box plots of MW 1, MW 2, MW 3, and MW 4 are shown in Fig. 10.19. Note the spread of data in the box plots for MW 3 and MW 4 whereas box plots of MW 1 and MW 2 show no spread. It may be mentioned that in many

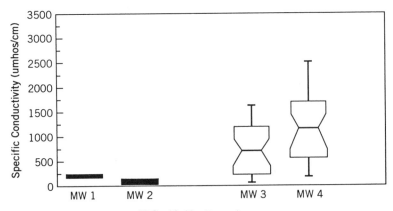

FIG. 10.19. Box plots.

TABLE 10.3. Specific Conductivity Data from a Landfill Monitoring Well

500	600	930
510	640	995
525	697	1030
537	750	1070
545	810	1130
557	850	1195

cases contamination of a well cannot be detected easily from a time vs concentration plot. From a detailed study on groundwater monitoring wells around landfills in Wisconsin, Fisher and Potter (1989) concluded that a box plot from a contaminated monitoring well will typically show a spread of data compared to the box plot(s) of uncontaminated background well(s), although natural variation with time may exist in both wells.

A box plot is a graphic display of the data that provides the summary information in the quartiles. In a box plot the first and third quartile data are connected by a box with the median indicated by a bar (Johnson and Bhattacharya, 1992). The highest and lowest data are connected to the box by a straight line. Example 10.1 shows how to draw a box plot. Box plots should be drawn for each well for each parameter. Study of box plots for each parameter for the monitoring wells around a landfill site will clearly show whether any well has been impacted.

Example 10.1

Calculate the first, second, and third quartiles for the specific conductivity data shown in Table 10.3 from a groundwater monitoring well down gradient of a landfill.

The first step is to arrange the data from the smallest to the largest value (already done in Table 10.3). The division of ordered data set into two equal halves is called median; when the division is in quarters it is called quartiles. The quartiles are 25th, 50th, and 75th percentiles. After arranging the data the quartile points are calculated by multiplying the quartile with the sample size (or number of data). Thus, for the example:

The first (or lower) quartile = $18 \times 0.25 = 4.5$. Since this is not an integer round it to the next integer. The value of the first quartile is the 5th data ($= 545$).

The second (or median) quantile = $18 \times 0.5 = 9$. Since this is an integer, the average of the 9th and the 10th data is the median $[= (697 + 750)/2 = 723.5]$

The third (or upper) quantile = $18 \times 0.75 = 13.5$. Since this is not an integer round it to the next integer. The value of the third quartile is the 14th data ($= 995$).

Use these numbers to draw the box plot as indicated in the text.

11 Landfill Operation

A simple, well-organized operating plan is the key to the successful operation of a landfill. An operating plan merely satisfying regulatory compliance is of little use. A good operating plan should provide guidance for day-to-day and year-to-year operation so that landfill volume is efficiently used, a safe working environment is created, and environmental nuisances are not created. No two landfills are alike. The following guidelines may be used to develop a site-specific plan:

1. The highest compaction is obtained by compacting from the base and up of the landfill.
2. End dumping from the side of the landfill should be avoided.
3. All-weather access roads should be built inside and outside of the landfill.
4. The active area should be as small as possible.
5. A maximum of three horizontal to one vertical waste slopes should be maintained on all internal waste faces.
6. The surface water should be diverted away from the landfill. Temporary surface water drainage swales should be constructed whenever possible.
7. Waste contact liquid must not be allowed to run off from the landfill.
8. Access to the landfill should be controlled by a fence and locking gate.
9. Proper care should be exercised during burning of waste (if allowed by the regulatory agency).
10. Clear and visible on-site directional signs should be posted for proper traffic routing. In municipal landfills burnable woods, white goods (refrigerators, etc.), recyclable material (paper, glass, etc.) should be deposited separately so that they may be salvaged.
11. Landfill operator(s) should know about fire control measures and emergency measures to be taken in case of accident leading to physical injury.
12. The operators should take necessary safety precautions when operating a landfill.
13. The operator(s) and other landfill personnel should be conscientious

about bacterial and chemical contamination while storing and eating food within the landfill office.

14. The operator(s) should know about monitoring and maintenance requirements for the landfill.
15. The operator(s) should have a basic knowledge about design and construction related issues of landfills.

The above guidelines are at best partial. Discussions of equipment used, litter and dust control measures, and routine maintenance issues follow.

11.1   EQUIPMENT USED FOR COMPACTION

Proper equipment should be chosen to operate a landfill. Proper attention should be given to the capability of the equipment and its need. In many instances backup equipment may be necessary if severe operational problems are envisioned during equipment downtime. Essentially three major operations are involved in landfilling a waste: spreading waste after dumping, compacting waste, and spreading and the compacting daily or intermediate cover (where necessary). Table 11.1 provides a list of landfill equipment and how each rates in performing the above operational needs. Although draglines are normally not used in landfills, they may be useful in disposing of sludge.

TABLE 11.1. Performance Characteristics of Landfill Equipment[a,b]

Equipment	Solid Waste		Cover Material			
	Spreading	Compacting	Excavating	Spreading	Compacting	Hauling
Crawler dozer	E	G	E	E	G	NA
Crawler loader	G	G	E	G	G	NA
Rubber-tired dozer	E	G	F	G	G	NA
Rubber-tired loader	G	G	F	G	G	NA
Landfill compactor	E	E	P	G	E	NA
Scraper	NA	NA	G	E	NA	E
Dragline	NA	NA	E	F	NA	NA

After Sorg and Bendixen (1975).

[a] Basis of evaluation: easily workable soil, and cover material haul distance greater than 1000 ft.

[b] Rating key: E, excellent; G, good; F, fair; P, poor; NA, not applicable.

Consultation with equipment salespersons regarding newly marketed equipment and its use should be undertaken before purchasing any equipment. A machine shed equipped to handle emergency repair and with a stock of accessories should be constructed for larger landfills. Backup equipment should be kept on site or arrangements should be made to borrow equipment whenever necessary. Routine maintenance should be performed as recommended by the manufacturer. Figure 11.1 shows a steel wheel landfill compactor and Fig. 11.2 shows a dozer.

11.2   PHASING PLAN

A phasing plan is important, both in terms of operating a landfill and for establishing financial proof for landfill closure (see Section 12.1.2 for details). A phasing plan is necessary for both natural attenuation and containment type landfills. An ideal phasing plan would be one in which each phase receives a final cover in the shortest possible time. This would mean disposing of waste in one phase until it reaches the final grade. Figure 11.3 shows a phasing plan for a landfill. However, such a simple phasing plan may not be applicable to all landfills. If the landfill height is more than 9 m (30 ft) from the base, usually an intermediate cover is used over parts of the landfill, at 3–4.5 m (10–15 ft) above ground (not above the base). In such cases an

FIG. 11.1. Steel wheel compactor.

FIG. 11.2. Dozer.

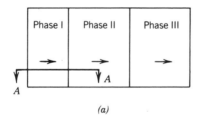

(a)

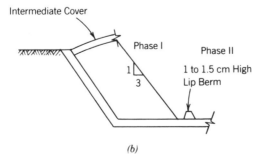

(b)

FIG. 11.3. Phasing plan for single stage landfill: (a) plan; (b) cross section A–A.

intermediate cover consisting of 60 cm (2 ft) of clayey soil and 15 cm (6 in.) of topsoil is used over the area. New phases are started on top of the lower phases after scraping the intermediate cover.

Figure 11.4 shows a phasing plan that follows the above concept. It may be noted that the soil used for constructing the intermediate cover should not be used for constructing the final cover because it gets contaminated with waste. The soil may be reused for daily cover or disposed of in the landfill. However, the topsoil may be reused in the final cover.

The direction of filling should be clearly indicated in the plan to avoid any confusion. The access road should not run over a closed phase. Usually a permanent all-weather access road is constructed parallel to the phases outside of the landfill area and branch roads, leading to the base of the landfill, are joined with it as necessary. Traffic routing should be planned carefully so that all waste is dumped in its final resting place.

11.3   COVERING WASTE

Three types of cover may be used in a landfill: daily cover, intermediate cover, and final cover. The design and construction of a final cover are discussed in Sections 8.8 and 9.9, respectively. Issues related to daily and intermediate covers are discussed in this section. Daily cover serves many

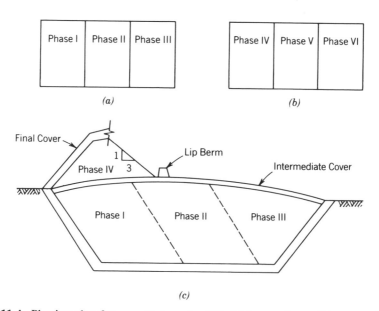

FIG. 11.4. Phasing plan for a multiphase landfill: (a) lower phase; (b) upper phase; (c) cross section.

functions that are considered essential for municipal landfills. For most municipal landfills 15 cm (6 in.) of daily cover is used, although its use may be warranted in some putrescible sludge landfills. Although a daily cover serves many functions it also uses up valuable landfill space. (Note: Daily cover occupies about one-fifth to one-sixth the volume of waste.) Depending on the geographical location of a landfill and considering other factors such as odor control, a weekly or monthly cover may also be used if permitted by the regulatory agency. Daily cover improves access, improves aesthetics, reduces windblown debris (e.g., paper, plastics), reduces risk of disease transmittal through vectors (e.g., birds, insects, and rats), reduces odors, reduces fire risk, and provides a media for partial attenuation of leachate. Historically, sandy soils have been the most commonly used daily cover material. If clay is used as daily cover then the daily cover should be scarified or removed prior to the placement of the next lift to avoid perching of leachate in the future. The other types of materials used for daily cover are: petrolium contaminated soil, synthetic textile, foundry sand, chemical foams and sluries, tire chips, bark and wood chips, and auto shredder fluff. Use of synthetic textiles saves valuable landfill space. The synthetic textile is stretched, either manually or using a dozer, over the lift of refuse at the end of a working day. The synthetic textile is removed prior to disposing waste next day. The same piece of synthetic textile can be reused for several days. If permitted by the regulatory agency permanent type daily cover (e.g., tire chips, soil) should be used every 7–10 days.

An intermediate cover is used when portions of a landfill are expected to remain open for a long period of time (2 years or more). Intermediate cover helps in reducing leachate production. The precipitation falling on the intermediate cover is allowed to run off the site. Surface runoff is not allowed from daily cover, so daily cover does not help in reducing leachate volume.

11.4 CODISPOSAL OF NONHAZARDOUS SLUDGE IN MUNICIPAL LANDFILLS

Codisposal of sludge should be done properly so that nuisance conditions do not develop in a landfill. Disposal of slurry type sludge should be avoided; the solids content of all sludge disposed in a landfill should be at least 40%. Mixing of sludge with municipal waste is preferred. Mixing can be done by spreading the sludge in thin layers and then covering it with municipal waste. The daily volume of sludge disposal should be restricted to a maximum of 25–30% of municipal waste. A 1:1 daily ratio of sludge to municipal waste may be used with due consideration toward fill stability and change in gas generation rate. It should be borne in mind that each landfill is designed for specific physical and chemical properties of waste. So design-related issues should be revisited prior to changing waste type in an active landfill. The interior access road should be carefully designed and maintained so hauling vehicles can maneuver easily. In some situations a separate disposal area of sludge is preferable for better operational control. Mixing of sludge with dry soil may be necessary to maintain trafficability within the landfill.

11.5   FIRE PROTECTION

Fire hazards exist in certain waste type landfills. Landfills, where open burning is practiced, are especially susceptible to such fires. Landfill fire can generate toxic fumes injurious to health. The following guidelines should be followed in landfills in which open burning is practiced. (Note: Permission from the regulatory agency may be necessary regarding open burning.)

1. An attendant should be on duty until the fire is extinguished.
2. A minimum of 3–4.5 m (10–15 ft.) of open space should be maintained around the burning area.
3. Asphalt, rubber, plastics and so on should not be burned because they will create air pollution problems.
4. A pit should be used to burn material. Ash generated from burning should be disposed in the landfill.
5. Public access to the burning pit should be restricted.

Usually open burning is allowed in small landfills located in remote places. Air curtain destructors may be used for a cleaner burning operation. A fire extinguisher, sand, and a water wagon should be readily available even in landfills in which open burning is not practiced. Operator(s) should be trained to extinguish small fires. The telephone number and location of the nearest fire station should be displayed in a conspicuous place in the landfill site office. In larger landfills a drill to train operator(s) in fire emergency handling procedures should be arranged once a year.

11.6   LITTER CONTROL

Paper and other lightweight material create littering problems around landfills. Dumping near the landfill base, especially during windy days, can help reduce littering. A small movable screen made of chicken wire may be placed near the active face to catch windblown papers. Regular manual pickup at the end of a working day may become necessary in some landfills. If windblown paper poses a serious problem, then very high fixed wire net may be placed down-gradient of the wind (direction prevailing in most days) or an earthen berm may be constructed up-gradient of the wind direction.

More than one row of trees or shrubs or an earthen berm may be used to screen a site from obvious public view. A well-kept landfill can improve public perception of landfills, which is beneficial to landfill siting both in the short and long term.

11.7   DUST CONTROL

Controlling dust in a landfill is somewhat difficult. The heavy traffic always creates dust problems within a landfill because of dry soil in the road. Although watering of the road can reduce dust, it may increase leachate volume. So judgment has to be used regarding the quantity of water added to the road inside a landfill to reduce dust. Dumping dusty loads in low areas of a landfill or spraying a small quantity of water to a dusty load before dumping will help in controlling dust.

11.8   ACCESS ROAD MAINTENANCE

The interior and exterior roads must be maintained properly at all times. Proper drainage of interior road beds is essential. The temporary interior roads are vital for proper operation of a landfill. Since heavy vehicles run to and from a landfill the exterior access roads deteriorate significantly. Sufficient funds should be allocated to maintain access roads. Landfill equipment that does not have rubber tires (e.g., steel wheel compactor or crawler tractor) should not be allowed on the paved portion of exterior roads. When necessary, this equipment may be allowed to move along unpaved portions of the road. A wide even surface, sufficiently wide to accommodate this equipment, should be maintained between the landfill and the maintenance shed.

11.9   LEACHATE COLLECTION SYSTEM MAINTENANCE

All items connected with leachate collection must be maintained properly. The items include mainly the leachate collection pipes, manhole(s), leachate collection tank and accessories, and pumps. The leachate lines should be cleaned once a year to clear out any organic growth. The manhole(s), tank(s), and pump(s) should be inspected visually once a year. Leachate can corrode metallic parts. An annual inspection and necessary repair will therefore prevent many future emergency type problems such as leachate overflow from the tank due to pump failure. Extreme care must be exercised when entering such confined space. Use of a harness and direct watch by personnel above ground are considered essential when entering manholes. All regulations regarding confined space entry must be followed to ensure the safety of the maintenance crews. A record of all repair activities should be maintained to assess (or claim) long-term warranties on pump(s) and other equipment.

11.10   FINAL COVER MAINTENANCE

The final cover should be maintained properly to reduce infiltration into the landfill. Erosion due to surface runoff, settlement, stress on vegetation due to gas in a landfill, and so on, damage the landfill final cover. Any damaged area should be repaired as soon as possible. The cause of the damage must be investigated so that proper repair measures are implemented. For instance, if the damage is due to stress on vegetation because of gas then proper gas venting measures should be undertaken. Otherwise, the problem will recur each year. On the other hand, if the damage is due to erosion, then the cause of the erosion must be investigated. The length and steepness of the slope, improper vegetation growth due to bad planting, uneven settlement of the solid waste, and so on, cause erosion. Protection matting should be used in long swales to reduce erosion. Once completed, access to the final cover area should be restricted to avoid damage by vehicular traffic. Every effort should be made to maintain a healthy vegetation because vegetation improves the stability of the slopes, reduces surface erosion (Gray and Leiser, 1982) and reduces leachate production by increasing evapotranspiration.

For the first 3–5 years the final cover area should be inspected twice a year: once during a season when the vegetative growth is minimum and once during a season when vegetative growth is maximum. Any problem of erosion can be detected easily when the surface is bare. Any problem of vegetative growth can be detected during the peak season inspection. The inspection frequencies may be reduced to once a year after the initial years and should be continued for another 10–15 years. Special attention should be given to protect the surface from burrowing animals. It is very likely that native vegetative species will invade the final cover in the long run. The root system of large trees may increase infiltration into the landfill. However, realistically very little can be done to stop the invasion of native species. No documentation of this phenomenon is available; this phenomenon is worth researching.

11.11   COMMENTS ON ENFORCEMENT RELATED ISSUES

It is normally the duty of the landfill manager to ensure that all conditions imposed by the regulatory agency are met. In general, the main areas of regulations are the health and safety of workers and environmental protection. In most cases two different government agencies are responsible for enforcing the relevant laws. It is essential that the landfill manager is familiar with both types of regulations. It is a good idea to meet with relevant officials at least prior to beginning the landfill operations and annually thereafter to discuss regulatory issues. The owner should include a lawyer in the group during the annual meeting who can help in understanding the legal aspects of regulatory requirements. Running an operation in accordance with government regulations costs a lot less than getting involved in litigation. However, if such litigation is unavoidable, then advice from an attorney should be sought at an early stage. The manager must read the conditions of permit and any changes thereof. All unrealistic permit conditions should be resolved at the beginning rather than trying to bypass them. A landfill manager should not knowingly force workers to engage in dangerous operations because the manager not the corporation (if privately owned) may be made personally liable for such action. The manager should be knowledgeable about the steps a regulatory agency takes prior to refering a case for legal action.

Although uniform legal procedures are not followed by all regulatory agencies, in many instances more than one notice is issued regarding a violation prior to taking legal action against a landfill owner. Each notice should be acted upon promptly and the lapse(s) should be rectified within the specified date.

11.12 LANDFILL MINING

Mining of landfill is undertaken to recover valuable landfill space and salvage materials. In landfill mining an old landfill is excavated and the materials are sorted into different categories. Big white goods (e.g., refrigerator) and other metallic items are salvaged and recycled whenever possible. The waste is sieved using trommels (approx 2.1 m diameter, 7 m long) that separates soil and burnable material. The landfill mining project in Lancaster county, Pennsylvania reported a 68% recovery of burnable materials and 28% of cover materials; only 4% of excavated material (by weight) was returned to the landfill (Flosdorf and Alexieff, 1993). Mining of a landfill in Florida has also been reported (U.S. Congress, 1989). According to Vasuki (1988), landfill mining is helpful for reusing of landfill space after a period of decomposition (which can extend the life of landfills), repairing of liners and leachate collection systems, and recovering materials of value. Although landfill mining is not widely practiced now, the idea merits consideration.

12 Economic Analysis

Enough money must be made available for not only constructing but also operating, maintaining, and monitoring a landfill. Proper cost analysis must be done to ensure cash flow for performing all the above tasks. Monitoring of a landfill may be required for 30–40 years after closing of the last phase. The total cost of long-term maintenance and monitoring (also known as long-term care) of a closed landfill could be higher than the cost of construction of a landfill. Enough money should be set aside during the active life of the landfill so that closure and long-term care can be performed. Methods for estimating these cost at a future date taking into account the inflation factor are discussed in Section 12.2.

Landfills may be either publicly (e.g., a municipality) or privately (e.g., industry) owned. The methods used to raise money for landfills differ for these two categories. For privately owned landfills the correct cash flow must be ensured by estimating costs for each activity discussed in subsequent sections.

Funds for a publicly owned landfill can be obtained from either general fund revenues or service charges. The disadvantages of using general fund revenues is that local officials are hard pressed to raise taxes. If service charges are used for funding purposes then updating these charges for inflation and so on must be carefully calculated. Municipalities usually draw from two basic sources for financing a project: capital borrowed funds and current revenue funds. Municipalities may raise money by issuing general obligation bonds, by revenue bonds, or by bank loans. It may also lease equipment necessary for operating the landfill. A detailed economic analysis of the suitability of both these methods of financing must be undertaken before a recommendation on funding is made. A municipality may also sell the operating permit to a private operator to operate the site. In such cases the private operator fixes the dumping fee.

12.1 COST ESTIMATES

As previously indicated, the life cycle cost can be categorized as follows: construction, operation (including all monitoring), closure, and long-term care (LTC). In many cases the cost of hauling is expected to be high because the landfills are located in remote areas. The cost of road construction for borrowed materials (e.g., clay) must be included in the unit cost.

12.1.1 Cost of Construction

Table 12.1 includes a list of typical construction items for a containment type landfill; the list is by no means complete. The basis for estimating unit costs is also included to provide some idea as to what activities should be included to arrive at unit costs. Items of construction for a natural attenuation landfill may also be arrived at easily from Table 12.1.

12.1.2 Cost of Closure

Table 12.2 includes a list of typical construction items necessary for closing a landfill and the basis of unit cost estimate. The unit cost should be a third party cost since it is possible that the owner (especially if private) may fail to close the landfill properly due to lack of funds. In that case the regulatory agency may have to close the site. In most cases the regulatory agency will have to hire an outside contractor to perform the job. So the cost of each item must reflect third party costs. If topsoil is available on site the cost of purchasing topsoil may not be included in the unit cost but a legal document

TABLE 12.1. Items for Estimating Site Construction Costs

Item	Basis of Cost
Clearing and grubbing	Clear and grub forest cover
Site access road	Place and compact crushed stone
Strip topsoil	Strip and stockpile topsoil
Drainage swale construction	Rough grade and erosion protection
Sedimentation basin construction	Berm, riprap, and discharge pipe
Excavation to subgrade	Cut for subgrade
Berm construction	Fill for perimeter berms (compaction only)
Collection lysimeter	Excavation, liner, bedding, piping, risers, and storage manhole
Clay liner construction	Hauling, placement, compaction, and borrow restoration
Drainage layer construction	Haul and place
Leachate collection piping	Trenching, bedding, backfill, pipe, and filter fabric
Leachate header pipe to manhole	Double encased, antiseep collars
Leachate header pipe in clay liner	Trench, piping, backfill, and compaction
Leachate cleanouts	Trench, pipe, backfill, and compaction
Leachate collection tank and loading station	Excavation, placement of existing tank, clay backfill and pump
Leachate pump manhole	Excavation, place MH, backfill, pump
Topsoil placement	Haul and place
Seed, fertilizer, and mulch	Using acreage, or similar criteria
Survey, documentation, and technical supervision	Inspection, testing, and report

TABLE 12.2. Items for Estimating Site Closure Costs

Item	Basis of Cost
Final cover construction	Construction of barrier layer and other layers, haul and place as required, restoration of borrow source(s), top soil placement, grading
Seed, fertilize, and mulch	Using acreage or similar criteria
Leachate head wells (if designed to be installed after completion)	Boring, well installation
Survey and documentation	Inspection, testing, report preparation, technical assistance, and supervision
Contingency	10–25% of total

providing access to the topsoil should be made available to the regulatory agency. A worst-case scenario (e.g., one phase has reached final grade but is not covered and waste is being disposed of in the next phase) should be assumed for estimating closure costs. Closure costs in the example case will include final cover construction in both phases and filling the second phase with soil up to the final grade.

12.1.3   Cost of Long-Term Care

Table 12.3 includes a list of typical items necessary for estimating long-term care costs for a containment type landfill. For an NA type landfill there will be no cost related to leachate management; all other cost items will remain the same. Here also the unit cost must be a third party cost. Damage of 10–20 m^2 of surface area per half a hectare (~100–200 ft^2/acre) of closed

TABLE 12.3. Items for Estimating Long-Term Care Cost

Item	Basis of Cost
Land surface care	Erosion damage repair, reseeding, mulching
Leachate and lysimeter pipeline cleaning	Cleaning of pipeline using proper techniques
Groundwater monitoring (includes lysimeter sample)	Sample collection, transportation, and laboratory and field testing
Gas monitoring	Field testing
Leachate monitoring	Sample collection, transportation, and laboratory and field testing
Leachate management	Hauling and treatment of leachate, leachate sump pump, and tank maintenance
Annual assessment	Site inspection and report preparation
Administration and contingency	10–25% (or as appropriate) of total

area per year may be assumed in most cases. Erosion loss is high in the first year and decreases steadily in a well-maintained landfill. The cost of land surface care can thus be reduced to 10% of the original value in a 7- to 10-year period. Leachate hauling and treatment are large cost items. The leachate production rate is reduced significantly a few years after closure. As mentioned in Section 3.4.2.6, the reduction in leachate production rate after landfill closure is difficult to estimate. A 20–30% infiltration of total precipitation may be used to estimate the leachate generation in the initial years (models mentioned in Section 3.4.2 may also be used to predict postclosure). The funds necessary for leachate management (hauling and treatment) can be adjusted in susequent years based on actual field data. Third party costs for hauling and treatment should be assumed even if the landfill owner owns a treatment plant. (Note: Most industries have waste water treatment plants that are capable of treating the landfill leachate.) Although leachate treatment can be continued in the owner's treatment plant, a third party cost is to be assumed only for estimating long-term care costs.

12.1.4   Cost of Operation

Table 12.4 includes a list of typical items for estimating the annual operating cost for a containment type landfill. Costs of leachate management and other irrelevant items (e.g., lysimeter monitoring) should be excluded when estimating the cost of operating an NA type landfill. A third party cost need not be used for estimating the annual operating cost. The waste fund is a fund administered in some states to which each landfill owner pays. The money from the fund may be used for closure, remedial action, or long-term care of abandoned landfills. Usually a fee is charged based on tonnage of waste disposed in a landfill. If such a fund does not exist then the fee need not be taken into consideration when estimating operational costs. In addition to costs of monitoring, necessary money for closure and long-term care should be invested in a fund during the active site life. In addition, money may also be saved for developing future landfill(s) during the active life of a landfill. Thus, once a landfill is constructed, the user fee may be estimated

TABLE 12.4. Items for Estimating Annual Operating Costs

Item	Basis of Cost
Waste placement	Manpower, equipment necessary for compaction of waste
Intermediate cover	Hauling, placement, and compaction of soil
Groundwater, leachate, and gas monitoring	See Table 12.3
Leachate management	See Table 12.3
Equipment	Purchase and maintenance
Annual payment to waste management	Disposal volume/weight

in such a way that it will cover all future activities as related to disposal of waste. In some instances the revenue generated from the user fee may be grossly underestimated if the disposal volume is reduced significantly due to increased recycling activities or wrong estimates of the waste volume generation rate. The chance of overestimating generation rate is high in small communities. So due care should be taken when estimating the waste generation rate. A discussion on cash flow estimates is included in Section 12.4, which will provide additional help in estimating operating costs.

12.2   ESTIMATE FOR PROOF OF FINANCIAL RESPONSIBILITY

Establishing proof of financial responsibility may be required by a regulatory agency. The purpose is to ensure that enough money is available for closure and long-term care of a landfill by the regulatory agency if the owner fails to perform these tasks due to a lack of funds. Not all regulatory agencies require this proof. The proof of financial responsibility (hereafter called "the proof") needs to be established prior to licensing the landfill site. There are two issues related to financial proof: methods of establishing the proof and estimation of the amount of proof. Each is discussed separately in the following sections.

12.2.1   Methods for Establishing Proof of Financial Responsibility

There are several methods by which proof can be established. The regulatory agency should be consulted to determine the acceptability of the chosen method. Since the cost of establishing the proof is different for different methods, the costs for each method should be calculated so that the owner can choose the method that is most suitable. Essentially there are three types of methods: putting money in an interest-bearing account, buying a payment guarantee from a insurance company or bank, and increasing company liability. If necessary, separate methods can be chosen to establish the proof to cover closure and LTC costs.

Money can be deposited in an interest-bearing account (e.g., escrow account) each year so that at the end of the active life of the landfill (hereafter called site life) enough money is available for closure and or LTC. Proof for LTC is required to be in place for several years (usually 30–40 years) after closure, whereas proof for closure is required to be in place until the site is closed. A trust fund may also be created for establishing the proof. The regulatory agency should be named as the joint operator of this account so that no cash can be withdrawn without written permission from the agency.

Forfeiture or a performance bond, a letter of credit issued by a financial institution, or risk insurance may also be used to establish the proof. The issuer of such bonds guarantees the payment for closure and or LTC if the owner fails to perform the task. The owner has to keep the bond in place for closure until closure is actually performed. Bonds for LTC must be kept in place until the LTC period (usually 30–40 years after closure) is over. The owner has to pay a fee to the bonding company for each year the bond is in place. The fee depends on the amount of the bond. The regulatory agency should be named as the beneficiary of the bond. The bond may be for cash payment to the agency or for performing the task by the issuing company to the satisfaction of the agency.

Large companies may guarantee closure and LTC by increasing their liability by cost of closure and LTC. A financial statement regarding economic soundness commonly known as "net worth of the company" is usually required to be submitted to the regulatory agency each year. Usually companies with a set minimum total asset (e.g., 5 million dollars) and a minimum net asset (total assets minus total liabilities) exceeding the total cost of closure and LTC by a factor (usually 6 or more) are allowed to use the "net worth" method. It is possible that a company's net assets in a year fall below the minimum required by the agency. In such cases the owner has to switch to a different method for establishing the proof. The cost of total LTC is reduced each year after the LTC is performed for the year. Money may be withdrawn, if allowed by the regulatory agency from an interest-bearing account after LTC is successfully performed in a year. If the money withdrawn from the interest-bearing account is saved, then it can be used for LTC in the current year; this means that cash flow is not needed every year for LTC. However, if the bond or net worth method is used for LTC then although the cost of the remaining LTC is reduced and the cost of buying the bond and so on is reduced, no money is really available for performing LTC in the current year. Thus, an interest-bearing account is the best choice for establishing the proof provided enough money is available initially.

12.2.2 Estimation of Financial Responsibility for the Amount of Proof

The costs of closure and LTC are estimated based on Section 12.1. These base costs are then used to calculate the costs of closure and LTC at a future date. These amounts must be available to perform the tasks at the specified time in the future. Inflation reduces the value of money. Thus, if $1000 is required for closure today, the amount required a year from today will be $(1000)(1 + f)$, where f is the percentage inflation rate for the 1-year period. (Note: Usually a constant inflation rate is assumed for each year.) At the end of year 2 the required amount will be $(1000)(1 + f)(1 + f)$. Similarly, the required amount at the end of year n is $(1000)(1 + f)^n$. Now, if the money is invested in an interest-bearing account then less money needs to be kept in that account because of the interest being earned. The money to be invested in an interest-bearing account to yield $(1000)(1 + f)$ at the end of year 1 is

$(1000)(1 + f)/(1 + i)$ $(=D$ say), where i is the percentage interest rate. Thus, the amount to be invested in an interest-bearing account to yield the necessary money at the end of year n will be $1000(1 + f)^n/(1 + i)^n$. This is the basic concept used to estimate the amount of financial proof required for closure and long-term care for a landfill. Proofs for both closure and LTC may be required to be established prior to disposing of waste in a landfill or within the active life (also known as site life) of the landfill. The regulatory agency should be contacted as to whether and when the proofs are to be established.

Formulas used for estimating the amount of proof depend on whether any interest is earned. Although interest increases the deposited amount, inflation reduces it. If a bond or net worth is used for proof then, only inflation needs to be accounted for because the additional money earned through interest on the deposited amount is not available to the bond account. The cost of closure and LTC should be increased each year by the inflation rate while using them for the net worth method so that these costs remain current throughout the LTC period. The official inflation factor is obtained only at the end of the year. If the trend indicates that the rate of inflation is increasing each year (e.g., year 1, 3%; year 2, 3.5%; year 3, 4.5%) for the last 2 years then a reasonable increase in inflation rate may be used (i.e., if the actual inflation rate is 4%, then a 5% inflation rate should be assumed). Similarly, if the trend indicates that interest rates will decrease each year then a reasonable reduction in interest rate may be used unless the bank has guaranteed a fixed rate for the entire LTC period. These suggestions regarding increases in the inflation rate and decreases in the interest rate are given to provide additional safeguards against shortages of the funds available to perform the closure and long-term care tasks. A discussion with the financial institution regarding available interest rates for an escrow account (if used) is helpful.

As mentioned in Section 12.1.3, LTC costs may vary over the LTC period (30–40 years after closure of a landfill). Therefore a variable LTC cost may need to be used to estimate the amount of financial proof. The approach for estimating the amount of financial proof for closure for an interest-bearing account is discussed at the beginning of this section. The approach for estimating the amount of financial proof for different cases are included in the following sections.

The Amount of Proof for Closure
For interest bearing accounts (C = cost of closure; SL = Site life)

$$D = C(1 + f)^{SL}/(1 + i)^{SL} \qquad (12.1)$$

For non-interest-bearing accounts (e.g., bonds) and net worth

$$D = C(1 + f)^{SL} \qquad (12.2)$$

The Amount of Proof for LTC

1. For non-interest-bearing accounts (e.g., bonds).

Equal Annual Out Payment. The account should be set up in such a way that enough money is available to pay all LTC-related bills at the end of each year of LTC. Thus, LTC money is needed for each year of the LTC period beginning 1 year after landfill closure. The total cost of LTC (A) for the entire LTC period (Fig. 12.1) is given by Eq. (12.4) in which L = the cost estimate of LTC calculated in year zero (i.e., in the year when waste was first disposed in the landfill).

$$A = L(1 + f)^{SL+1} + L(1 + f)^{SL+2} + \cdots + L(1 + f)^{SL+LTC}$$

or

$$A = L(1 + f)^{SL+1}[1 + (1 + f) + (1 + f)^2 + \cdots +(1 + f)^{LTC-1}]$$
(12.3)

$1 + (1 + f) + (1 + f)^2 + \cdots +(1 + f)^{LTC-1}$ is a geometric series of the form $1 + a + a^2 + \cdots + a^{n-1}$, the sum of which is expressed as (Tuma, 1979) $(a^n - 1)/(a - 1)$. Thus, Eq. (12.3) can be written as

$$A = L(1 + f)^{SL+1}\left[\frac{(1 + f)^{LTC} - 1}{f}\right]$$
(12.4)

Example 12.1

Find the amount of financial proof required when a bond is used for a constant LTC cost of \$5000/year estimated 2 years after the landfill was opened. The

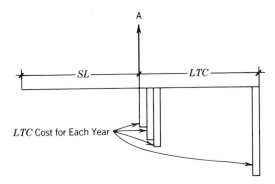

FIG. 12.1. Cash flow pattern used for estimating proof of financial responsibility for long-term care.

estimated SL for the landfill is 10 years, the LTC period is 20 years, and the constant rate of inflation rate is found to be 5%.

Note that the LTC estimate was made 2 years after the landfill was opened, which has an SL of 10 years. Thus, the SL to be used in Eq. (12.4) shall be $10 - 2 = 8$ years rather than 10 years.

$$A = 5000 (1 + 0.05)^{8+1} \left[\frac{(1 + 0.05)^{20} - 1}{0.05} \right]$$

$$= \$256,480.74$$

Unequal Annual Out Payments. For estimating the amount of proof for variable LTC L_x for year x, A is given by

$$A = L_1 (1 + f)^{SL+1} \sum_{\substack{x=2 \\ r=1}}^{\substack{r=LTC-1 \\ x=n}} L_x (1 + f)^r \tag{12.5}$$

2. For interest-bearing accounts.

Equal Annual Out Payments. In deriving this formula it is assumed that all the money needed for LTC is deposited in an interest-bearing bank account at the end of the site life but the cost estimate for LTC was made at the beginning of the site life. Therefore, the money earns interest for 1 year before payment is needed for LTC. The money that should be deposited in an interest-bearing account is given by

$$
\begin{aligned}
A &= L(1 + f)^{SL+1}/(1 + i) + L(1 + f)^{SL+2}/(1 + i)^2 + \cdots \\
&\quad + L(1 + f)^{SL+LTC}/(1 + i)^{LTC} \\
&= L(1 + f)^{SL} [(1 + f)/(1 + i) + (1 + f)^2/(1 + i)^2 + \cdots \\
&\quad + (1 + f)^{LTC}/(1 + i)^{LTC}]
\end{aligned}
\tag{12.6}
$$

The series within the brackets is a geometric series. Thus, Eq. (12.6) can be written as

$$A = L(1 + f)^{SL} \left[\frac{\left(\dfrac{1 + f}{1 + i} \right)^{LTC+1} - 1}{\left(\dfrac{f - i}{1 + i} \right)} - 1 \right] \tag{12.7}$$

Note that the formula is not valid when $f = i$.

Example 12.2

Estimate the amount to be deposited in an interest-bearing bank account for long-term care. Use the following values of various variables: $L + \$10,000$, $SL = 2, f + 3\%, i = 6\%, LTC$ period $= 5$.

$$A = 10,000(1 + 0.03)^2 \left[\frac{\left(\frac{1 + 0.03}{1 + 0.06} \right)^{5+1} - 1}{\left(\frac{0.03 - 0.06}{1 + 0.06} \right)} - 1 \right]$$

$$= \$48,707.56$$

The interest earned and the expenditure for LTC for each year is included in Table 12.5; Fig. 12.2 depicts partial cash flow.

Unequal Out Payments. To estimate the amount of proof (A) for variable LTC costs of L_x for year x, Eq. (12.8) is to be used:

$$
\begin{aligned}
A &= L_1(1 + f)^{SL+1}/(1 + i) + L_2(1 + f)^{SL+2}/(1 + i)^2 + \cdots \\
&\quad + L_n(1 + f)^n/(1 + i)^n \\
&= \sum_{x=1}^{x=LTC} L_x(1 + f)^{SL+x}/(1 + i)^x
\end{aligned}
\tag{12.8}
$$

3. For net worth. (Note: Net worth must be evaluated each year.)

$$A = L(1 + f) \tag{12.9}$$

The following symbols are used in the above formulas: A = the total in-payment for long-term care per year of site life, i = the estimated annual rate of interest, expressed as a decimal, f = the estimated annual rate of inflation, expressed as a decimal, SL = the estimated life of the facility in

TABLE 12.5. Long-Term Care Cash Flow for Example 12.2

Year	Opening Balance ($)	Interest Earned ($)	LTC Cost ($)	Balance ($)
1	48,707.57	2,922.45	10,927.27	40,702.75
2	40,702.75	2,442.17	11,255.09	31,889.83
3	31,889.83	1,913.39	11,592.74	22,210.49
4	22,210.40	1,332.63	11,940.52	11,602.51
5	11,602.51	696.15	12,298.74	≈0.00

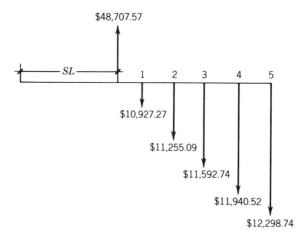

FIG. 12.2. Partial cash flow for Example 12.2.

years rounded to the lower nonzero whole number, L = the estimated annual long-term care costs, L_x = the estimated unequal annual costs, x = the year of long-term care, and LTC = the period of long-term care.

12.3   USER FEE ESTIMATE

User fees can be calculated by dividing the total of the annual operating cost, the average annual cost for a future landfill, and the average annual cost for establishing financial proof, by the estimated annual disposal weight (or volume). The fee should be updated for inflation and revenue shortfall due to decreased disposal volume and so on. The following example enumerates how to estimate user fees.

Example 12.3

The following items are estimated for construction, operation, and long-term care of a landfill. The site life of the landfill is 3 years. It is estimated that the landfill will receive 25,000 tons of garbage each year. It was decided to raise funds for a future landfill of the same volume when the current landfill is closed. An escrow account will be used to establish proof of financial responsibility for closure and long-term care. The proof for closure is to be established in year 1 of the landfill site life and the proof for LTC is to be established by the end of the site life of the landfill. Cost estimates are included in Table 12.6. Estimate the user fee for the landfill. Assume an inflation rate of 4%, an interest rate of 5%, and a long-term care period of 20 years.

TABLE 12.6. Cost Estimates for the Landfill in Example 12.3

Items		Cost ($)
a. Construction of the landfill		70,000.00
b. Site investigation, report preparation, etc.		25,000.00
c. Annual operating cost		
1. Waste placement		20,000.00
2. Intermediate cover construction		2,000.00
3. Groundwater monitoring		5,000.00
4. Lysimeter monitoring		1,000.00
5. Gas monitoring		500.00
6. Leachate management		50,000.00
7. Equipment purchase and maintenance		20,000.00
8. Waste fund payment (at 5¢/ton)		1250.00
	Subtotal	99,750.00
d. Estimated cost of closure at the beginning of site life		10,000.00
e. Estimated cost of land surface care each year		3,000.00

Average Cost for a Future Landfill

Cost for the landfill at the end the 3-year period

$$= \text{(item a + item b of Table 12.6)} (1 + f)^3$$

$$= (70,000 + 25,000)(1 + 0.04)^3 = \$106,862.00$$

The average approximate amount to be made available each year is

$$106,862/3 = \$35,621.00$$

The amount kept in an interest-bearing account (interest rate: 5% per year) would accrue interest and hence the amount actually needed to be deposited in the first year would be

$$35,621/(1.05)^3 = \$30,770.76$$

Similarly the cost for the two subsequent years can be estimated as follows:

$$\text{Second year deposit} = 35,621/(1.05)^2 = \$32,309.30$$

$$\text{Third year deposit} = 35,621/(1.05) = \$33,924.76$$

To estimate user fees one may use an average of these three amounts [$(30,770.76 + 32,309.30 + 33,924.76)/3 = \$32,334.94$] or the maximum annual payment ($33,924.76). In this example the average amount is used.

Average Annual Cost for Establishing Proof of Financial Responsibility

Closure

$$D = C(1 + f)^{SL}/(1 + i)^{SL}$$
$$= 10,000 \ (1 + 0.04)^3/(1 + 0.05)^3 + \$9716.99$$

The average amount to be made available each year for closure is $9716.99/3 = \$3238.99$

Long-Term Care. The total cost for long-term care is obtained by adding items c3, c4, c5, c6, and e in Table 12.6.

$$\text{Long-term care cost} = (5000 + 1000 + 500 + 50,000 + 3000)$$
$$= \$59,500.00$$

$$\text{Total amount of proof} = 59,500 \ (1 + 0.04)^3 \left[\frac{\left(\dfrac{1 + 0.04}{1 + 0.05}\right)^{20+1} - 1}{\left(\dfrac{0.05 - 0.04}{1 + 0.05}\right)} - 1 \right]$$

$$= \$1,212,468.26$$

The average amount to be made available each year is $1,212,468.26/3 = \$404,156.09$.

Average Annual Cost. The total amount to be made available each year for closure and long-term care is $3238.99 + \$404,156.09 = \$407,395.08$. The amounts to be saved each year are: year 1, $407,395.08/(1 + 0.05)^3 = \$351,923.19$; year 2, $407,395.08/(1 + 0.05)^2 = \$369,519.35$; and year 3, $407,395.08/(1 + 0.05) = \$387,995.31$. The average amount to be used for user fee is

$$(\$351,923.19 + \$369,519.35 + \$387,995.31)/3 = \$369,812.62$$

Money must be accumulated during the site life of a landfill to perform LTC of the landfill for 1 year after closure. Usually money is released from the LTC proof at the end of an LTC period. Therefore from the second year onward money will be available from the financial proof withholding if an escrow account is used. However, if a non-escrow type method is used for LTC then money to perform LTC for the entire LTC period must be accumulated in a fund. Money for closure will not be available at the time of performing closure. However, the proof may be released by the regulatory agency once the closure document is accepted. Thus, if an escrow account is maintained for closure then a lump sum amount may be available in the second year of the LTC period that may be used for LTC.

Thus, one should estimate which of the following will be the cheapest way of establishing proof: depositing money in an interest-bearing account, a bond, net worth, or a combination of two methods. The total average annual cost to be used for estimating user fee is (note: the annual operating cost for the landfill is obtained from item c of Table 12.6)

$$\$32,334.94 + \$99,750.00 + \$369,812.62 = \$501,897.56$$

Thus the user fee is

$$\$501,897.56/25,000 = \$20.08/\text{ton}$$

The user fee may be raised by 10–15% for contingency. For privately owned landfills, the user fee should be increased to include profit.

12.4   CASH FLOW ESTIMATE

Cash flow estimate is important for the smooth running of a landfill. As previously indicated, the annual cost for running a landfill should include both the cost of the daily operation and the long-term cost.

The above example will provide some ideas for estimating cash flow for a landfill. The money to be borrowed for running the landfill for the first month can be somewhat high because of equipment purchase and so on, and this should be taken into account. Additional cash is needed at the end of the landfill site life for closure of the landfill and to perform long-term care. A careful study of the cash flow needed should be undertaken for the smooth running of a landfill. It is important to keep a record of waste actually disposed of in a landfill for future planning purposes. A faulty cash flow estimate, especially underestimates, can significantly increase the cost of operation because of emergency borrowing of money.

LIST OF SYMBOLS

D	= required deposit for closure proof
f	= inflation rate
i	= interest rate.
SL	= site life
A	= Total cost of long-term care
L_x	= the estimated unequal annual costs of long-term care

x = the year of long-term care
LTC = the period of long-term care
Σ = the sum from year 1 through the last year of LTC
C = cost of closure
L = cost of long-term care per year

APPENDIX I
Hazardous Constituents

Acetonitrile (ethanenitrile)

Acetatophenone (ethanone, 1-phenyl)

3-(α-Acetonylbenzyl)-4-hydroxycoumarin and salts (warfarin)

2-Acetylaminofluorene (acetamide, N-(9H-fluoren-2-yl)-)

Acetylchloride (ethanoyl chloride)

1-Acetyl-2-thiourea (acetamide, N-(aminothioxomethyl)-)

Acrolein (2-propenal)

Acrylamide (2-propenamide)

Acrylonitrile (2-propenenitrile)

Aflatoxins

Aldrin   (1,2,3,4,10,10-hexachloro-1,4,4a,5,8,8a,8b-hexahydro-endo,   exo
    (1,4,5,8-dimethanonaphthalene)

Allyl alcohol (2-propen-1-ol)

Aluminum phosphide

4-Aminobiphenyl([1,1F-biphenyl]-4-amine)

6-Amino-1,1a,2,8,8a,8b-hexahydro-8-(hydroxymethyl)-8a-methoxy-5-
    methyl-carbamate azirino [2F,3F: 3,4] pyrrolo[1,2-a]alindole-4,7-dione
    (ester) (mitomycin C) azirino[2F,3F:3,4]pyrrolo(1,2-a)indole-4,7-one, 6-
    amino-8-[((amino-carbonyl)oxy)methyl]-1,1a,2,8,8a,8b-hexahydro-8a-
    methoxy-5-methyl-)

5-(Aminomethyl)-3-isoxazolol   (3(2H)-isoxazolone,   5-(aminomethyl)-)   4-
    amino-pyridine (4-pyridinamine)

Amitrole (1H-1,2,4-triazol-3-amine)

Aniline (benzenamine)

Antimony and compounds, N.O.S.

Aramite (sulfurous acid, 2-chloroethyl-,2-[4-(1,1-dimethylethyl)phenoxyl]-1-
    methyl-ethyl ester)

Arsenic and compounds, N.O.S.

Arsenic acid (orthoarsenic acid)

Arsenic pentoxide (arsenic(V) oxide)

Arsenic trioxide

(Arsenic(III) oxide)

Auramine (benzenamine, 4,4F-carbonimidoylbis]N,N-dimethyl-, monohy-
    drochloride)

Azaserine (L-serine, diazoacetate (ester))

Barium and compounds, N.O.S.

Barium cyanide

Benz[c]acridine (3,4-benzacridine)

Benz[a]anthracene (1,2-benzanthracene)

Benzene (cyclohexatriene)

Benzene, 2-amino-1-methyl (o-toluidine)

Benzene, 4-amino-1-methyl (p-toluidine)

Benzenearsonic acid (arsonic acid, phenyl-)

Benzene, dichloromethyl-(benzal chloride)

Benzenethiol (thiophenol)

Benzidine ([1,1F-biphenyl]-4,4F-diamine)

Benzo[b]fluoranthene (2,3-benxofluoranthene)

Benzo[j]fluoranthene (7,8-benzofluoranthene)

Benzol[a]pyrene (3,4-benzopyrene)

p-Benzoquinone (1,4-cyclohexadienedione)

Benzotrichloride (benzene, trichloromethyl-)

Benzyl chloride (benzene, (chloromethyl)-)

Beryllium and compounds, N.O.S.

Bis(2-chloroethoxy)methane (ethane, 1,1F-[methylenebis(oxy)]bis[2-chloro-])

Bis(2-chloroethyl) ether [ethane, 1,1F-oxybis[2-chloro-]) N

N-Bis(2-chloroethyl)-2-naphthylamine (chlornaphazine)

Bis(2-chloroisopropyl) ether (propane,2,2F-oxybis[2-chloro-]

Bis(chloromethyl) ether (methane, oxybis[chloro-])

Bis(2-ethylhexyl) phthalate (1,2-benzenedicarboxylic acid, bis(2-ethylhexyl)ester)

Bromoacetone (2-propanone, 1-bromo-)

Bromomethane (methyl bromide)

4-Bromophenyl phenyl ether

(Benzene, 1-bromo-4-phenoxy-)

Brucine (strychnidin-10-one, 2,3-dimethoxy-)

2-Butanone peroxide (methyl ethyl ketone, peroxide)

Butyl benzyl phthalate (1,2-benzenedicarboxylic acid, butyl phenyl-methyl ester)

2-sec-Butyl,-4, 6-dinitrophenol (DNBP) (phenol, 2,4-dinitro-6-(1-methylpropyl)-)

Cadmium and compounds, N.O.S.

Calcium chromate (chromic acid, calcium salt)

Calcium cyanide carbon disulfide (carbon bisulfide)

Carbon oxyfluoride (carbonyl fluoride)

Chloral (acetaldehyde, trichloro-)

Chlorambucil (butanoic acid, 4-[bis(2-chloroethyl)amino]benzene-)

Chlordane (α and γ isomers) (4,7-methanoindan, 1,2,4,5,6,7,8,8-octachloro-3,4,7,7a-tetrahydro-) (α and γ isomers)

Chlorinated benzenes, N.O.S.

Chlorinated ethane, N.O.S.
Chlorinated fluorocarbons, N.O.S.
Chlorinated naphthalene, N.O.S.
Chlorinated phenol, N.O.S.
Chloroacetaldehyde (acetaldehyde, chloro-)
Chloroalkyl ethers, N.O.S. *p*-chloroaniline (benzenamine, 4-chloro-)
Chlorobenzene (benzene, chloro-)
Chlorobenzilate (benzeneacetic acid, 4-chloro-α-(4-chlorophenyl)-α-hydroxy-, ethyl ester) 2-chloro-1, 3-butadiene (chloroprene)*p*-chloro-*m*-cresol (phenol, 4-chloro-3-methyl)
1-Chloro-2,3-expoxypropane (oxirane, 2-(chloromethyl)-)
2-Chloroethyl vinyl ether (ethane, (2-chloroethoxy)-)
Chloroform (methane, trichloro-)
Chloromethane (methyl chloride)
Chloromethyl methyl ether (methane, chloromethoxy-)
2-Chloronaphthalene (naphathalene, β-chloro-)
2-Chlorophenol (phenol-*o*-chloro-) 1-(*o*-chlorophenyl)thiourea (thiourea, (2-chloro-phenyl)-)
3-Chloropropene (allylchloride)
3-Chloropropionitrile (propanenitrile, 3-chloro-)
Chromium and compounds, N.O.S.
Chrysene (1,2-benzphenanthrene)
Citrus red No. 2 (2-naphthol, 1-[(2,5-dimethoxyphenyl)azol-)
Coal tars
Copper cyanide
Creosote (creosote, wood) cresols (cresylic acid) (phenol, methyl-) crotonaldehyde (2-butenal)
Cyanides (soluble salts and complexes), N.O.S.
Cyanogen (ethanedinitrile)
Cyanogen bromide (bromine cyanide)
Cyanogen chloride (chlorine cyanide)
Cycasin (β-D-glucopyranoside, (methyl-ONN-azoxy)methyl-)
2-Cychlohexyl-4, 6-dinitrophenol (phenol, 2-cyclohexyl,-4,6-dinitro-)
Cyclophosphamide (2*H*-1,3,2,-oxazaphosphorine, [bis(2-chloroethyl) amino]-tetrahydro-, 2-oxide
Daunomycin (5,12-naphthacenedione, (8S-*cis*)-8-acetyl-10-[(3-amino-2,3,6-trideoxy)-α-L-lyxo-hexo pyranosyl)oxy]7,8,9,10-tetrahydro-6,8,11-trihydroxy-1-methoxy-)
DDD (dichlorodiphenyldichloroethane) (ethane, 1,1-dichloro-2,2-bis(*p*-chloro-phenyl)-)
DDE (ethylene, 1,1-dichloro-2,2-bis(4-chlorophenyl)-)
DDT (dichlorodiphenyltrichloroethane) (ethane, 1,1,1-trichloro-2,2-bis(*p*-chloro-phenyl)-)
Diallate (*S*-(2,3-dichloroallyl)diisopropylthiocarbamate)
Dibenz[*a,h*]acridine (1,2,5,6-dibenzacridine)

Dibenz[*a,j*]acridine (1,2,7,8-dibenzacridine)
Dibenz[*a,h*]anthracene (1,2,5,6-dibenzanthracene)
7*H*-Dibenzo[*c,g*]carbazole (3,4,5,6-dibenzcarbazole)
Dibenzo[*a,e*]pyrene (1,2,4,5-dibenzpyrene)
Dibenzo[*a,h*]pyrene (1,2,5,6-dibenzpyrene)
Dibenzo[*a,i*]pyrene (1,2,7,8-dibenzpyrene) 1,2-dibromo-3-chloropropane
    (propane,1,2-dibromo-3-chloro-)
1,2-Dibromoethane (ethylene dibromide)
Dibromomethane (methylene bromide)
Di-*n*-butyl phthalate (1,2-benzenedicarboxylic acid, dibutyl ester)
o-Dichlorobenzene (benzene, 1,2-dichloro-)
m-Dichlorobenzene (benzene, 1,3-dichloro-)
p-Dichlorobenzene (benzene, 1,4-dichloro-)
Dichlorobenzene, N.O.S. (Benzene, dichloro-, N.O.S.) 3,3*F*-dichloroben-
    zidine ([1,1*F*-biphenyl]-4-4*F*-diamine, 3,3*F*-dichloro-) 1,4-dichloro-2-bu-
    tene (2-butene, 1,4-dichloro-)
Dichlorodifluoromethane (methane, dichlorodifluoro-)
1,1-Dichloroethane (ethylidene dichloride) 1,2-Dichloroethane (Ethylene di-
    chloride) trans-1,
2-Dichloroethene (1,2-dichloroethylene)
Dichloroethylene, N.O.S. (ethene, dichloro-, N.O.S.) 1
1-Dichloroethylene (ethene, 1,1-dichloro-)
Dichloromethane (methylene chloride)
2,4-Dichlorophenol (phenol, 2,4-dichloro-)
2,6-Dichlorophenol (phenol, 2,6-dichloro-)
2,4-Dichlorophenoxyacetic acid (2,4-D), salts and esters (acetic acid)
    2,4-dichlorophen-oxy-, salts and esters)
Dichlorophenylarsine (phenyl dichloroarsine)
Dichloropropane, N.O.S. (propane, dichloro-, N.O.S.)
1,2-Dichloropropane (propylene dichloride)
Dichloropropanol, N.O.S. (propanol, dichloro-, N.O.S.)
Dichloropropene, N.O.S. (propene, dichloro-, N.O.S.)
1,3-Dichloropropene (1-propene, 1,3-dichloro-)
Dieldrin (1,2,3,4,10,10-hexachloro-6,7-epoxy-1,4,4a,5,6,7,8,8a-octa-hydro-
    endo, exo-1,4:5,8-dimethanonaphthalene)
1,2:3,4-Diepoxybutane (2,2*F*-bioxirane)
Diethylarsine (arsine, diethyl-)
N,N-Diethylydrazine (hydrazine, 1,2-diethyl)
O,O-Diethyl *S*-methyl ester of phosphoros dithioic acid (phosphorodithioic
    acid,
O,O-Diethyl *S*-methyl ester
O,O-Diethylphosphoric acid, *O-p*-nitrophenyl ester (phosphoric acid, diethyl
    p-nitrophenyl ester)
Diethyl phthalate (1,2,-benzenedicarboxylic acid, diethyl ester) *O,O*-diethyl
    O-2,pyrazinyl phosphorothioate (phosphorothioic acid, *O,O*-diethyl *O*-
    pyrazinyl ester

Diethylstilbesterol (4,4F-stilbenediol, α,α-diethyl, bis(dihydrogen phosphate, (E)-)

Dihydrosafrole (benzene, 1,2-methylene-dioxy-4-propyl-)

3,4-Dihydroxy-α-(methylamino)methyl benzyl alcohol (1,2-benzenediol, 4-[1-hydroxy-2(methylamino)ethyl]-)

Diisopropylfluorophosphate (DFP) (phosphorofluoridic acid, bis(1-methylethyl) ester)

Dimethoate (phosphorodithioic acid, O,O-dimethyl S-[2-(methylamino)-2-oxoethyl] ester

3,3F-Dimethoxybenzidine ([1,1F-biphenyl]-4,4F-diamine, 3-3F-dimethoxy-)

p-Dimethylaminoazobenzene (benzenamine, N,N-dimethyl-4-(phenylazo)-)

7,12-Dimethylbenz[a]anthracene (1,2-benzanthracene, 7,12-dimethyl-)

3,3F-Dimethylbenzidine ([1,1F-biphenyl]-4,4F-diamine, 3,3F-dimethyl-)

Dimethylcarbamoyl chloride (carbamoyl chloride, dimethyl-)

1,1-Dimethylhydrazine (hydrazine, 1,1-dimethyl-)

1,2-Dimethylhydrazine (hydrazine, 1,2-dimethyl-)

3,3-Dimethyl-1-(methylthio)-2-butanone)O-[(methylamino) carbonyl]oxime (thiofanox) α,α-dimethylphenethylamine (ethanamine, 1,1-dimethyl-2-phenyl-)

2,4-Dimethylphenol (phenol, 2,4-dimethyl-)

Dimethyl phthalate (1,2-benzendicarboxylic acid, dimethyl ester)

Dimethyl sulfate (sulfuric acid, dimethyl ester)

Dinitrobenzene, N.O.S. (benzene, dinitro-, N.O.S.)

4,6-Dinitro-o-cresol and salts (phenol, 2,4-dinitro-6-methyl-, and salts)

2,4-Dinitrophenol (phenol, 2,4-dinitro-)

2,4-Dinitrotoluene (benzene, 1-methyl-2,4-dinitro-)

2,6-Dinitrotoluene (benzene, 1-methyl-2,6-dinitro-)

Di-n-octyl phthalate (1,2-benzenedicarboxylic acid, dioctyl ester)

1,4-Dioxane (1,4-diethylene oxide)

Diphenylamine (benzenamine, N-phenyl-)

1,2-Diphenylhydrazine (hydrazine, 1,2-diphenyl-)

Di-n-propylnitrosamine (N-nitroso-di-n-pro pylamine)

Disulfoton (O,O-diethyl S-[2-(ethylthio)ethyl]phosphorodithioate)

2,4-Dithiobiuret (thioimidodicarbonic diamide)

Endosulfan (5-norbornene, 2,3-dimethanol, 1,4,5,6,7,7-hexachloro-, cyclic sulfite)

Endrin and metabolites (1,2,3,4,10,10-hexachloro-6,7-epoxy-1,4,4a,5,6, 7,8,8a-octahydroendo, endo-1,4:5,8-dimethanonaphthalene, and metabolites)

Ethyl carbamate (urethan) (carbamic acid, ethyl ester)

Ethyl cyanide (propanenitrile)

Ethylenebisdithiocarbamic acid, salts and esters (1,2-ethanediylbiscarbamodithioic acid, salts and esters)

Ethylene, glycol monoethyl ether (ethanol, 2-ethoxy)

Ethyleneimine (aziridine)

Ethylene oxide (oxirane)

Ethylenethiourea (2-imidazolidinethione)
Ethyl methacrylate (2-propenoic acid, 2-methyl-ethyl ester)
Ethyl methanesulfonate (methanesulfonic acid, ethyl ester)
Fluoranthene (benzo[*j,k*]fluorene)
Fluorine 2-fluoroacetamide (acetamide, 2-fluoro-)
Fluoroacetic acid, sodium salt (acetic acid, fluoro-, sodium salt)
Formaldehyde (methylene oxide)
Formic acid (methanoic acid)
Glycidylaldehyde (1-propanol-2,3-epoxy)
Halomethane, N.O.S.
Heptachlor (4,7-methano-1*H*-indene, 1,4,5,6,7,8,8-heptachloro-3a,4,7,7a-tetrahydro-)
Heptachlor epoxide (α, β, and γ isomers) (4,7-methano-1*H*-indene, 1,4,5,6,7,8,8-heptachloro-2,3-epoxy-3a,4,7,7-tetrahydro-, α, β, γ isomers)
Hexachlorobenzene (benzene, hexachloro-)
Hexachlorobutadiene (1,3-butadiene, 1,1,2,3,4,4-hexachloro-)
Hexachlorocyclohexane (all isomers) (lindane and isomers)
Hexachlorocyclopentadiene (1,3-cyclopentadiene, 1,2,3,4,5,5-hexachloro-)
Hexachloroethane (ethane, 1,1,1,2,2,2-hexachloro-) 1,2,3,4,10,10-hexachloro-1,4,4a,5,8,8a-hexahydro-1,4:5,8-endo, endo-dimethanonaphthalene (hexachloro-hexahydro-endo-endo-dimethanonaphthalene)
Hexachlorophene (2,2*F*-methylenebis(3,4,6-trichlorophenol)
Hexachlororpropene (1-propene, 1,1,2,3,3,3-hexachloro-)
Hexaethyl tetraphosphate (tetraphosphoric acid, hexaethyl ester)
Hydrazine (diamine)
Hydrocyanic acid (hydrogen cyanide)
Hydrofluoric acid (hydrogen fluoride)
Hydrogen sulfide (sulfur hydride)
Hydroxydimethylarsine oxide (cacodylic acid)
Indeno (1,2,3-*c,d*)pyrene (1,10-(1,2-phenylene)pyrene)
Iodomethane (methyl iodide)
Iron dextran (ferric dextran)
Isocyanic acid, methyl ester (methyl isocyanate)
Isobutyl alcohol (1-propanol, 2-methyl-)
Isosafrole (benzene, 1,2-methylenedioxy-4-allyl-)
Kepone (decachloroctahydro-1,3,4-methano-2*H*-cyclobuta[*c,d*]pentalen-2-one)
Lasiocarpine (2-butenoic acid, 2-methyl-, 7-[2,3-dihydroxy-2-2(1-methoxyethyl)-3-methyl-1-oxobutoxy)methyl] [-2,3,5,7a-tetrahydro-1*H*-pyrrolizin-1-yl-ester)
Lead and compounds, N.O.S.
Lead acetate (acetic acid, lead salt)
Lead phosphate (phosphoric acid, lead salt)
Lead subacetate (lead, bis(acetato-*O*)tetrahydroxytri-)
Maleic anhydride (2,5-furandione)

Maleic hydrazide (1,2-dihydro-3,6-pyridazinedione)

Malononitrile (propanedinitrile)

Melphalan (alanine, 3-[p-bis(2-chloroethyl)amino]phenyl-, L-)

Mercury fulminate (fulminic acid, mercury salt)

Mercury and compounds, N.O.S.

Methacrylonitrile (2-propenenitrile, 2-methyl-)

Methanethiol (thiomethanol)

Methapyrilene (pyridine, 2-[dimethylamino)ethyl]-2-thenylamino-)

Methomyl (acetimidic acid, N-[(methylcarbamoyl)oxy]thio-, methyl ester

Methoxychlor (ethane, 1,1,1-trichloro-2,2F-bis)p-methoxyphenyl)-)

2-Methylaziridine (1,2-propylenimine)

3-Methylcholanthrene (benz[j]aceanthrylene, 1,2-dihydro-3-methyl-)

Methyl cholocarbonate (carbonochloridic acid, methyl ester)

4,4F-Methylenebis(2-chloroaniline) (benzenamine, 4,4F-methylenebis-(2-chloro-)

Methyl ethyl ketone (MEK) (2-butanone)

Methyl hydrazine (hydrazine, methyl-)

2-Methyllactonitrile (propanenitrile, 2-hydroxy-2-methyl-)

Methyl methacrylate (2-propenoic acid, 2-methyl-, methyl ester)

Methyl methanesulfonate (methanesulfonic acid, methyl ester)

2-Methyl-2-(methylthio)propionaldehydro-(methylcarbonyl) oxime (propanal, 2-o-methyl-2-(methylthio)-,

O-[(Methylamino)carbonyl]oxime

N-Methyl-NF-nitro-N-nitrosoguanidine (guanidine, N-nitroso-N-methyl-NF-nitro-)

Methyl parathion (O,O-dimethyl O-(4-nitrophenyl)phosphorothioate)

Methylthiouracil (4-IH-pyrimidinone, 2,3-dihydro-6-methyl-2-thioxo-)

Mustard gas (sulfide, bis(2-chloroethyl)-)

Naphthalene

1,4-Naphthoquinone (1,4-naphthalene-dione)

1-Naphthylamine (α-naphthylamine)

2-Naphthylamine (β-naphthylamine)

1-Naphthyl-2-thiourea (thiourea, 1-naphthalenyl-)

Nickel and compounds, N.O.S.

Nickel carbonyl(nickel) tetracarbonyl

Nickel cyanide (nickel(II) cyanide)

Nicotine and salts (pyridine, (S)-3-(1-methyl-2-pyrrolidinyl)-, and salts

Nitric oxide (nitrogen(II) oxide)

p-Nitroaniline (benzenamine, 4-nitro-)

Nitrobenzene (benzene, nitro-)

Nitrogen dioxide (nitrogen(IV) oxide)

Nitrogen mustard and hydrochloride salt (ethanamine, 2-chloro-,N-(2-chloroethyl)-N-methyl-, and hydrochloride salt)

Nitrogen mustard N-oxide and hydrochloride salt (ethanamine, 2-chloro, N-(2-chloroethyl)-N-methyl-, and hydrochloride salt

Nitroglycerine (1,2,3-propanetriol, trinitrate)
4-Nitrophenol (phenol, 4-nitro-)
2-Nitropropane (propane, 2-nitro-)
4-Nitroquinoline-1-oxide (quinoline, 4-nitro-1-oxide-)
Nitrosamine, N.O.S.
N-Nitrosodi-n-butylamine (1-butanamine, N-butyl-N-nitroso-)
N-Nitrosodiethanolamine (ethanol, 2,2F-(nitrosoimino)bis-)
N-Nitrosodiethylamine (ethanamine, N-ethyl-N-nitroso-)
N-Nitrolsodimethylamine (dimethylnitrosamine)
N-Nitroso-N-methylurea (carbamide, N-ethyl-N-nitroso-)
N-Nitrosodiphenylamine (ethanamine, N-methyl-N-nitroso-)
N-Nitroso-N-methylurea (carbaminde, N-methyl-N-nitroso-)
N-Nitroso-N-methylurethane (carbamic acid, methylnitroso-, ethyl ester)
N-Nitrosomethylvinylamine (ethenamine, N-methyl-N-nitroso-)
N-Nitrosomorpholine (morpholine, N-nitroso-)
N-Nitrosonornicotine (nornicotine, N-nitroso-)
N-Nitrosopiperidine (pyridine, hexahydro-, N-nitroso-)
N-Nitrosopyrrolidine (pyrrole, tetrahydro-, N-nitroso-)
N-Nitrososarcosine (Sarcosine, N-nitroso-)
5-Nitro-o-toluidine (benzenamine, 2-methyl-5-nitro-)
Octamethylpyrophosphoramide (diphosphoramide, octamethyl-)
Osmium tetroxide (osium(VIII)oxide)
7-Oxabicyclo[2.2.1]heptane-2,3-dicarboxylic acid (endothal)
Paraldehyde (1,3,5-trioxane, 2,4,6-trimethyl-)
Parathion (Phosphorothioic acid, O,O-diethyl O-(p-nitrophenyl)ester)
Pentachlorobenzene (benzene, pentachloro-)
Pentachloroethane (ethane, pentachloro-)
Pentachloronitrobenzene (PCNB) (benzene, pentachloronitro-)
Pentachlorophenol (phenol, pentachloro-)
Phenacetin (acetamide, N-(4-ethoxy-phenyl)-)
Phenol (benzene, hydroxy-)
Phenylenediamine (benzenediamine)
Phenylmercury acetate (mercury, acetatophenyl-)
N-Phenylthiourea (thiourea, phenyl-)
Phosgene (carbonyl chloride)
Phosphine (hydrogen phosphide)
Phosphorothioic acid, O,O-diethyl S-[(ethylthio)methyl]ester(phorate)
Phosphorotioic acid, O,O-dimethyl O-[p-((dimethylamino)sulfonyl) phenyl] ester (famphur)
Phthalic acid esters, N.O.S. (benzene, 1,2-dicarboxylic acid, esters, N.O.S.)
Phthalic anhydride (1,2-benzenedicarboxylic acid anhydride)
2-Picoline (pyridine, 2-methyl-)
Polychlorinated biphenyl, N.O.S.
Potassium cyanide
Potassium silver cyanide (argentate(1-), dicyano-, potassium)

Pronamide (3,5-dichloro-*N*-(1,1-dimethyl-2-propynyl) benzamide

1,3-Propane sultone (1,2-oxathiolane, 2,2-dioxide)*n*-propylamine (1-propanamine)

Propylthiouracil (undecamethylenediamine, *N*, *NF*-bis(2-chlorobenzyl)-, dihydrochloride 2-propyn-1-ol (propargyl alcohol)

Pryidine reserpine (yohimban-16-carboxylic acid, 11,17-dimethoxy-18-[3,4,5-trimethoxybenzoyl)oxy], methyl ester)

Resorcinol (1,3-Benzenediol)

Saccharin and salts (1,2-benzoisothiazolin-3-one, 1,1-dioxide, and salts)

Safrole (benzene, 1,2-methylenedioxy-4-allyl-)

Selenious acid (selenium dioxide)

Selenium and compounds, N.O.S.

Selenium sulfide (sulfur selenide)

Selenourea (carbamimidoselenoic acid)

Silver and compounds, N.O.S.

Silver cyanide

Sodium cyanide

Streptozotocin (D-glucopyranose, 2-dioxy-2-(3-methyl-3-nitrosoureido)-)

Strontium sulfide

Strychnine and salts (strychnidin-10-one, and salts)

1,2,4,5-Tetrachlorobenzene (benzene, 1,2,4,5-tetrachloro-)

2,3,7,8-Tetrachlorodibenzo-*p*-dioxin (TCDD) (dibenzo-*p*-dioxin, 2,3,7,8-tetrachloro-)

Tetrachloroethane, N.O.S. (ethane, tetrachloro-, N.O.S.)

1,1,1,2-Tetrachloroethane (ethane, 1,1,1,2-tetrachloro-)

1,1,2,2-Tetrachloroethane (ethane, 1,1,2,2-tetrachloro-)

Tetrachloroethene (ethene, 1,1,2,2,-tetrachloro-)

Tetrachloromethane (carbon tetrachloride)

2,3,4,6-Tetrachlorophenol (phenol, 2,3,4,6-tetrachloro-)

Tetraethylidithiopyrosphosphate (dithiopyrophosphoric acid, tetraethyl-ester)

Tetraethyl lead (plumbane, tetraethyl-)

Tetraethylpyrophosphate (pyrophosphoric acid, tetraethyl ester)

Tetranitromethane (methane, tetranitro-)

Thallium and compounds, N.O.S.

Thallic oxide (thallium(III)oxide)

Thallium(I) acetate (acetic acid, thallium(I)salt)

Thallium(I) carbonate (carbonic acid, dithallium(I)salt)

Thallium(I) chloride

Thallium(I) nitrate (nitric acid, thallium(I)salt)

Thallium selenite

Thallium(I) sulfate (sulfuric acid, thallium(I)salt)

Thioacetamide (ethanethiomide)

Thiosemicarbazide (hydrazinecarbothioamide)

Thiourea (carbamide thio-)

Thiuram (bis(dimethylthiocarbamoyl) disulfide)

Toluene (benzene, methyl-)

2,4-Toluenediamine

2,6-Toluenediamine

3,4-Toluenediamine

Toluenediamine, N.O.S. (diaminotoluene, N.O.S.) *o*-toluidine hydrochloride (benzenamine, 2-methyl-, hydrochloride)

Tolylene diisocyanate (benzene, 1,3-diisocyanatomethyl-)

Toxaphene (camphene, octachloro-)

Tribromomethane (bromoform)

1,2,4-Trichlorobenzene (benzene, 1,2,4-trichloro-)

1,1,1-Trichloroethane (methyl chloroform)

1,1,2-Trichloroethane (ethane, 1,1,2-trichloro-)

Trichloroethene (trichloroethylene)

Trichloromethanethiol (methanethiol, trichloro-)

Trichloromonofluoromethane (methane, trichlorofluoro-)

2,4,5-Trichlorophenol (phenol, 2,4,5-trichloro-)

2,4,6-Trichlorophenol (phenol, 2,4,6-trichloro-)

2,4,5-Trichlorophenoxyacetic acid (2,4,5-T)(acetic acid, 2,4,5-Trichlorophe-noxy-)

2,4,5-Trichlorophenoxypropionic acid (2,4,5-TP) (silvex) (propionoic acid, 2-(2,4,5-trichlorophenoxy)-)

Trichloropropane, N.O.S. (propane, trichloro-, N.O.S.)

1,2,3-Trichloropropane (propane, 1,2,3-trichloro-)

O,O,O-Triethyl phosphorothioate (phosphorothioic acid *O,O,O*-triethyl ester)

sym-Trinitrobenzene (benzene, 1,3,5-trinitro-)

Tris(1-azridinyl) phosphine sulfide (phosphine sulfide, tris(1-aziridnyl)-)

Tris(2,3-dibromopropyl) phosphate (1-propanol, 2,3-dibromo-,phosphate)

Trypan blue (2,7-naphthalenedisulfonic acid, 3,3F-dimethyl(1,1F-biphenyl)-4,4F-diyl)bis(azo)]bis(5-amino-4-hydroxy-,tetrasodium salt)

Uracil mustard

Urcail 5-bis(2-chloroethyl)amino]-

Vanadic acid, ammonium salt (ammonium vandadate)

Vanadium pentoxide (vanadaium(V)oxide)

Vinyl chloride (ethene, chloro-)

Zinc cyanide

Zinc phosphide

APPENDIX II
Conversion of U.S. Customary Units to International System of Units (SI)

SI units are derived from seven base units [length in meters (m); mass in kilograms (kg); time in seconds (s); electic current in amperes (A); thermodynamic temperature in Kelvin (K); amount of substance in moles (mol); luminous intensity in candelas (cd)] and two supplementary units [plane angle in radians (rad); solid angle in steradians (sr)]. Force is expressed as $kg\text{-}m/s^2$ and is termed a newton (N) and pressure is expressed as N/m^2 and is termed a pascal (Pa). A list of commonly used factors and their names used to form unit multiples and submultiples is given in Table IIA. Prefixes are used closed up to the unit symbol. For example, millimeter is written as mm, kilonewton is written as kN, and so on. Table IIB lists conversion factors for commonly used engineering units. The following references may be consulted for a detailed guide on the SI system.

1. "ASTM E-380," American Society for Testing Materials, 1916 Race Street, Philadelphia, PA 19103.
2. T. Goldman and R. J. Bell, eds., "The International System of Units (SI)," NBS Publ. No. 330. National Bureau of Standards, Washington, D.C., 1981.

TABLE IIA. Name and Factors of Commonly Used SI Units for the Formation of Multiples and Submultiples

Name (Symbol)	Factor
mega (M)	10^6
kilo (k)	10^3
hecto (h)	10^2
deka (da)	10
deci (d)	10^{-1}
centi (c)	10^{-2}
milli (m)	10^{-3}
micro (μ)	10^{-6}

TABLE IIB. Conversion Factors

To Convert From	To	Multiply By
inches	centimeters	2.54
feet	centimeters	30.48
yard	meters	0.914
miles	kilometers	1.609
mils	inches	1×10^{-3}
mils	millimeters	0.0254
square inches	square centimeters	6.452
square feet	square meters	0.0929
square feet	hectares	9.29×10^{-6}
square yards	square meters	0.836
square yards	hectares	8.361×10^{-5}
acres	hectares	0.405
acres	square kilometers	4.047×10^{-3}
acre-feet	cubic meters	1.23×10^{3}
cubic inches	cubic millimeters	16.387×10^{3}
cubic feet	cubic meters	28.317×10^{-3}
cubic feet	U.S. gallons	7.48
cubic feet	liters	28.317
cubic feet	acre-feet	22.957×10^{-6}
cubic yards	cubic meters	0.765
U.S. gallon	liters	3.785
pounds mass	kilograms	0.454
short tons (2000 lb)	kilograms	907.185
slugs	kilograms	14.594
ounces	grams	28.3495
pounds force	newtons	4.448
kilogram force	newtons	9.81
dynes	newtons	1.0×10^{-5}
feet per second	meters per second	0.305
feet per second squared	meters per second squared	0.305
standard gravitational acceleration	meters per second squared	9.807
cubic feet per second	liters per second	28.317
U.S. gallons per minute	cubic meters per second	0.631×10^{-4}
U.S. gallons per minute	liters per second	0.063
acre-feet per day	cubic meters per day	1233.79
pounds force per square inch	kilopascal	6.895
pounds per square foot	kilopascal	0.048
kilopascals	newtons per square meter	1.0×10^{-3}
pounds per foot-hour	pascal second	4.134×10^{-4}
pounds per foot-hour	centipoise	0.413
centipoise	pascal second	1.0×10^{-3}
pounds per foot-second	centipoise	1.488×10^{3}
pounds per cubic foot	kilograms per cubic meter	16.018
pounds per U.S. gallon	kilograms per cubic meter	119.826

$$°C = \frac{°F - 32}{1.8} = °K - 273.15$$

Bibliography

Acar, Y. B., and Ghosh, A. (1986). Role of activity in hydraulic conductivity of compacted soils permeated with acetone. *Proc. Int. Symp. Environ. Geotechnol.* **1,** 403–412.

Acar, Y. B., and Seals, R. K. (1984). Clay barrier technology for shallow land waste disposal facilities. *Hazard. Waste* **1**(2), 167–181.

Alexander, M. (1981). Biodegradation of chemicals of environmental concern. *Science* **211,** 132–138.

Alkire, B. D., and Morrison, J. M. (1982). Changes in soil structure due to freeze–thaw and repeated loading. *Transport. Res. Rec.* **918,** 15–22.

Allison, F. E., and Klein, C. J. (1962). Rates of immobilization and release of nitrogen following additions of carbonaceous materials and nitrogen to soils. *Soil Sci.* **93,** 383.

Alther, G. R. (1982). The role of bentonite in soil sealing applications. *Bull. Assoc. Eng. Geol.* **19**(4), 401–409.

Alther, G. R. (1983). The methylene blue test for bentonite liner quality control. *Geotech. Test. J.* **6**(3), 128–132.

American Conference of Governmental Industrial Hygienists (1987). "Threshold Limit Values and Biological Exposure Indices for 1987–1988." ACGIH, Cincinnati, Ohio.

American Public Works Association, (APWA) (1966). "Municipal Refuse Disposal." Public Administration Service, Chicago, Illinois.

American Society of Civil Engineers (ASCE) (1960). "Design and Construction of Sanitary and Storm Sewers," Manual of Engineering Practice, No. 37. ASCE, New York.

American Society of Civil Engineers (ASCE) (1976). "Sanitary Landfill," ASCE Manuals and Reports on Engineering Practice, No. 39. ASCE, New York.

American Society of Civil Engineers/Water Pollution Control Federation (ASCE/ WPCF) (1977). "Waste Water Treatment Plant Design," WPCF Manual of Practice No. 8. WPCF, Washington, D.C.

American Society for Testing and Materials (ASTM) (1986). "Annual Book of ASTM Standards," ASTM, Philadelphia, Pennsylvania.

American Society for Testing and Materials (ASTM) (1990). *In:* "Geotecchnics of Waste Fills—Theory and Practice, STP. 1070" (A. O. Landva and G. D. Knowles, eds.). ASTM, Philadelphia.

Andersen, J. R., and Dornbush, J. N. (1967). Influence of sanitary landfill on groundwater quality. *J. Am. Water Works Assoc.* **59**(4), 457–470.

Andersland, O. B., and Anderson, D. M. (1978). "Geotechnical Engineering for Cold Regions." McGraw-Hill, New York.

Andersland, O. B., and Mathew, P. W. (1973). Consolidation of high ash paper mill sludges. *J. Soil Mech. Found. Div., Am. Soc. Civ. Eng.* **99**, No. SMS-5, 365–374.

Anderson, D. G., Husmand, B., and Martin, G. R. (1992). Seismic response of landfill slopes. *Proc Stability Perform. Slopes Embankments II,* Geotech. Sp. Publ. 31 (Am. Soc. Civ. Eng.) 973–989.

Anderson, D. M., Push, R., and Penner, E. P. (1978). Physical and thermal properties of frozen ground. *In* "Geotechnical Engineering for Cold Regions" (O. B. Andersland and D. M. Anderson, eds.), pp. 37–102. McGraw-Hill, New York.

Apgar, M. A., and Langmuir, D. G. (1971). Groundwater pollution potential of a landfill above the water table. *Groundwater* **9**(6), 76–96.

Axelrod, H. D., and Lodge, J. P., Jr. (1977). Sampling and calibration of gaseous pollutants. *In* "Air Pollution" (A. C. Stern, ed.), 3rd ed., Vol. 3, 145–177. Academic Press, New York.

Babcock, J. D., and Mossien, C. P. (1993). Leachate cleaning test for the Mill Seat landfill. *Proc. 16th Int. Madison Waste Conf.* 384–405.

Bagchi, A. (1980). Discussion of leachate generation from sludge disposal area by Charlie et al. (1979). *J. Environ. Eng. Div. (Am. Soc. Civ. Eng.)* **106**, No. EE-5, 1005.

Bagchi, A. (1983). Design of natural attentuation landfills. *J. Environ. Eng. Div. (Am. Soc. Civ. Eng.)* **109**, No. EE-4, 800–811.

Bagchi, A. (1984). Closure to "Design of natural attentuation landfills" by Bagchi (1983). *J. Environ. Eng. Div. (Am. Soc. Civ. Eng.)* **110**, No. EE-6, 1211–1212.

Bagchi, A. (1986a). Landfill geostructure construction using blended soil. *Proc. Int. Symp. Environ. Geotechnol.* **1**, 43–52.

Bagchi, A. (1986b). Simplified analysis of clay liner blow out. *J. Civ. Eng. Pract. Des. Eng.* **5**(7), 533–551.

Bagchi, A. (1987a). Improving stability of a paper mill sludge. *Proc. Purdue Ind. Waste Conf.* **42**, 137–141.

Bagchi, A. (1987b). "Natural Attenuation Mechanisms of Landfill Leachate and Effects of Various Factors on the Mechanisms," Waste Management and Research, Denmark.

Bagchi, A. (1987c). Discussion on "Hydraulic conductivity of two prototype clay liners" by S. R. Day and D. E. Daniel (1985). *J. Geotech. Eng. Div. (Am. Soc. Civ. Eng.)* **113** (7), 796–799.

Bagchi, A. (1987d). Discussion of Overburden pressures exerted on clay liners, by J. J. Peirce (1986). *J. Environ. Engl. Div. (Am. Soc. Civ. Eng.)* **113** (EE-5), 1180–1182.

Bagchi, A. (1993). Effect of freeze-thaw on hydraulic conductivity of compacted clay liners in Wisconsin. *Proc. 16th Int. Madison Waste Conf.* 583–592.

Bagchi, A., and Carey, D. (1986). More effective methane gas monitoring at landfill. *Public Works* **117**(12), 44–45.

Bagchi, A., and Ganguly, A. (1990). Leachate apportionment in active landfills. *Proc. 13th Madison Waste Conf.* 14–27.

Bagchi, A., and Sopcich, D. (1989). Characterization of MSW incinerator ash. *J. Environ. Eng. Div. (Am. Soc. Civ. Eng.)* **115** (EE-2), 447–452.

Bagchi, A., Dodge, R. L., and Mitchell, G. R. (1980). Application of two attenuation mechanism theories to a sanitary landfill. *Proc. 3rd Annu. Madison Conf. Appl. Res. Pract. Munic. Ind. Waste* 201–213.

Barber, E. S. (1946). Application of triaxial test results to the calculations of flexible pavement thickness. *Proc. 26th Annu. Res. Meet., Highway Res. Board* 26–39.

Barden, L., and Sides, G. R. (1970). Engineering behavior and structure of compacted clay. *J. Soil Mech. Found. Div., Am. Soc. Civ. Eng.* **96** (SM-4), 1171–1200.

Barker, J. F., Tessman, J. S., Poltz, P. E., and Reinhard, M. (1986). The organic geochemistry of a sanitary landfill leachate plume. *J. Contam. Hydrol.* **1**(2), 171–189.

Barlaz, M. A., and Ham, R. K. (1993). Leachate and Gas Generation, *In* "Geotechnical Practice For Waste Disposal" (D. E. Daniel, ed.), Chapman & Hall, London, 113–136.

Barvenic, M. D., Hadge, W. E., and Goldberg, D. T. (1985). Quality control of hydraulic conductivity and bentonite content during soil bentonite cutoff wall construction, *In* "Land Disposal of Hazardous Waste" (N. P. Barkley, proj. officer), 11th Annu. Res. Symp. EPA-600/9-85/013, pp. 66–79. U.S. Environ. Prot. Agency, Cincinnati, Ohio.

Bass, J. (1986). "Avoiding Failure of Leachate Collection and Cap Drainage Systems," EPA-600/2-86/058. U.S. Environ. Prot. Agency, Cincinnati, Ohio.

Bass, J. M. (1985). Avoiding failure of leachate collection systems. *Waste Manage. Res.* **3**, 233–243.

Bass, J. M., Cornish, R. M., Ehrenfield, J. F., Spennenburg, S. P., and Vallentine, J. R. (1983). Potential mechanisms for clogging of leachate drain systems. *In* "Land Disposal of Hazardous Waste" (D. W. Shultz, ed.), 9th Annu. Res. Symp., EPA-600/9-83/018, pp. 148–156. U.S. Environ. Prot. Agency, Cincinnati, Ohio.

Basu, A. N., Mikherjee, D. C., and Mikherjee, S. K. (1964). Interaction between humic acid fraction of soil and trace element cations. *J. Indian Soc. Soil Sci.* **12,** 311–318.

Battista, J., and Connelly, J. P. (1988). VOC contamination at selected municipal and industrial landfills in Wisconsin—Sampling results and policy implications. *Proc. 11th Annu. Madison Waste Conf. Cations.*

Bear, J. (1969). Hydrodynamic dispersion. *In* "Flow Through Porous Media" (R. J. M. De Wiest, ed.), pp. 109–199. Academic Press, New York.

Bear, J. (1972). "Dynamics of Fluids in Porous Media." Am. Elsevier, New York.

Benson, C. H. (1990). Minimum thickness of compacted soil liners. *13th Annu. Madison Waste Conf.* 395–422.

Benson, C. H., and Daniel, D. E. (1990). Influence of clods on hydraulic conductivity of compacted clay. *J. Geotech. Eng.* (Am. Soc. Civ. Eng.) 116(8), 1231–1248.

Benson, N. (1980). "Waste Characterization of Pulp and Papermill Sludges," unpublished report. Bureau of Solid & Hazardous Waste Management, Wisconsin Dept. of Natural Resources, Madison.

Bishop, A. W. (1955). Use of slip circle in slope stability analysis. *Geotechnique* **5**(1), 7–17.

Bishop, A. W., and Morgenstern, N. R. (1960). Stability co-efficients for earth slopes. *Geotechnique* **10**(4), 129–150.

Blackwell, R. J. (1959). "Laboratory Studies of Microscopic Dispersion Phenomena in Porous Media," Jt. Symp. Fundam. Concepts Miscible Fluid Displacement, Part II. Am. Inst. Chem. Eng., San Francisco, California.

Bolt, G. H. (1955). Analysis of the Validity of Gouy-Chapman theory of the electric double layer. *J. Colloid Sci.* **10,** 206.

Bonaparte, R., Giroud, J. P., and Gross, B. A. (1989). Rates of leakage through landfill liners. *Proc. Geosynthetics,* San Diego, California, **89,** 18–29.

Bosscher, P. J. (1987). "P.C. Disk of N.Y. State Slope Stability Program." University of Wisconsin, Geotech. Eng. Div., Madison.

Bouwer, H. (1979). "Groundwater Hydrology," pp. 268–276. McGraw-Hill, New York.

Bowders, J. J., Daniel, D. E., Broderick G. P., and Liljestrand, M. M. (1984). "Methods for Testing the Compatibility of Clay Liners with Landfill and Leachate," 4th Annu. Symp. Hazard. Ind. Solid Waste Test., Am. Soc. Test. Mater., Arlington, Virginia.

Bower, C. A., and Hatcher, J. T. (1967). Adsorption of fluoride by soils and minerals. *Soil Sci.* **103,** 151–54.

Boyle, W. C., and Ham, R. K. (1974). Biological treatability of landfill leachate. *J. Water Pollut. Control Fed.* **46**(5), 860–872.

Bramble, G. M. (1973). Spray irrigation of sanitary landfill leachate. Unpublished Master's Thesis, Department of Civil Engineering, University of Cincinnati, Cincinnati, Ohio.

Broadbent, F. E., and Ott, J. B. (1957). Soil organic matter-metal complexes. 1. Factors affecting retention of various cations. *Soil Sci.* **83,** 419.

Brookman, R. S., Kamp, L. C., and Weintraub, L. (1984). PVC/EVA grafts: A new polymer for geomembranes. *Proc. Int. Conf. Geomembr.* **1,** 25–28.

Brown, K. W., and Anderson, D. C. (1980). Effect of organic chemicals on clay liner permeability: A review of the literature. *In* "Land Disposal of Municipal Solid Waste" (D. W. Schultz, ed.), EPA-600/9-80/010. U.S. Environ. Port. Agency, Cincinnati, Ohio.

Brown, K. W., and Thomas, J. C. (1985). Influence of concentrations of organic chemicals on the colloidal structure and hydraulic conductivity of clay soils. *In* "Land Disposal of Hazardous Waste," (N. P. Barkley, proj. officer), 11th Annu. Res. Symp., EPA-600/9085/013, p. 272, (abstr.). U.S. Environ. Prot. Agency, Cincinnati, Ohio.

Brown, K. W., Green, J. W., and Thomas, J. C. (1983). The influence of selected organic liquids on the permeability of clay liners. *In* "Land Disposal of Hazardous Waste," (D. W. Schultz, ed.), 9th Annu. Res. Symp., EPA-600/9-83/018, pp. 114–125. U.S. Environ. Prot. Agency, Cincinnati, Ohio.

Bruch, J. C., and Street, R. L. (1967). Two dimensional dispersion. *J. Sanit. Eng. Div., Am. Soc. Civ. Eng.* **93,** (SA-6), 17–39.

Bumb, A. C., McKee, C. R., Evans, R. B., and Eccles, L. A. (1988). Design of lysimeter leak detector networks for surface impoundments and landfills. *Ground Water Monit. Rev.* **8**(2), 102–114.

Bureau, B. (1981). "Design of Impervious Clay Liners by Unsaturated Flow Principles," Proc. 4th Symp. Uranium Mill Tailings Manage., pp. 647–664. Colorado State University, Ft. Collins.

Bureau of Reclamation (1978). "Drainage Manual," 1st ed. Bureau of Reclamation, Eng. Res. Cent., Denver Federal Center, Denver, Colorado.

Bureau of Solid Waste Management (1985). "Guidelines for Monitoring Well Installation." Wisconsin Dept. of Natural Resources, Madison.

Burton, P. E. (1987). An evaluation of three vadose zone monitoring techniques. Unpublished M. S. Thesis, University of Texas at Austin.

Byrne, R. J., Kendall, J., and Brown, S. (1992). Cause and mechanism of failure, Kettleman Hills landfill B-19, Unit IA. *Proc. Stability Perform. Slopes and Embankment,* Geotech. Sp. Publ. 31 (Am. Soc. Civ. Eng.) 1188–1215.

Cadena, F., and Jeffers, S. W. (1987). Use of tailored clays for selective adsorption of hazardous pollutants. *Proc. Purdue Ind. Waste Conf.* **42**, 113–119.

Cadwallader, M. W. (1986). Selection and specification criteria for flexible membrane liner of the high density polyethylene variety. *Proc. Int. Symp. Environ. Geotechnol.* **1**, 323–333.

Caffrey, R. P., and Ham, R. K. (1974). The role of evaporation in determining leachate production from milled refuse landfills. *Compost Sci.* **15**(2), 11–15.

California Water Pollution Control Board (CWPCB) (1954). "Report on the Investigation of Leaching for a Sanitary Landfill," Publ. No. 10. Resources Agency of California, Sacramento.

California Water Pollution Control Board (CWPCB) (1961). "Effects of Refuse Dumps on Ground Water Quality," Publ. No. 24. Resources Agency of California, Sacramento.

Callahan, M. A., Slimak, M. W., Gabel, N. W., May, I. P., and Fowler, C. F. (1979). "Water Related Environmental Fate of 129 Priority Pollutants," EPA-440/4-79-029a and b. U.S. Environ. Prot. Agency, Washington, D.C.

Cameron, R. D., and McDonald, E. C. (1982). Toxicity of landfill leachates. *J. Water Pollut. Control Fed.* **52**(4), 760–769.

Campbell, G. S. (1974). A simple method for determining unsaturated hydraulic conductivity from moisture retention data. *Soil Sci.* **117**(6), 311–314.

Carlson, D. A., and Johansen, O. J. (1975). Aerobic treatment of leachates from sanitary landfills. *In* "Waste Management Technology and Resource and Energy Recovery," pp. 359–382. U.S. Environ. Prot. Agency, Washington, D.C.

Carman, P. C. (1956). "Flow of Gases Through Porous Media." Academic Press, New York.

Carpenter, G. W., and Stephenson, R. W. (1986). Permeability testing in the triaxial cell. *Geotech. Test. J.* **9**(1), 3–9.

Carroll, R. G., Jr. (1983). "Geotextile Filter Criteria, Engineering Fabrics in Transportation Construction," Transp. Res. Rec. No. 916. National Academy of Sciences, Washington, D.C.

Carter, J. N., and Allison, F. E. (1961). The effect of rates of application of ammonium sulphate and gaseous losses of nitrogen from soils. *Soil Sci. Soc. Am. Proc.* **25,** 484.

Cedergren, H. R. (1977). "Seepage, Drainage, and Flow Nets," 2nd ed. Wiley, New York.

Chamberlain, E. J., and Bouin, S. E. (1976). Freeze–thaw enhancement of the drainage of consolidation of fine grained dredged material in confined disposal areas. CRREL Final Report, U.S. Army Cold Regions Research and Engineering Laboratory, Hanover, N.H.

Chamberlain, E. J., and Blouin, S. E. (1977). "Frost Action as a Factor in Enhancement of the Drainage and Consolidation of Fine Grained Dredged Material," Tech. Rep. No. D-77-16. Department of the United States Army Engineer Water Ways Experiment Station, Dredged Material Research Program, Hanover, New Hampshire.

Chamberlain, E. J., and Gow, A. J. (1978). "Effect of Freezing and Thawing on the Permeability and Structure of Soils," Int. Symp. Ground Freez. Department of the United States Army, Hanover, New Hampshire.

Chamberlain, E. J., and Gow, A. J. (1979. Effect of freezing and thawing on the permeability and structure of soils. *Eng. Geol.* **13,** 73–92.

Chamberlain, E. J., Iskander, I., and Hansiker, S. E. (1990). Effect of freeze-thaw on the permeability and microstructure of soils. *Proc. Int. Symp. Frozen Soil Impact on Agricultural Range and Forest Lands,* March 21–22, Spokane, WA, 145–155.

Chan, K. Y., Davey, B. G., and Geering, H. R. (1978). Interaction of treated sanitary landfill leachate with soil. *J. Environ. Qual.* **7**(3), 306–310.

Chao, T. T., Harward, M. E., and Fang, S. C. (1962). Soil constituents and properties in the adsorption of sulphate ions. *Soil Sci.* **94,** 276.

Charlie, W. A., Wardwell, R. E. (1979). Leachate generation from sludge disposal area. *J. Environ. Eng. Div. Am. Soc. Civ. Eng.* **105** EE-5), 947–960.

Chazin, J. D., Allen, M., and Pippin, D. D. (1987). "Measurement, Assessment and Control of Hazardous (Toxic) Air Contaminants in Landfill Gas Emmissions in Wisconsin," EPA/APCA Symp. Meas. Toxic Other Relat. Air Pollut. Research Triangle Park, North Carolina.

Chen, H. W., and Yamamoto, L. O. (1987). *In* "Permeability Tests for Hazardous Waste Management Unit Clay Liners," Geotechnical and Geohydrological Aspects of Waste Management (D. J. A. VanZyl et al., eds.), pp. 229–243. Lewis Publishing Inc., Chelsea, MI.

Chen, Y. H., Simons, D. B., and Demery, P. M. (1981). Hydraulic testing of plastic filters. *J. Irrig. Drain. Div., Am. Soc. Civ. Eng.* **107**(IR-3), 307–324.

Cherry, J. A., Barker, J. F., Buszka, P., and Hewetson, J. P. (1981). Contaminant occurrence in an unconfined sand aquifer at a municipal landfill. *Proc. 4th Annu. Madison Conf. Appl. Res. Pract. Munic. Ind. Waste* 393–411.

Chian, E. S., and Dewalle, F. B. (1975). "Compilation of Methodology for Measuring Pollution Parameters of Landfill Leachate," EPA-600:3075-011. U.S. Environ. Prot. Agency, Washington, D.C.

Chian, E. S., and Dewalle, F. B. (1977). "Evaluation of Leachate Treatment," Vols. 1 and 2, EPA-600/2-77/186a and b. U.S. Environ. Prot. Agency. Cincinnati, Ohio.

Chow, V. T. (1959). "Open-Channel Hydraulics." McGraw-Hill, New York.

Chow, V. T., ed. (1964). "Handbook of Applied Hydrology." McGraw-Hill, New York.

Christel, B. J., Rehm, B. W., and Lowery, B. (1985). Field performance of pressure-transducer equipped tensionmeters in fly ash. *Proc. Natl. Water Well Assoc. Conf. Charact. Monit. Vadose (Unsaturated) Zone* 182–197.

Collins, K., and McGown, A. (1974). The form and functions of microfabric features in a variety of natural soils. *Geotechnique* **24**(2), 223–254.

Connelly, J. (1994). Monitoring Well Sampling. (Personal communication). Wisconsin Dept. of Natural Resources, Madison.

Connolly, R. A. (1972). Soil burial tests: Soil burial of materials and structures. *Bell Syst. Tech. J.* **51**(1), 1–21.

Cook, E. N., and Foree, E. G. (1974). Aerobic biostabilization of sanitary landfill leachate. *J. Water Pollut. Control Fed.* **46**(2), 380–392.

Corps of Engineers (1977). "Civil Works Construction Guide Specification CW 02215." U.S. Department of the Army, Washington, D.C.

Daniel, D. E. (1981). "Problems in Predicting the Permeability of Compacted Clay Liners," Proc. 4th Symp. Uranium Mill Tailings Mange., pp. 665–675. Colorado State University, Fort Collins.

Daniel, D. E. (1984). Predicting the hydraulic conductivity of compacted clay liners. *J. Geotech. Eng.* (Am. Soc. Civ. Eng.) **110**(2), 285–300.

Daniel, D. E. (1989). In situ hydraulic conductivity tests for compacted clay. *Geotech. Eng. Div.* (Am. Soc. Civ. Eng.) **115**(9), 1205–1226.

Daniel, D. E., and Benson, C. H. (1990). Water content–density criteria for compacted soil liners. *J. Geotech. Eng.* (Am. Soc. Civ. Eng.) **116**(12), 1811–1830.

Daniel, D. E., and Kurtovich, M. (1987). Monitoring for hazardous waste leaks. *Civ. Eng.* (*N.Y.*) **57**(2), 48–51.

Daniel, D. E., and Shackelford, C. D. (1988). Disposal barriers that release contaminants only by molecular diffusion. *Nucl. Chem. Waste Manage.* **8**, 299–305.

Daniel, E. E., and Trantwein, S. T. (1986). Field permeability test for earthen liners. *Proc. Am. Soc. Civ. Eng. Spec. Conf. Use In-Situ Tests Geotech. Eng.* 146–160.

Daniel, D. E., Hamilton, J. M., and Olson, R. E. (1981). Suitability of thermocouple psychrometers for studying moisture movement in unsaturated soils. *ASTM Spec. Tech. Publ.* **STP 746**, 84–100.

Daniel, D. E., Trantwein, S. J., Boyton, S. S., and Foreman, D. E. (1984). Permeability testing with flexible wall permeameters. *Geotech. Test. J.* **7**(3), 113–122.

D'Appolonia, D. (1980). Soil-bentonite slurry trench cutoffs. *Geotech. Eng. Div. Am. Soc. Civ. Eng.* **106** (GT-4), 399–417.

Dass, P., Tamke, G. R., and Stoffel, C. M. (1977). Leachate production at sanitary landfill sites. *J. Environ. Eng. Div., Am. Soc. Civ. Eng.* **103** (EE-6), 981–988.

Davis, J. L., Sing, R., Stegman, B. G., and Waller, M. J. (1984). "Innovative Concepts for Detecting and Locating Leaks in Waste Impoundment Liner Systems: Acoustic Emission Monitoring and Time Domain Reflectrometry," EPA-600/52-84/058. U.S. Environ. Prot. Agency, Cincinnati, Ohio.

Dawson, G. W., and Mercer, B. W. (1986). "Hazardous Waste Management," pp. 43–105. Wiley, New York.

Day, S. R., and Daniel, D. E. (1985). Hydraulic conductivity of two prototype liners. *J. Geotech. Eng. Div., Am. Soc. Civ. Eng.* **111** (8), 957–970.

Demetracopoulos, A. C., and Korfiatis, G. P. (1984). Design considerations for landfill bottom collection systems. *J. Civ. Eng. Pract. Des. Eng.* **3**(10), 967–984.

Demetracopoulos, A. C., Korfiatis, G. P., Bourodimos, E. L., and Nawy, E. G. (1984). Modeling for design of landfill bottom liners. *J. Environ. Eng. Div., Am. Soc. Civ. Eng.* **110**(6), 1084–1098.

Demetracopoulos, A. C., Sehayek, L., and Erdogan, H. (1986). Modeling leachate production from municipal landfills. *J. Environ. Eng. Div. Am. Soc. Civ. Eng.* **112** (EE-5), 849–866.

Department of the Navy (1971). "Design Manual, Soil Mechanics, Foundations, and Earth Structures," NAVFAC DM-7, pp. 7–9–8 to 7–9–9. U.S. Dept. of the Navy, Washington, D.C.

Dirksen, C., and Miller, R. D. (1966). Closed system freezing of unsaturated soil. *J. Am. Soil. Sci. Soc.* **50,** 1114–1112.

Druschel, S. J., and Underwood, E. R. (1993). Design of lining and cover system sideslopes. *Proc. Geosynthetics 93 Conf.* Vancouver, B.C., Canada, 1341–1356.

Dunn, R. J. (1986). Clay liners and barriers—considerations of compacted clay structure. *Proc. Int. Symp. Environ. Geotechnol.* **1,** 293–302.

Duranceau, P. E. (1987). US EPA's new leaching test: The toxicity characteristic leaching procedure (TCLP). *Proc. 10th Annu. Madison Waste Conf.* 547–560.

Dvinoff, A. H., and Munion, D. W. (1986). Stability failure of a sanitary landfill. *Proc. Int. Symp. Env. Geotechnol.* **1,** 25–36.

Edil, T. B., Park, J. K., and Berthouex, P. M. (1992). Attenuation and transport of volatile organic compounds in clay liners. *Proc. Mediterranean Conf. Env. Geotech.,* Izmir, Turkey.

Ehrig, H-J. (1983). Quality and quantity of sanitary landfill leachate. *Waste Manage. Res.* **1**(1), 53–68.

Ehrig, H. J. (1984). Treatment of sanitary landfill leachate biological treatment. *Waste Manage. Res.* **2**(2), 131–152.

Ellis, B. G., and Knezek, B. D. (1972). "Adsorption Reactions of Micro-Nutrients in Soils," Micro-Nutr. Agric., p. 59. Soil Sci. Soc. Am., Madison, Wisconsin.

EMCON Associates (1975). "Sonoma County Solid Waste Stabilization Study," EPA-530/SW-65dl. U.S. Environ. Prot. Agency, Cincinnati, Ohio.

EMCON Associates (1980). "Methane Generation and Recovery from Landfills," Ann Arbor Science, Ann Arbor, Michigan.

Everett, L. G. (1981). Monitoring in the vadose zone. *Groundwater Monit. Rev.* **1**(2), 44–51.

Everett, L. G., Wilson, L. G., and McMillion, L. G. (1982). Vadose zone monitoring concepts for hazardous waste sites. *Groundwater* **20**(3), 312–324.

Fair, G. M., and Geyer, J. C. (1954). "Water Supply and Waste Water Disposal." Wiley, New York.

Fang, H. Y., and Evans, J. C. (1988). Long-term permeability tests using leachate on a compacted clayey liner material. *ASTM Spec. Tech. Publ.* **STP 936,** 397–404.

Farquhar, G. K. (1977). Leachate treatment by soils methods. *In* "Management of Gas and Leachate in Landfills" (S. K. Banerjee, ed.), EPA-600/9-77-026, pp. 187–207. U.S. Environ. Prot. Agency, Cincinnati, Ohio.

Farquhar, G. J., and Rovers, F. A. (1976). Leachate attentuation in undisturbed and remoulded soils. In "Gas and Leachate from Landfills" (E. J. Genetelli and J. Cirello, eds.), EPA-600/9-76-004, p. 55. U.S. Environ. Prot. Agency, Cincinnati, Ohio.

Fattah, D. N. (1974). Investigation and verification of a model for the dispersion co-efficient tensor in flow thru anisotropic homogeneous porous media with application to flow from a recharge well through a confined aquifer. Unpublished Ph.D. Thesis, University of Wisconsin, Madison.

Fenn, D. G., Hanley, K. J., and Degeare, T. V. (1975). "Use of the Water Balance Method for Predicting Leachate Generation from Solid Waste Disposal Sites," EPA-530/SW-168. U.S. Environ. Prot. Agency, Cincinnati, Ohio.

Fernandez, F., and Quigley, M. R. (1985). Hydraulic conductivity of natural clays permeated with simple liquid hydrocarbons. *Can. Geotech. J.* **22**(2), 205–214.

Fisher, E. A. (1927). Some factors affecting the evaporation of water from soil. II. *J. Agric. Sci.* **17**, 407–419.

Fisher, S. R., and Potter, K. W. (1989). "Methods for Determining Compliance with Groundwater Quality Regulations at Waste Disposal Facilities." Wisc. Dept. of Natl. Res, BSHWM, Madison, Wisconsin.

Follett, R. H., and Lindsay, W. L. (1971). Changes in DPTA-extractable zinc, iron, manganese and copper in soils following fertilization. *Soil Sci. Soc. Am. Proc.* **35**(4), 600–602.

Fong, M. A., and Haxo, H. E., Jr. (1981). Assessment of liner materials for municipal solid waste landfills. *In* "Land Disposal of Municipal Solid Waste" (D. W. Shultz, ed.), 7th Annu. Res. Symp., EPA/600/9-81/002a, pp. 148–162. U.S. Environ. Prot. Agency, Cincinnati, Ohio.

Ford, H. W. (1974). Low pressure jet cleaning of plastic drains in sandy soil. *Trans. ASAE* **17**(5), 895–897.

Foreman, D. E., and Daniel, D. E. (1984). Effects of hydraulic gradient and method of testing on the hydraulic conductivity of compacted clay to water, methanol, and heptane. *In* "Land Disposal of Hazardous Waste" (N. P. Barkley, proj. officer), 10th Annu. Res. Symp., EPA-600/9-84-007, 138–144. U.S. Environ. Prot. Agency, Cincinnati, Ohio.

Forseth, J. M., and Kmet, P. (1983). Flexible membrane liners for solid and hazardous waste landfills—a state of the art review. *Proc. 6th Annu. Madison Waste Conf.* 138–166.

Foster, W. S., and Sullivan, R. H. (1977). "Sewer Infiltration and Inflow Control Product and Equipment Guide," EPA-600/12-77/107C. U.S. Environ. Prot. Agency, Cincinnati, Ohio.

Foth, H. D., and Turk, L. M. (1943). "Fundamentals of Soil Science," 5th ed., pp. 84–85. Wiley, New York.

Freeze, R. A., and Cherry, J. A. (1979). "Groundwater," pp. 384–438. Prentice-Hall, Englewood Cliffs, New Jersey.

Friedman, M. A. (1988). "Volatile Organic Compounds in Ground Water and

Leachate at Wisconsin Landfills," PUBL-WR-192-88. Wisconsin Dept. of Natural Resources, Madison.

Fuller, W. H. (1977). "Movement of Selected Metals, Asbestos and Cyanide in Soil, Applications to Waste Disposal Problem," EPA 600/2-77/020. U.S. Environ. Prot. Agency, Cincinnati, Ohio.

Fuller, W. H., and Korte, N. (1976). Attenuation mechanisms of pollutants through soils. *In* "Gas and Leachate from Landfills" (E. J. Genetelli and J. Cirello, eds.), EPA 600/9-76-004, pp. 111–122. U.S. Environ. Prot. Agency, Cincinnati, Ohio.

Fungaroli, A. A., and Steiner, R. L. (1979). "Investigation of Sanitary Landfill Behavior," Vol. 1, Final rep. EPA-600/2-79/053a, p. 331. U.S. Environ. Prot. Agency, Cincinnati, Ohio.

Garland, G. A., and Mosher, D. C. (1975). Leachate effects of improper land disposal. *Waste Age* **6**(3), 42–48.

Gear, B. (1988). "Leachate Head Well Design" (personal communication). Wisconsin Dept. of Natural Resources, Madison.

Gebhard, A. (1978). The potential of leachate attenuation in soils. *In* "Disposal of Residuals by Landfilling" (A. Gebhard, proj. manager), pp. vi-1–vi-110. Minnesota Pollut. Control Agency, Minneapolis.

Gee, G. W., and Dodson, M. E. (1981). Soil water content by microwave drying: A routine procedure. *Soil Sci. Soc. Am.* **45**(6), 1234–1237.

Gee, J. R. (1986). Predicting percolation at solid waste disposal sites—A direct method. *Proc. 9th Annu. Madison Conf. Appl. Res. Pract. Munic. Ind. Waste* 623–645.

Gemmell, R. S. (1971). Mixing and sedimentation. *In* "Water Quality and Treatment" (P. D. Haney, Chair.), 3rd ed., pp. 123–157. McGraw-Hill, New York.

Gerhardt, R. A. (1977). Leachate attentuation in the unsaturated zone beneath three sanitary landfills in Wisconsin. *Inf. Circ.—Wis. Geol. Nat. Hist. Surv.* **35**, 93.

Ghassemi, M., Haro, M., Metzgar, J., et al. (1983). 'Assessment of Technology for Constructing Cover and Bottom Liner Systems for Hazardous Waste Facilities," EPA/68-02/3174. U.S. Environ. Prot. Agency, Cincinnati, Ohio.

Giroud, J. P. (1982). Filter criteria for geotextiles. *Proc. Int. Conf. Geotext. 2nd, 1982* **1**, 103–108.

Giroud, J. P. (1984). Impermeability: The myth and a rational approach. *Proc. Int. Conf. Geomembr.* **1**, 157–162.

Girvin, D. C., and Sklarew, D. S. (1986). "Attenuation of Polychlorinated Biphenyls in Soil: Literature Review." CS 4396 EPRI, 3412 Hillview Ave., Palo Atlo, CA 94304.

Glaub, J. C., Savage, G. M., Tuck, J. K., and Henderson, T. M. (1983). Waste characterization for North Santa Clara County, California. *Proc. 6th Annu. Madison Conf. Appl. Res. Pract. Munic. Ind. Waste* 74–96.

Godish, T. (1987). "Air Quality," Lewis Publishers, Chelsea, Michigan.

Goldman, L. J., Truesdale, R. W., Kingsbury, G. L., Northeim, C. M., and Damle, A. S. (1986). "Design Construction and Evaluation of Clay Liners for Waste Management Facilities," EPA/530-SW-86/007. U.S. Environ. Prot. Agency, Cincinnati, Ohio.

Gordon, M. E., Huebbner, P. M., and Miazga, T. J. (1989). Hydraulic conductivity of three landfill clay liners. *J. Geotech. Eng. Div., Am. soc. Civ. Eng.* **115**(8), 1148–1160.

Gray, D. H., and Leiser, A. T. (1982). "Biotechnical Slope Protection and Erosion Control," p. 271. Van Nostrand-Reinhold, New York.

Green, W. J., Lee, G. F., and Jones, R. A. (1981). Clay-soils permeability and hazardous waste storage. *J. Water Pollut. Control Fed.* **54**(8), 1347–1354.

Greenland, D. J. (1970). Sorption of organic compounds by clays and soils. *SCI Monogr.* **37**, 79.

Griffin, R. A. (1977). "Geochemical Considerations Bearing a Disposal of Industrial Chemicals at Willsanuil, Macoupin City, Illinois," Doc. SS-2-2, Courthouse No. 77-CH-10 and 77-CH-13. Illinois Environ. Prot. Agency, Macoupin City.

Griffin, R. A., and Chian, E. S. K. (1980). "Attenuation of Water-Soluble Polychlorinated Biphenyls by Earth Materials," final Report, EPA600/2-80-027. U.S. Environ. Prot. Agency, Washington, D.C.

Griffin, R. A., and Shimp, N. F. (1976). Leachate migration through selected clays. *In* "Gas and Leachate from Landfills" (E. J. Genetelli and J. Cirello, eds.), EPA-600/9-76/004. U.S. Environ. Prot. Agency, Cincinnati, Ohio.

Griffin, R. A., Cartwright, K., Shimp, N. F., Steele, J. D., Ruch, R. R., White, W. A., Hughes, G. M., and Gilkenbon, R. H. (1976a). "Attenuation of Pollutants in Municipal Landfill Leachate by Clay Minerals." Illinois State Geol. Surv., Urbana.

Griffin, R. A., Frost, R. R., and Shimp, N. F. (1976b). Effect of pH on removal of heavy metal from leachate by clay minerals. *In* "Residual Management by Land Disposal" (W. H. Fuller, Ed.), EPA-600/9-76/015, pp. 259–268. U.S. Environ. Prot. Agency, Cincinnati, Ohio.

Grim, R. E. (1968). "Clay Minerology," 2nd ed. McGraw-Hill, New York.

Grube, W. E., Jr., Roulier, M. H., and Hermann, J. G. (1987). Implications of current soil liner permeability research results. *In* "Land Disposal of Hazardous Waste," 13th Annu. Res. Symp. EPA-600/9-87/015. U.S. Environ. Prot. Agency, Cincinnati, Ohio.

Hagerty, D. J., Ulrich, C. R., and Thacker, B. K. (1977). Enginering properties of FGD sludges. *Proc. Conf. Geotech. Pract. Disposal Solid Waste Mater., 1977,* Am. Soc. Civ. Eng., 23–40.

Ham, R. K. (1980). "Decomposition of Residential and Light Commercial Solid Waste in Test Lysimeters," EPA/SW-190c. U.S. Environ. Prot. Agency, Cincinnati, Ohio.

Ham, R. K., and Anderson, C. R. (1974). Pollutant production by refuse degradation in test lysimeter. *Waste Age* **5**(9), 33–39.

Ham, R. K., and Anderson, C. R. (1975a). Pollutant production by refuse degradation in test lysimeter. *Waste Age* **6**(1), 30–36.

Ham, R. K., and Anderson, C. R. (1975b). Pollutant production by refuse degradation in test lysimeter. *Waste Age* **6**(2), 38–48.

Ham, R. K., Anderson, M. A., Stegmann, R., and Stanforth, R. (1979). "Background Study on the Development of a Standard Leaching Test," EPA-600/2-79/109. U.S. Environ. Prot. Agency, Cincinnati, Ohio.

Hamilton, N. F. (1988). Antimicrobial controls effects of bioslime. *Mod. Plast.* **65**(5), 166–170.

Hammer, M. J. (1977). "Water and Waste-Water Technology." Wiley, New York.

Hannapel, R. J., Fuller, W. H., and Fox, R. H. (1964). Phosphorus movement in a calcareous soil: Soil microbial activity and organic phosphorus movement. *Soil Sci.* **97**(6), 421–427.

Hansen, B. P. (1980). Reconnaissance of the effect of landfill leachate on the water quality of Marshall Brook, Southwest Harbor, Hancock County, Maine. *Geol. Surv. Open-File Rep. (U.S.)* **80-1120.**

Harleman, D. R. F., Mehlhorn, P. F., and Rummer, R. R., Jr. (1963). Dispersion permeability co-relation in porous media. *J. Hydraul. Div., Am. Soc. Civ. Eng.* **89,** No. HY-2, 67–85.

Harper, S. R., and Pohland, F. G. (1988). Landfills: Lessening environmental impacts. *Civil Eng.* **58,** 66–69.

Haxo, H. E., Jr. (1982). Effects on liner materials of long term exposure in waste environments. *In* "Land Disposal of Municipal Solid Waste" (D. W. Schultz, ed.). 8th Annu. Res. Symp., EPA-600/9-82/002. U.S. Environ. Prot. Agency, Cincinnati, Ohio.

Haxo, H. E., Miedema, J. A., and Nelson, N. A. (1984). Permeability of polymeric membrane lining materials. *Proc. Int. Conf. Geomembranes,* Denver, Colorado **VI,** 151–156.

Haxo, H. E., Jr., Haxo, R. S., Nelson, N. A., Haxo, P. D., White, R. M., and Dakessian, S. (1985). "Liner Materials Exposed to Hazardous and Toxic Wastes," EPA/600/2-84/169. U.S. Environ. Prot. Agency, Cincinnati, Ohio.

Hentges, G. T., Thies, F., and Lemar, T. S. (1993). Leachate extraction well assessments Des Moines, Iowa Metropolitan Park sanitary landfill, Hamilton County Iowa Sanitary Landfill. *Proc. 16th Int. Madison Waste Conf.* 406–441.

Hickey, M. E. (1969). "Investigation of Plastic Films for Canal Linings," Res. Rep. No. 19. U.S. Dept. of Interior, Bureau of Reclamation, Washington, D.C.

Hilf, J. W. (1975). Compacted fill. *In* "Foundation Engineering Handbook" (H. F. Winterkorn and H. Y. Fang, eds.), pp. 244–341. Van Nostrand-Reinhold, New York.

Hoekstra, P. (1966). Moisture movement in soils under temperature gradients with the cold-side temperature below freezing. *Water. Res. Res.* **2**(2), 241–250.

Hoffman, M. C., and Oettinger, T. P. (1987). Landfill leachate treatment with the PACT system. Tech. Report No. HT 602, Zimpro Inc., Rothschild, WI 54474.

Horz, R. C. (1984). "Geotextiles for Drainage and Erosion Control at Hazardous Waste Landfills," EPA Interagency Agreement No. AD-96-F-1-400-1. U.S. Environ. Prot. Agency, Cincinnati, Ohio.

Houle, M. (1976). Industrial hazard waste migration potential. *In* "Residual Management by Land Disposal" (W. H. Fuller, ed.), EPA-600/9-76/015, pp. 76–85. U.S. Environ. Prot. Agency, Cincinnati, Ohio.

Howland, J. D., and Landva, A. O. (1992). Stability analysis of a municipal solid waste landfills. *Proc. Stability and Perform. Slopes and Embankments II,* Geotech. Sp. Publ. 31 (Am. Soc. Civ. Eng.) 1216–1231.

Huang, C. P., Elliot, H. A. and Ashmead, R. M. (1977). Interfacial reactions and the fate of heavy metals in soil-water systems. *J. Water Pollu. Control Fed.* **49**(5), 745.

Hwang, S. (1987). Four types of probes for monitoring the vadose zone. Unpublished M.S. Thesis, University of Texas at Austin.

Inyang, H. I. (1994). A weibull based reliability analysis of waste containment systems. *Proc. 1st. Int. Congress Env. Geotech.* Edmonton, Alberta, Canada.

Inyang, H. I., and Myers, V. B. (1993). "Geotechnical Systems for Structures on Contaminated Sites," EPA 530-R-93-002, PB93-209.419. U.S. Env. Prot. Agency, Washington, D.C.

Iwata, V., Westlake, W. E., and Gunther, F. A. (1973). Varying persistence of polychlorinated biphenyis in six California soils under laboratory conditions. *Bull. Environ. Contain. Toxicol.* **9**, 204–211.

James Clem Corporation (1992). Sales documents for geosynthetic clay liner (GCL), 444 North Michigan, suite 1610, Chicago, IL 60611.

Jenne, E. A. (1968). "Controls on Mn, Fe, Co, Ni, Cu and Zn Concentrations in Soils and Water: The Significant Role of Hydrous Mn and Fe Oxides," Adv. Chem. Ser. No. 73. Am. Chem. Soc., Washington, D.C.

Johnson, A. I. (1954). Symposium on permeability. *ASTM, Spec. Tech. Publ.* **STP 163,** 98–114.

Johnson, A. W., and Sollberg, J. R. (1960). "Factors That Influence Field Compaction of Soils," Bull. No. 272. Highway Res. Board, Washington, D.C.

Johnson, R. A., and Bhattacharya, G. K. (1992). "Statistics Principles and Methods," 2nd ed., p. 686. Wiley, New York.

Johnson, T. M., and Cartwright, K. (1980). Monitoring of leachate migration in the unsaturated zone in the vicinity of sanitary landfills. *Circ.—Ill. State Geol. Surv.* **514,** 82.

Jumikis, A. R. (1962). "Soil Mechanics." Van Nostrand, Princeton, New Jersey.

Kalfka, S. (1986). "Control Measures for Vinyl Chloride Contained in Landfill Gas," unpublished Air Management Section Memo dated May 30. Wisconsin Dept. of Natural Resources, Madison.

Kansas State Highway Commission (1947). "Design of Flexible Pavement Using the Triaxial Compression Test," Bull. No. 8. Highway Research Board, Washington, D.C.

Katsman, K. H. (1984). Hazardous waste landfill geomembrane: Design, installation and monitoring. *Proc. Int. Conf. Geomembr.* **1,** 215–220.

Kayhanian, M., Lindenauer, K., Hardy, S., and Tchobanoglous, G. (1991). "The Recovery of Energy and Production of Compost from the Biodegradable Organic Fraction of MSW Using the High-Solids Anaerobic Digestion/Aerobic Biodrying Process." Dept of Civil and Environmental Eng., Univ. of California, Davis, CA 95616, 2-1

Keen, B. H., Crowther, E. M., and Coutts, J. R. H. (1926). The evaporation of water from soil. III. A critical study of the techniques. *J. Agric. Sci.* **16**, 105–122.

Kester, R. A., and Van Slyke, S. M. (1987). Air toxic emissions from landfill gas flares. Presented at the Wisconsin Chapter APCA Meeting, April, Madison.

Keswich, B. H., and Gerba, C. P. (1980). Viruses in ground water. *Environ. Sci. Technol.* **14**, 1290–1297.

Kim, W. H., and Daniel, D. E. (1992). "Effects of Freezing on Hydraulic Conductivity of Compacted Clay." *J. Geotech. Eng. (Am. Soc. Civ. Eng.)* **118**(7).

Kimmel, G. E., and Braids, O. C. (1974). Leachate plume in a highly permeable aquifer. *Groundwater* **12**(6), 388–393.

Klausmeir, R., and Andrews, C. (1981). Microbial biodeterioration. *In* "Economic Microbiology" (A. H. Rose, ed.), Vol. 6, pp. 432–472. Academic Press, New York.

Klute, A., and Peters, D. B. (1962). A recording tensiometer with a short response time. *Soil Sci. Soc. Am. Proc.* **26**(1), 87–88.

Kmet, P., and Lindorff, D. E. (1983). Use of collection lysimeters in monitoring sanitary landfill performance. *Proc. Natl. Water Well Assoc. Conf. Charact. Monit. Vadose (Unsaturated) Zone.*

Kmet, P., Quinn, K. J., and Slavic, D. (1981). Analysis of design parameters effecting the collection efficiency of clay lined landfills. *Proc. 4th Annu. Madison Conf. Appl. Res. Pract. Munic. Ind. Waste* 250–265.

Kmet, P., Mitchell, G., and Gordon, M. (1988). Leachate collection system design and performance—Wisconsin's experience. Presented at the ASTSWMO National Solid Waste Forum on Intregrated Municipal Waste Management, July, Lake Buena Vista, Florida.

Koerner, R. M. (1986). "Designing with Geosynthetics." Prentice-Hall, Englewood Cliffs, New Jersey.

Konrad, J. M., and Morgenstern, N. R. (1980). Mechanistic theory of ice lens formations in fine-grained soils. *J. Can. Geotech.* **17**, 473–486.

Korfiatis, G. P., and Demetracopoulos, A. C. (1986). Flow characteristics for landfill leachate collection systems and liners. *J. Environ. Eng. Div. (Am. Soc. Civ. Eng.)* **112** (EE-3), 538–550; errata: **113** (EE-6), 1393.

Korfiatis, G. P., Demetracopoulos, A. C., Bourodimas, E. L., and Nawy, E. G. (1984). Moisture transport in a solid waste column. *J. Environ. Eng. Div., Am. Soc. Civ. Eng.* **110** (EE-4), 780–796.

Korte, N. E., Skopp, J., Fuller, W. H., Niebla, E. E., and Alesii, B. A. (1976). Trace element movement in soils, influence of soil physical and chemical properties. *Soil Sci.* **122**(6), 350–359.

Kremer, J. G., Lo, M. P., Martyn, P. C., and Directo, L. S. (1987). Regulation of toxic organics in industrial sewer discharges at the sanitation districts of Los Angeles County. *Proc. Purdue Ind. Waste Conf.* **42**, 137–141.

Kristiansen, R. (1981). Sand filter trenches for purification of septic tank effluent. I. The clogging mechanism and soil physical environment. *J. Environ. Qual.* **10**, 53–64.

Kulhawy, F. H., Sangrey, D. A., and Grove, C. S., Jr. Geotechnical behavior of solvay process wastes. *Proc. Conf. Geotech. Pract. Disposal Solid Waste Mater., 1977* 118–135.

Kuster, E., and Azadi-Bokhsh, A. (1973). Studies on Microbial Degradation of Plastic Films, Proc. Conf. Degrad. Poly. Plast. Inst. Electr. Eng. Plast. Inst., London.

Lahti, L. R., King, K. S., Readers, D. W., and Bacopoulos, A. (1987). Quality assurance monitoring of large clay liner. *Proc. Geotech. Practice Waste Disp.* 87, Geotech. Sp. Publ. 13 (Am. Soc. Civil. Eng.) 640–654.

Lambe, T. W. (1954). The permeability of fine grained soils. *ASTM Spec. Tech. Publ.* **STP 163,** 56–67.

Lambe, T. W., and Silva-Tulla, F. (1992). Stability analysis of an earth slope. *Proc. Stability Perform. Slopes and Embankments* II, Geotech. Sp. Publ. 31 (Am. Soc. Civ. Eng.) 27–69.

Lange, C. R., Hartman, J. R., Chong, N. M., Weber, A. S., and Matsumoto, M. R. (1987). Constraints of bioaugmentation in enhancing biological treatment process performance. *Proc. Purdue Ind. Waste Conf.* **42,** 275–284.

Larsen, S., and Widowsen, A. E. (1971). Soil fluoride. *J. Soil Sci.* **22,** 210.

Lawson, C. R. (1982). Filter criteria for geotextile: Relevance and use. *J. Geotech. Eng. Div. Am. Soc. Civ. Eng.* **108** (GT-10), 1300–1317.

Lechner, P., Lahner, T., and Binner, E. (1993). Reactor landfill experiences gained at the Breitenau Research Landfill in Austria. *Proc. 16th Int. Madison Waste Conf.* 169–180.

Leckie, J. O., Pacey, J. G., and Halvadakis, C. (1979). Landfill management with moisture control. *J. Environ. Eng. Div., Am. Soc. Civ. Eng.* **105** (EE-2), 337–355.

Lee, G. F., and Jones, R. A. (1990). Managed fermentation and leaching: An alternative to MSW landfills. *Biocycle* May, 78–83.

Lee, J. (1974). Selecting membrane pond liners. *Pollut. Eng.* **6**(1), 33–40.

Lentz, J. J. (1981). Apportionment of net recharge in landfill covering layer into separate components of vertical leakage and horizontal seepage. *Water Resour. Res.* **17**(4), 1231–1234.

Lindorff, D. E., Feld, J., and Connelly, J. (1987). "Groundwater Sampling Procedures Guidelines," PUBL WR-153-87. Wisconsin Dept. of Natural Resources, Madison.

Lindsay, W. L. (1972). "Inorganic Phase Equilibria of Micro-Nutrients in Soils," Micro-Nutr. Agric., p. 41. Soil Sci. Soc. Am., Madison, Wisconsin.

Linell, K. A., and Kaplar, C. W. (1959). "The Factor of Soil and Material Type in Frost Action," Bull. No. 225, pp. 81–128.. Highway Res. Board, Washington, D.C.

Linsley, R. E., and Franzini, J. B. (1972). "Water Resources Engineering." McGraw-Hill, New York.

Little, A. L., and Price, V. E. (1958). The use of an electronic computer for slope stability analysis. *Geotechnique* **8**(3), 113–120.

Long, F. L. (1982). A new solid state device for reading tensiometers. *Soil Sci.* **133**(2), 131–132.

Longman, F. P. (1990). "A Treatment Concept for Leachate from Sanitary Landfills" (Personal Communication), Stork Friesland B.V., P. O. Box. 13, 8400 AA Gorredijk, Netherlands.

Lord, A. E., Jr., and Koerner, R. M. (1984). Fundamental aspects of chemical degradation of geomembranes. *Proc. Int. Conf. Geomembr.* **1,** 293–298.

Low, P. F. (1961). Physical chemistry of clay water interaction. *Adv. Agron.* **13,** 269–327.

Lowe, R. K., and Andersland, O. B. (1981). Decomposition effects on shear strength of paper mill sludge. *Proc. Tech. Assoc. Pulp Pap. Ind. Environ. Conf.*, New Orleans, *1981*, 239–244.

Lu, J. C. S., Morrison, R. D., and Stearns, R. J. (1981). Leachate production and management from municipal landfills: Summary and assessment. *In* "Land Disposal of Municipal Solid Waste" (D. W. Shultz, ed.), 7th Annu. Res. Symp., EPA-600/9-81/002a. 1–17. U.S. Environ. Prot. Agency, Cincinnati, Ohio.

Lu, J. C. S., Eichenberger, B., and Stearns, R. J. (1985). "Leachate from Municipal Landfills, Production and Management," pp. 109–121. Noyes Publ., Park Ridge, New Jersey.

Lucas, R. E., and Knezek, B. D. (1972). "Climatic and Soil Conditions Promoting Micro-Nutrient Deficiencies in Plants," Micro-Nutr. Agri., p. 265. Soil Sci. Soc. Am., Madison.

Lundgren, T. A. (1981). Some bentonite sealants in soil mixed blankets. *Proc. Int. Conf. Soil Mech. Found. Eng. 10th, 1981* **2**, 349–354.

Lustiger, A., and Corneliussen, R. D. (1988). Microscopy shows way to better service in underground PE pipe. *Mod. Plast.* **63**(3), 74–82.

Lutton, R. J., Regan, G. L., and Jones, L. W. (1977). "Design and Construction of Covers for Solid Waste Landfills," EPA-600/2-79/165, p. 249. U.S. Environ. Prot. Agency, Cincinnati, Ohio.

Makdisi, F. I., and Seed, H. B. (1978). Simplified procedure for estimating dam and embankment earthquake induced deformations. *J. Geotech. Eng. Div.* (Am. Soc. Civ. Eng.), 104(GT7), 849–867.

Mar, B. W. (1980). Discussion on leachate generation from sludge disposal area by Charlie and Wardwell (1979). *J. Environ. Eng. Div., Am. Soc. Civ. Eng.* **106** (EE-3), 677–678.

Marcuson, W. F., III, Hynes, M. E., and Franklin, A. G. (1992). Seismic stability and permanent deformation analysis: The last 25 years. *Proc. Stability Perform. Slopes Embankments II*, Geotech. Sp. Publ. 31 (Am. Soc. Civ. Eng.) 552–592.

Marthaler, H. P., Vogelsanger, W., Richard, F., and Wierenga, P. J. (1983). A pressure transducer for field tensiometers. *Soil Sci. Soc. Am. J.* **47**(4), 624–627.

Martin, R. T. (1958). "Rhythmic Ice Banding in Soil," Bull. No. 218, pp. 11–23. Highway Res. Board, Washington, D.C.

Maser, K. R., and Kaelin, J. J. (1986). Leakage detection of liners using seismic boundry waves. *Barrier Technol.* 362–368.

Mbela, K. K., Sridharan, L., O'Leary, P., Bagchi, A., Mack, D. P., and Mitchell, G. R. (1991). Leachate generation patterns in MSW landfills in Wisconsin and HELP model assessment. *Proc. Waste Tech'91*, Toronto, Canada.

McBean, E. A., Poland, R., Rovers, F., and Crutcher, A. J. (1982). Leachate collection design for containment landfills. *J. Environ. Eng. Div., Am. Soc. Civ. Eng.* **108** (EE-1), 204–209.

McGinley, P. M., and Kmet, P. (1984). "Formation, Characteristics, Treatment and Disposal of Leachate from Municipal Solid Waste Landfills." Bur. Solid Waste Manage., Wisconsin Dept. of Natural Resources, Madison.

McKee, C. R., and Bumb, A. C. (1988). A three dimensional analytical model to aid in selecting monitoring locations in the vadose zone. *Ground Water Monit. Rev.* **8**(2), 124–136.

Mcshane, S. F., Montgomery, J. M., Lebel, A., Pollock, T. E., and Stirrat, B. A. (1986). Biophysical treatment of landfill leachate containing organic compounds. *Proc. 41st Purdue Industrial Waste Conf.* 167–177.

Means, R. E., and Parcher, J. V. (1963). "Physical Properties of Soils." Charles E. Merrill Book Inc., Columbus, Ohio.

Meidl, J. A., and Peterson, R. L. (1987). The treatment of contaminated groundwater and RCRA wastewater at Bofors—Nobel Inc. *Proc. 4th Natl. RCRA Conf. Haz. Waste Haz. Materials.*

Menser, H. A., and Winant, W. M. (1980). Landfill leachate as nutrient source for vegetable crops. *Compost Sci./Land Util.* **21**(4), 48–53.

Merill, S. D., and Rawlins, S. C. (1972). Field measurement of soil water potential with thermocouple psychrometers. *Soil Sci.* **113**(2), 102–109.

Metcalf & Eddy, Inc. (1979). "Wastewater Engineering: Treatment, Disposal, Reuse," 2nd ed. McGraw-Hill, New York.

Meyer, C. F., ed. (1973). "Polluted Groundwater Source: Causes, Effects, Controls and Monitoring," EPA-600/4-73/001b. U.S. Environ. Prot. Agency, Cincinnati, Ohio.

Michaels, A. S., and Lin, C. S. (1954). The permeability of kaolinite. *Ind. Eng. Chem.* **46**, 1239–1246.

Miller, C. J., and Mishra, M. (1989). Discussion of "Field Verification of HELP Model for Landfills" by Peyton and Schroedder (1988). *J. Environ. Eng. Div.* (Am. Soc. Civ. Eng.), **115**(4), 882–884.

Minnesota Pollution Control Agency (1978): "Disposal of Residuals by Landfilling." Minnesota PCA, Minneapolis.

Mitchell, J. K. (1956). The fabric of natural clays and its relation to engineering properties. *Proc. Highway Res. Board* **35**, 693–713.

Mitchell, J. K. (1976). "Fundamentals of Soil Behavior." Wiley, New York.

Mitchell, J. K., and Madsen, F. T. (1987). Chemical effects on clay hydraulic conductivity. *Proc. Conf. Geotech. Pract. Waste Disposal, 1987* Geotech. Spec. Publ. No. 13, Am. Soc. Civ. Eng., 87–116.

Mitchell, J. K., Hooper, D. R., and Campanella, R. G. (1965). Permeability of compacted clay. *J. Soil Mech. Found. Div., Am. Soc. Civ. Eng.* **91**(SM-4), 41–65.

Mitchell, R. A., and Mitchell, J. K. (1992). Stability evaluation of waste landfills. *Proc. Stability Perform. Slopes Embankments II,* Geotech. Sp. Publ. 31 (Am. Soc. Civ. Eng.) 1152–1187.

Morgenstern, N. R. (1992). The evaluation of slope stability—A 25 year perspective. *Proc. Stability Perform. Slopes Embankments II,* Geotech. Sp. Publ. 31 (Am. Soc. Civ. Eng.) 1–26.

Morrison, W. R., and Parkhill, L. D. (1985). "Evaluation of Flexible Membrane Liner Seams after Chemical Exposure and Simulated Weathering," Interagency Agreement No. DW 14930547-01-2. Hazard. Waste Eng. Res. Lab., U.S. Environ. Prot. Agency, Cincinnati, Ohio.

Mundell, J. A. (1984). Discussion of "Design of natural attenuation landfills" by Bagchi (1983). *J. Environ. Eng. Div., Am. Soc. Civ. Eng.* **110** (EE-6), 1207–1210.

Nasim, K., Meyer, M. C., and Autian, J. (1972). Permeation of aromatic organic compounds from aqueous solutions through polyethylene. *J. Pharm. Sci.* **61**(11), 1775–1780.

Newmark, N. M. (1965). Effects of earthquakes on dams and embankments. *Geotechnique* **29**(3), 215–263.

Nicholson, R. V., Cherry, J. A., and Reardon, E. J. (1980). "Hydrogeologic Studies on an Aquifer at an Abandoned Landfill," Part 6. University of Waterloo, Dept. of Earth Science, Ontario, Canada.

Norcell, W. A. (1972). "Equilibria of Metal Chelates in Soil Solution." Micro-Nutr. Agric. Soil Sci. Soc. Am., Madison, Wisconsin.

Norman, L. E. J. (1958). A comparison of values of liquid limit determined with apparatus with bases of different hardness. *Geotechnique* **8**(1), 70–91.

Norstorm, J. M., Willams, C. E., and Pabor, P. A. (1991) Properties of leachate from construction/demolition waste landfills. *Proc. 14th Annu. Madison Waste Conf.* 357–366.

Nutall, P. M. (1973). The effects of refuse tip liquor upon stream biology. *Environ. Pollut.* **4**, 215–222.

Oakley, R. E., III (1987). Design and performance of earth lined containment systems. *Proc. Conf. Geotech. Pract. Waste* Disposal, '87. Geotech. Spec. Publ. No. 13, Am. Soc. Civ. Eng. 117–136.

Oden, J. T., and Ripperger, E. A. (1981). "Mechanics of Elastic Structures," p. 460. Hemisphere Publishing, New York.

Olsen, H. W. (1962). Hydraulic flow through saturated clays. *Proc. Natl. Conf. Clays Clay Miner., 9th* 131–161.

Olson, R. E., and Daniel, D. E. (1979). Field and laboratory measurements of the permeability of saturated and partially saturated fine-grained soils. *ASTM Spec. Tech. Publ.* **STP 746**, 67.

Olson, R. E., and Daniel, D. E. (1981). Measurement of the hydraulic conductivity of fine-ground soils. ASTM, STP 746, ASTM, Philadelphia, Pennsylvania, 18–64.

Orr, W. R., and Finch, M. O. (1990). Solid waste landfill performance during the Loma Prieta earthquake. *In:* "Geotechnics of Waste Fills Theory and Practice, STP 1070" (A. O. Landra and G. D. Knowles, eds.). ASTM, Philadelphia.

Osantowski, R. A., Kormanik, R. A., and Huibregtse, G. L. (1989). Leachate treatment system design—flexibility is a must. *Proc. 12th Annu. Madison Waste Conf.* 542–555.

Oster, J. D., and Ingualson, R. D. (1967). In situ measurement of soil salinity with a sensor. *Soil Sci. Soc. Am. Proc.* **31**, 572–574.

Othman, M. A., and Benson, C. A. (1991). Influence of freeze–thaw on the hydraulic conductivity of a compacted clay. *Proc. 14th Annu. Madison Waste Conf.* 296–312.

Owein, I. S., and Khera, R. P. (1990). Geotechnology of Waste Management, pp 261–263, Butterworths, London.

Pal, D., Weber, J. B., and Overcash, M. R. (1980). Fate of polychlorinated biphenyls (PCB'S) in soil-plant system. *Residue Rev.* **74**, 45–98.

Palmer, L. A., and Barber, E. S. (1940). Soil displacement under circular loaded area. *Proc. 20th Annu. Res. Meet., Highway Res. Board* 279–286.

Parizek, R., and Lane, B. E. (1970). Soil-water sampling using pan and deep pressure-vacuum lysimeters. *J. Hydrol.* **9**(1), 1–21.

Paruvakat, N. (1993). Discussion of ''Effects of Freezing on Hydraulic Conductivity of Compacted Clay'' by Kim and Daniel (1992). *J. Geotech. Eng. (Am. Soc. Civ. Eng.)* **119**(11), 1862–1864.

Paruvakat, N. (1993). Deflection analysis of polyethylene leachate collection pipes. *Proc. Geosynthetics-1993,* Vancouver, B.C., Canada, 1413–1424.

Paruvakat, N., Sevick, G. W., and Buechel, L. J. (1990). Freeze–thaw effects on landfill clay liners. *Proc. 13th Annu. Madison Waste Conf.* 452–469.

Paschke, N. W. (1982). Mean behavior of buoyant contaminant plumes in groundwater. Unpublished M. S. Thesis, Department of Civil and Environmental Engineering, University of Wisconsin, Madison.

Paschke, N. W., and Hoppes, J. A. (1984). Buoyant contaminant plumes in groundwater. *Water Resour. Res.* **20**(2), 1183–1192.

Patrick, W. H., Jr., and Mahapatra, I. C. (1968). Transformation and availability to rice of nitrogen and phosphorus in waterlogged soils. *Adv. Agron.* **29,** 323.

Peirce, J. J., Sallfors, G., and Murray, L. (1986). Overburden pressure exerted on clay liners. *J. Environ. Eng. Div., Am. Soc. Civ. Eng.* **112** (EE-2), 280–291.

Penman (1948). Natural evaporation from open water, bare soil and grass. *Proc. R. Soc. London, Ser. A* **193,** 120–145.

Perkins, T. K., and Johnston, O. C. (1963). A review of diffusion and dispersion in porous media. *Soc. Pet. Eng. J.* **3**(1), 70–84.

Perrier, E. R., and Gibson, A. C. (1980). ''Hydrologic Simulation on Solid Waste Disposal Sites,'' EPA/SW-868. U.S. Environ. Prot. Agency, Cincinnati, Ohio.

Perry, R. H. (1976). ''Engineering Manual,'' 3rd ed. McGraw-Hill, New York.

Pervaiz, M. A., and Lewis, K. H. (1987). Geotechnical properties of industrial sludge. *Proc. Int. Symp. Environ. Geotechnol.* **2,** 57–76.

Petrasek, A. C., Kugelman, I. J., Austern, B. M., Pressley, T. A., Winslow, L. A., and Wise, R. H. (1983). Fate of toxic organic compounds in wastewater treatment plants. *J. Water Pollut. Control Fed.* **55**(10), 1286–1295.

Peyton, L., and Shroeder, P. R. (1988). Field verification of HELP model for landfills. *J. Environ. Eng. Div., Am. Soc. Civ. Eng.* **114** (EE-2), 247–269.

Peyton, R. L., and Schroeder, P. R. (1989). Closure to ''Field verification of HELP Model for Landfills'' by Peyton and Schroeder (1988). *J. Environ. Eng. Div. (Am. Soc. Civ. Eng.)* **115**(4), 884–886.

Phene, C. J., Hoffman, G. J., and Rawlins, S. L. (1971a). Measuring soil matric potential in situ by sensing heat dissipation within a porous body: I. Theory and sensor calibration. *Soil Sci. Soc. Am. Proc.* **35**(1), 27–33.

Phene, C. J., Rawlins, S. L., and Hoffman, G. J. (1971b). Measuring soil matric potential in situ by sensing heat dissipation within a porous body. II. Experimental results. *Soil Sci. Soc. Am. Proc.* **35**(2), 225–229.

Phillips, C. R., and Nathwani, J. (1976). ''Soil Waste Interactions, A State-of-the-Art Review,'' Solid Waste Manage. Rep., EPS 3-EC-76-14m. Environmental Conservation Directorate, Toronto, Ontario, Canada.

Pickens, J. F., and Lennox, W. C. (1976). Numerical simulation of waste movement in steady groundwater flow system. *Water Resour. Res.* **12**(2), 171–180.

Pohland, F. G. (1975). "Sanitary Landfill Stabilization with Leachate Recycle and Residual Treatment," EPA-600/2-75/043. U.S. Environ. Prot. Agency, Cincinnati, Ohio.

Polkowski, L. B., and Boyle, W. C. (1970). "Groundwater Quality Adjacent to Septic Tank Soil Adsorption System." Wisconsin Dept. of Natural Resources, p. 85. Madison.

Ponnamperuma, F. N. (1972). The chemistry of submerged soils. *Adv. Agron.* **24**, 29–96.

Ponnamperuma, F. N. (1973). "Oxidation-Reduction Reactions, Soil Chem., Vol. 2. Decker, New York.

Potts, J. E., Clendinning, R. A., and Ackart, W. B. (1973). "The Effects of Chemical Structure on the Biodegradability of Plastics," Proc. Conf. Degrad. Poly. Plast., Inst. Electr. Eng. Plast. Inst., London.

Preul, H. C. (1964). Travel of nitrogen compounds in soils. Unpublished Ph.D. Thesis, Department of Civil Engineering, No. MN U-D, pp. 64–115. University of Minnesota, Minneapolis.

Proctor, R. R. (1933). Fundamental principles of soil compaction. *Eng. News Rec.* **3** (Aug. 31, Sept. 7, 21, and 28).

Quasim, S. R., and Burchinal, J. C. (1970a). Leaching from simulated landfills. *J. Water Pollut. Control Fed.* **43**(3), 371–379.

Quasim, S. R., and Burchinal, J. C. (1970b). Leaching of pollutants from refuse beds. *J. Sanit. Eng. Div., Am. Soc. Civ. Eng.* **96** (SA-1), 49–58.

Quastel, J. H., and Scholefield, P. G. (1953). Arsenite oxidation in soil. *J. Soil Sci.* **75**, 279–285.

Raimondi, P., Gardner, G. H., and Petric, C. B. (1959). "Effect of Pore Structure and Molecular Diffusion on the Mixing of Miscible Liquids Flowing in Porous Media," Jt. Symp. Fundam. Concepts Miscible Fluid Displacement, Part II. Am. Inst. Chem. Eng., San Francisco, California.

Rao, S. K., Moulton, L. K., and Seals, R. K. (1977). Settlement of refuse landfills. *Proc. Conf. Geotech. Pract. Disp. Solid Waste Mater. Am. Soc. Civ. Eng.* 574–598.

Raudkivi, A. J., and Callandar, R. A. (1976). "Analysis of Groundwater Flow." Wiley, New York.

Rebhun, M., and Galil, N. (1987). Biotreatment inhibition by hazardous compounds in an integrated oil refinery. *Proc. Purdue Ind. Waste Conf.* **42**, 163–174.

Reeve, R. C., and Tamaddoni, G. H. (1965). Effect of electrolyte concentration on laboratory permeability and field intake rate of a sodic soil. *Soil Sci.* **99**(4), 261–266.

Revah, A., and Avnimeleih, Y. (1979). Leaching of pollutants from sanitary landfill models. *J. Water Pollut. Control Fed.* **51**(11), 2705–2716.

Reynolds, W. D., and Elrick, D. E. (1985). In situ measurement of field-saturated hydraulic conductivity, sorptivity and the a-parameter using Guelph permeameter. *Soil Sci.* **140**(4), 292–302.

Rhoades, J. D. (1979). Inexpensive four-electrode probe for monitoring soil salinity. *Soil Sci. Soc. Am. J.* **43**(4), 817–818.

Rhoades, J. D., and Van Schilfgaarde, J. (1976). An electrical conductivity probe for determining soil salinity. *Soil Sci. Soc. Am. J.* **40,** 647–651.

Richards, D. J., and Shieh, W. K. (1986). Biological fate of organic priority pollutants in the aquatic environment. *Water Res.* **20**(9), 1077–1090.

Richards, L. A., and Gardner, W. O. (1936). Tensiometers for measuring the capilary tension of soil water. *J. Am. Soc. Agron.* **28,** 352–358.

Ridley, W. P., Dizikes, L. P., and Wood, J. M. (1977). Biomethylation of toxic elements in the environment. *Science* **197,** 329–332.

Roberts, K. J., and Sangrey, D. A. (1977). "Attenuation of Inorganic Landfill Leachate Constituents in Soils of New York," Res. Rep. Contract No. C110367. Presented to N.Y. State Department of Environmental Conservation, Division of Solid Waste Management.

Robinson, H. D., and Maris, P. J. (1985). The treatment of leachates from domestic waste in landfill sites. *J. Water Pollut. Control Fed.* **57**(1), 30–38.

Rogowski, A. S., and Richie, E. B. (1984). Relationship of laboratory and field determined hydraulic conductivity in compacted clay soils. *Ind. Waste Proc., Mid-Atl. Conf., 16th, 1984* 520–533.

Rogowski, A. S., Weinrich, B. E., and Simmons, D. E. (1985). Permeability assessment of a compacted clay liner. *Proc. 8th Annu. Madison Conf. Appl. Res. Pract. Munic. Ind. Waste* 315–337.

Rovers, F. A., and Farquhar, G. J. (1973). Infiltration and landfill behavior. *J. Environ. Eng. Div., Am. Soc. Civ. Eng.* **99** (EE-5), 671–690.

Rowe, P. W. (1972). The relevance of soil fabric to site investigation practice. *Geotechnique* **22**(2), 195–300.

Sai, J. O., and Anderson, D. C. (1991). *In:* "State-of-the-Art Field Hydraulic Conductivity Testing of Compacted Soils," EPA/600/S2-91/022. U.S. Environ. Prot. Agency, Cincinnati, Ohio.

Salame, M. (1961). An empirical method for the prediction of liquid permeation in polyethylene and related polymers. *SpE Trans.* (Poly. Eng and Science), October 1961 153–163.

Salvato, J. A., Witkie, W. G., and Mead, B. E. (1971). Sanitary landfill leaching prevention and control. *J. Water Pollut. Control Fed.* **43**(10), 2084–2100.

Sangrey, D. A., Noonan, D. K., and Webb, G. S. (1976). Variation in Atterberg limits of soil due to hydration history and specimen preparation. *ASTM Spec. Tech. Publ.* **STP 599,** 158–168.

Santillan-Medrano, J. and Jarinak, J. J. (1975). The chemistry of lead and cadmium in solid phase formation. *Soil Sci. Soc. Am. Proc.* **39,** 851.

Scharch, J. F., Huebner, P. M., Mack, D. P. (1985). An improved analytic method for estimating leachate heads within clay lined landfills. *Proc. 8th Annu. Madison Conf. App. Res. Prod. Munic. Ind. Waste* 338–354.

Schmugge, T. J., Jackson, T. J., and McKim, H. L. (1980). Survey of methods for soil moisture determination. *Water Resour. Res.* **16**(6), 961–979.

Schnabel, P., Lysmer, J., and Seed, H. B. (1972). SHAKE: A computer program for earthquake response analysis of horizontally layered sites. Report No. EERC 72-2, Earthquake Eng. Res. Center. Univ, California, Berkeley, CA.

Schroeder, P. R. (1985). Personal communication, with the author regarding HELP

Model experiments, dated August 5, 1985. Waterways Exp. Stn., Corps of Engineers, Vicksburg, Mississippi.

Schroeder, P. R., Gibson, A. C., and Smolen, M. D. (1984). "The Hydrologic Evaluation of Landfill Performance (HELP) Model," Vol. II, Doc. for Version 1, EPA/ 530-SW-84-101. U.S. Environ. Prot. Agency, Cincinnati, Ohio.

Schultz, D. W., and McKias, M. P., Jr. (1980). Assessment of liner installation procedures. *In* "Land Disposal of Municipal Solid Waste" (P. W. Shultz, ed.), 6th Annu. Res. Symp., EPA 600/9-80/010. U.S. Environ. Prot. Agency, Cincinnati, Ohio.

Schwab, G. O., and Manson, P. W. (1957). Engineering aspects of land drainage. *In* "Drainage of Agricultural Land" (J. N. Luthin, ed.), Chapter 3, pp. 287–344, Am. Soc. Agron., Madison, Wisconsin.

Seed, H. B., and Bonaparte, R. (1992). Seismic analysis and design of lined waste fills: Current practice. *Proc. Stability Perform. Slopes Embankments II*, Geotech. Sp. Publ. 31 (Am. Soc. Civ. Eng.) 1521–1545.

Seed, H. B., and Chan, C. K. (1959). Structure and strength characteristics of compacted clays. *J. Soil Mech. Found. Div., Am. Soc. Civ. Eng.* **85** (SM-5), 87–128.

Seed, H. B., Woodward, R. J., and Lundgren, R. (1962). Prediction of swelling potential for compacted clays. *J. Soil Mech. Found. Div., Am. Soc. Civ. Eng.* **90** (SM-4), 107–131.

Seed, H. B., Woodward, R. J., and Lundgren, R. (1964a). Clay minerological aspects of the Atterberg limits. *J. Soil Mech. Found. Div., Am. Soc. Civ. Eng.* **90** (SM-4), 107–131.

Seed, H. B., Woodward, R. J., and Lundgren, R. (1964b). Fundamental aspects of the Atterberg limits. *J. Soil Mech. Found. Div., Am. Soc. Civ. Eng.* **90** (SM-6), 75–105.

Sherard, J. L., Dunnigan, L. P., and Talbot, J. R. (1984a), Basic properties of sand and gravel filters. *J. Geotech. Eng. Div., Am. Soc. Civ. Eng.* **110** (GT-6), 684–700.

Sherard, J. L., Dunnigan, L. P., Talbot, J. R. (1984b). Filters for silts and clays. *J. Geotech. Eng. Div., Am. Soc. Civ. Eng.* **110** (GT-6), 701–717.

Silka, L. R. (1988). Simulation of vapor transport through the unsaturated zone— interpretation of soil-gas surveys. *Groundwater Monit. Rev.* **8**(21), 115–123.

Singh, S. (1992). "Response of Clay Liner System to Seismic Loading." CE Report No. 92-5, Dept. Civ. Eng., Santa Clara Univ., Santa Clara, California.

Singh, S., and Murphy, B. (1990). Evaluation of stability of sanitary landfills. *In:* "Geotechnics of Wastefills—Theory and Practice STP 1070" (A. D. Landva and G. D. Knowles, eds.), pp. 240–258. ASTM, Philadelphia.

Skempton, A. W. (1953). The colloidal activity of clay. *Proc. Int. Conf. Soil Mech. Found. Eng., 3rd, 1953* **1**, 57–61.

Skempton, A. W. (1964). Long-term stability of clay slopes, *Geotechnique* **14**, 77–101.

Slane, K. O. (1987). An evaluation of groundwater models to predict groundwater mounding beneath proposed groundwater gradient control systems for sanitary landfill design. Unpublished M.S. Thesis, University of Wisconsin, Dept. of Civil and Environmental Engineering, Madison.

Soil Conservation Service (SCS) (1971). "National Engineering Handbook, Section 16—Drainage of Agricultural Land." U.S. Dept. of Agriculture, Eng. Div., Washington, D.C.

Soil Conservation Service (SCS) (1972). "Procedure for Computing Sheet and Rill Erosion on Project Areas," Release No. 51. U.S. Dept. of Agriculture, Eng. Div., Washington, D.C.

Soil Conservation Service (SCS) (1975). "Urban Hydrology for Small Watersheds," Tech. Release No. 55. U.S. Dept. of Agriculture, Eng. Div., Washington, D.C.

SCS Engineers (1976). "The Selection and Monitoring of Land Disposal Case Study Sites," Vol. I, Project Description and Findings: Contract No. 68-01-2973. U.S. Environ. Prot. Agency, Washington, D.C.

Solseng, P. B. (1978). Determining the moisture content of residual waste. In "Disposal of Residuals by Landfilling" (A. Gebhard, proj. manager). Minnesota Pollut. Control Agency, Minneapolis. I-1 to I-40

Somogyi, F., and Gray, D. H. (1977). Engineering properties affecting disposal of red muds. *Proc. Conf. Geotech. Pract. Disposal Solid Waste Mater., 1977* 1–22.

Sopcich, D., and Bagchi, A. (1988). Discussion on mass burning of MSW with energy recovery, by Beckman and Dragovich (1986), *J. Environ. Eng. Div., Am. Soc. Civ. Eng.* **114** (EE-1), 235–236.

Sorg, T. J., and Bendixen, T. W. (1975). Sanitary landfill. *In* "Solid Wastes: Origin, Collection, Processing, and Disposal" (C. L. Mantell, ed.), pp. 71–113. Wiley, New York.

Sowers, G. B., and Sowers, G. F. (1970). "Introductory Soil Mechanics and Foundations," 3rd ed. Macmillan, New York.

Spangler, M. G., and Handy, R. L. (1982). "Soil Engineering," 4th ed. Harper & Row, New York.

Spitzglass, J. M. (1912). Flow of gas formula, derived, analyzed and checked by experimental data with diagrams for figuring the flow of gas in street mains and services. *J. Am. Gas Light* **96,** 269–315.

Sridharan, L., and Didier, P. (1988). Leachate quality from containment landfills in Wisconsin. *Proc. Int. Solid Waste Conf. 5th* **2,** 133–138.

Stanczyk, T. F. (1987). The development of treatment alternatives through an understanding of waste chemistry. *Proc. Purdue Ind. Waste Conf.* **42,** 309–320.

Starke, J. O. (1989). Effect of freeze/thaw weather conditions on compacted clay liners. *Proc. 12th Annu. Madison Waste Conf.* 412–420.

Stegmann, R. (1979). Leachate treatment at the sanitary landfill of Lingen, West Germany—Experiences with the design operation of the aerated lagoons. *Proc. 2nd Annu. Madison Conf. Appl. Res. Pract. Munic. Ind. Waste* 456–471.

Steiner, R. L., Keenan, J. D., and Fungaroli, A. A. (1979). "Demonstrating Leachate Treatment," Report on a Full-scale Operating Plant, EPA SW-758. U.S. Environ. Prot. Agency, Cincinnati, Ohio.

Stephens, D. B., Unruh, M., Havelana, J., Knowlton, R. G., Jr, Mattson, E., and Cox, W. (1988). Vadose zone characterization of low permeability sediments using field permeameters. *Ground water Monitor. Rev.* **8**(2), 59–66.

Stone, R. (1974). "Disposal of Sewage Sludge into a Sanitary Landfill," EPA-SW-71d. U.S. Environ. Prot. Agency, Cincinnati, Ohio.

Stone, R. (1975). Aerobic landfill stabilization. *In* "Solid Waste: Origin, Collection, Processing, and Disposal," (C. L. Mantell, ed.) pp. 153–183. Wiley, New York.

Straub, W. A., and Lynch, D. R. (1982a). Models of landfill and leaching: Moisture flow and inorganic strength. *J. Environ. Eng. Div., Am. Soc. Civ. Eng.* **108** (EE-2), 231–250.

Straub, W. A., ,and Lynch, D. R. (1982b). Models of landfill leaching: Organic strength. *J. Environ. Eng. Div., Am. Soc. Civ. Eng.* **108** (EE-2), 251–268.

Streng, D. R. (1976). The effects of the disposal of industrial waste within a sanitary landfill environment. *In* "Residual Management by Land Disposal" (W. H. Fuller, ed.), EPA-600/9-76/015, pp. 51–70. U.S. Environ. Prot. Agency, Cincinnati, Ohio.

Stumm, W., and Morgan, J. J. (1970). "Aquatic Chemistry." Wiley (Interscience), New York.

Suflita, J. M., Gerba, C. P., Ham, R. K., Palmisano, A. C., Rathje, W. L., and Robinson, J. A. (1992). The world's largest landfill—a multidisciplinary investigation. *Environ. Sci. Technol.* **26**(8), 1486–1494.

Swaine, D. J., and Mitchell, R. L. (1960). Trace element distribution in soil profiles. *J. Soil Sci.* **11**, 347–368.

Sykes, J. F., Soyupak, S., and Farquhar, G. J. (1969). "Modeling of Leachate Migration and Attenuation in Groundwater below Sanitary Landfill." University of Waterloo, Dept. of Civil Engineering, Ontario, Canada.

Tabak, H. H., Quave, S. A., Mashni, C. I., and Barth, E. R. (1981). Biodegradibility studies with organic priority pollutant compounds. *J. Water Pollut. Control Fed.* **53**(10), 1503–1518.

Taylor, D. W. (1948). "Fundamentals of Soil Mechanics." Wiley, New York.

Taylor, G. S., and Luthin, J. N. (1978). A model for couple head and moisture transfer during soil freezing. *J. Can. Geotech.* **15**, 548–555.

Technical Enforcement Guidance Document (1986). "RCRA Ground-Water Monitoring Technical Enforcement Guidance Document," OSWER-9950. 1. U.S. Environ. Prot. Agency, Office of Solid Waste and Emergency Response, Cincinnati, Ohio.

Telles, R. W., Unger, S. L., and Lubowitz, H. R. (1988). "Technical Considerations for De Minimis Pollutant Transport Through Polymeric Liners," EPA/600/2-88/042. U.S. Environ. Prot. Agency, Cincinnati, Ohio.

Terzaghi, K. (1936). The shearing resistance of saturated soils. *Proc. Int. Conf. Soil Mech. Found. Eng., 1st, 1936* **1**, 54–56.

Tharp, L. (1991). Leachate characteristics for Missouri sanitary landfills. *Proc 14th Annu. Madison Waste Conf.* 313–326.

Thiel, T. J., Focuss, J. C., and Leech, A. P. (1963). Electrical water pressure transducers for field and laboratory use. *Soil Sci. Soc. Am. Proc.* **27**(5), 601–602.

Thornthwaite, C. W., and Mather, J. R. (1955). "The Water Balance," Publ. Climatol. Lab. Climatol. Drexel Institute of Technology, Lab. Climatol. Centerton, New Jersey.

Thornthwaite, C. W., and Mather, J. R. (1957). "Instructions and Tables for Computing Potential Evapotranspiration and the Water Balance," Publ. Climatol. Lab. Climatol. Drexel Institute of Technology, Lab. Climatol., Centerton, New Jersey.

Tisdale, S. L., and Nelson, W. L. (1975). "Soil Fertility and Fertilizer." Macmillan, New York.

Topp, G. C., Davis, J. L., and Annan, A. P. (1980). Electromagnetic determination of soil water content: Measurements in coaxial transmission lines. *Water Resour. Res.* **16**, No. 3.

Torrance, J. K. (1974). A laboratory investigation of the effect of leaching on the compressibility and shear strength of Norwegian marine clays. *Geotechnique* **24**(2), 155–173.

Tortensson, B. A. (1984). A new system for groundwater monitoring. *Groundwater Monit. Rev.* **4**(4), 131–138.

Trainor, D. P. (1986). Moisture and saturation effects on hydraulic conductivity testing. *Proc. 9th Annu. Madison Conf. App. Res. Pract. Munic. Ind. Waste* 646–657.

Tuma, J. J. (1978). "Engineering Mathematics Handbook," p. 394. McGraw-Hill, New York.

Uloth, V. C., and Mavinic, D. S. (1977). Aerobic bio-treatment of a high strength leachate. *J. Environ. Eng. Div., Am. Soc. Civ. Eng.* **103**(EE-4), 647–661.

Uni-Bell Plastic Pipe Association (Uni-Bell) (1979). "Handbook of PVC Pipe Design and Construction. Uni-Bell, Dallas, Texas.

U.S. Congress (1989). "Facing America's Trash: What's Next for Municipal Solid Waste?" OTA-0-424, Office of Technology Assessment, U.S. Govt. Printing Office, Washington, D.C.

U.S. Environmental Protection Agency (USEPA) (1986). "Test Methods for the Evaluation of Solid Waste," EPA SW-846. USEPA, Washington D.C.

Vallejo, L. E. (1980). A new approach to the stability analysis of thawing slopes. *Can. Geotech. J.* **17**, 607–612.

Vallejo, L. E., and Edil, T. B. (1981). Stability of thawing slopes: Field and theoretical investigation. *Proc. Int. Conf. Soil Mech. Found. Eng., 10th, 1981* **3**, 545–548.

Van Olphen, H. (1963). "An Introduction to Clay Colloid Chemistry." Wiley (Interscience), New York.

Varshney, R. S. (1979). "Engineering Hydrology," pp. 368–386. NemChand & Bros. Roorkee, U.P, India.

Vasuki, N. C. (1988). Why not recycle the landfill. *Waste Age* November, 165–170.

Veihmeyer, F. J. (1964). Evapotranspiration. *In* "Handbook of applied Hydrology" (V. T. Chow, ed.), pp. 11–1 to 11–38. McGraw-Hill, New York.

Veihmeyer, F. J., and Henderickson, A. H. (1955). Rates of evaporation from wet and dry soils and their significance. *Soil Sci.* **80**, 61–67.

Viessman, W., Jr., Knapp, J. W., Lewis, G. L., and Harbaugh, T. E. (1977). "Introduction to Hydrology," pp. 43–87. Harper & Row, New York.

Walker, W. H. (1969). Illinois groundwater pollution. *J. Am. Water Works Assoc.* **61**, 31–40.

Walsh, J. J., and Kinman, R. N. (1979). Leachate and gas production. Under controlled moisture conditions. *In* "Land Disposal: Municipal Solid Waste" (D. W. Shultz, ed.), 5th Annu. Res. Symp., EPA-600/9-79/023, pp. 41–57. U.S. Environ. Prot. Agency, Cincinnati, Ohio.

Walsh, J. J., and Kinman, R. N. (1981). Leachate and gas from municipal solid waste landfill simulators. *In* "Land Disposal: Municipal Solid Waste" (D. W. Shultz, ed.), 7th Annu. Res. Symp., EPA-600/9-81/002, pp. 67–93. U.S. Environ. Prot. Agency, Cincinnati, Ohio.

Wardwell, R. E., and Charlie, W. A. (1981). Effects of fiber decomposition on the compressibility and leachate generation at a combined paper mill sludge landfill area. *Proc. Tech. Assoc. Pulp Pap. Ind. Environ. Conf.* New Orleans 233–238.

Water Pollution Control Federation (WPCF) (1980). "Operation and Maintenance of Wastewater Collection Systems." Moore & Moore Lithographers, Washington, D.C.

Water Pollution Control Federation (WPCF) (1981). Standard Methods for the Examination of Water and Wastewater." WPCF, Washington, D.C.

Watkins, R. K. (1990). Plastic pipes under high landfills. *In:* "Buried Plastic Pipe Technology, ASTM STP 1093" (G. S. Buczala and M. J. Cassady, eds.). ASTM, Philadelphia.

Weiler, W. A., Ozarowski, P. P., and Soydemir, C. (1993). Earthquake engineering for landfills. *Waste Age* **24**(8), 53–62.

Wellings, F. M., Lewis, A. L., and Mountain, C. W. (1974). Virus survival following wastewater spray irrigation of sandy soils. *In* "Virus Survival in Water and Wastewater Systems" (J. F. Mallina and B. P. Sagic, eds.), Water Resour. Symp., No. 7, pp. 253–260. University of Texas, Austin.

Wenger, R. B., and Rhyner, C. R. (1984). Optional service regions for solid waste facilities. *Waste Manage. Res.* **2**(1), 1–15.

Weston, A. F. (1984). Obtaining reliable priority-pollutant analyses. *Chem. Eng. (N.Y.)* **9**(9), 54–60.

Whitman, R. V., and Bailey, W. A. (1967). Use of computers for slope stability analysis. *J. Soil. Mech. Found. Div., Am. Soc. Civ. Eng.* **93** (SM-4), 475–498.

Wigh, R. J. (1979). "Boone County Field Site Interim Report," EPA-600/2-79/058. U.S. Environ. Prot. Agency, Cincinnati, Ohio.

Wigh, R. J., and Brunner, R. D. (1981). Summary of landfill research, Boone County field site. *In* "Land Disposal of Municipal Solid Waste" (D. W. Schultz, ed.), 7th Annu. Res. Symp., EPA-600/9-81/002a, pp. 209–242. U.S. Environ. Prot. Agency, Cincinnati, Ohio.

Williams, N. D., and Houlihan, M. (1986). Evaluation of friction coefficients between geomembranes, geotextiles and related products. *Proc. Int. Conf. Geotext., 3rd, 1986* 891–896.

Williams, P. J. (1986). Pore pressure at a penetrating frost line and their prediction. *Geotechnique* **16**(3), 187–208.

Wilson, L. G. (1980). "Monitoring in the Vadose Zone: A Review of Technical Elements and Methods," EPA-600/7-80/134. U.S. Environ. Prot. Agency, Cincinnati, Ohio.

Wilson, L. G. (1981). Monitoring in the vadose zone. Part I. Storage changes. *Ground Water Monit. Rev.* **1**(3), 32–41.

Wilson, L. G. (1982). Monitoring in the vadose zone. Part II. *Ground Water Monit. Rev.* **2**(1), 31–42.

Wilson, L. G. (1983). Monitoring in the vadose zone. Part III. *Ground Water Monit. Rev.* **3**, 155–166.

Wollak, H. F. (1984). PVC/ethylene interpolymer alloys for plasticizer-free membranes. *Proc. Int. Conf. Geomembr.* **1**, 29–32.

Wong, J. (1977). The design of a system for collecting leachate from a lined landfill site. *Water Resour. Res.* **13**(2), 404–410.

Wood, J. A., and Porter, M. L. (1987). Hazardous pollutants in class II landfills. *J. Air Pollut. Control Assoc.* **37**(5), 609–615.

Wood, J. M., Segal, H. J., Ridley, W. P., Cheh, A., Chudyk, W., and Thayer, J. S. (1975). Metabololic cycles for toxic elements in the environment. *Proc. Int. Conf. Heavy Met. Environ.,* Toronto, Ontario, Canada, *1975* **1**, 49–68.

Woodley, R. M. (1988). How to select, install and prevent damage to membrane liners used in settling ponds. *Pulp Paper* **52**(13), 28–32.

Ying, W., Bonk, R. R., and Sojka, S. A. (1987). Treatment of a landfill leachate in powdered activated carbon enhanced sequencing batch bioreactors. *Env. Progress* **6**(1), 1–8.

Yoder, E. J. (1967). "Principles of Pavement Design." Wiley, New York.

Yong, R. N. (1986). Selective leaching effects on some mechanical properties of a sensitive clay. *Proc. Int. Symp. Environ. Geotechnol.* **1**, 349–362.

Yong, R. N., and Warkentin, B. P. (1966). "Introduction to Soil Behavior." Macmillan, New York.

Yong, R. N., Sethi, A. J., Booy, E., and Dascal, O. (1979). Basic characterization and effect of some chemicals on a sensitive clay from outardes 2. *Eng. Geol.* **14**, 83–107.

Ziezel, A. J., Walton, W. C., Sasman, R. T., and Prickett, T. A. (1962). Groundwater resources of Dupage County, Illinois. *Ill. State Geol. Surv. Ill. State Water Surv. Rep.* No. 2.

Zimmerman, R. E., and Perpich, W. M. (1978). "Design and construction of pulp and paper mill residue disposal sites. *Proc. 1st Annu. Conf. Appl. Res. Pract. Munic. Ind. Waste* 495–508.

Zimmerman, R. E., Chen, W. W. H., and Franklin, A. G. (1977). Mathematical model for solid waste settlement. *Proc. Conf. Geotech. Pract. Disposal Solid Waste Mater., 1977* 210–226.

Zimmie, T. F., and Laplante, C. (1990). The effect of freeze–thaw cycles on the permeability of a fine grained soil. *Proc. 22nd Mid Atlantic Industrial Waste Conf.,* Philadelphia, PA, July 24–27, 580–593.

Zimmie, T. F., Doynow, J. S., and Wardwell, J. T. (1981). Permeability testing of soils for hazardous waste disposal sites. *Proc. Int. Conf. Soil Mech. Found. Eng., 10th, 1981* **2,** 403–406.

Index